Materials Horizons: From Nature to Nanomaterials

Series Editor

Materials are an indispensable part of human civilization since the inception of life on earth. With the passage of time, innumerable new materials have been explored as well as developed and the search for new innovative materials continues briskly. Keeping in mind the immense perspectives of various classes of materials, this series aims at providing a comprehensive collection of works across the breadth of materials research at cutting-edge interface of materials science with physics, chemistry, biology and engineering.

This series covers a galaxy of materials ranging from natural materials to nanomaterials. Some of the topics include but not limited to: biological materials, biomimetic materials, ceramics, composites, coatings, functional materials, glasses, inorganic materials, inorganic-organic hybrids, metals, membranes, magnetic materials, manufacturing of materials, nanomaterials, organic materials and pigments to name a few. The series provides most timely and comprehensive information on advanced synthesis, processing, characterization, manufacturing and applications in a broad range of interdisciplinary fields in science, engineering and technology.

This series accepts both authored and edited works, including textbooks, monographs, reference works, and professional books. The books in this series will provide a deep insight into the state-of-art of Materials Horizons and serve students, academic, government and industrial scientists involved in all aspects of materials research.

Review Process

The proposal for each volume is reviewed by the following:

1. Responsible (in-house) editor
2. One external subject expert
3. One of the editorial board members.

The chapters in each volume are individually reviewed single blind by expert reviewers and the volume editor.

Arunachalam M. Rajendran ·
Srinivasan Gopalakrishnan · Stephan J. Bless

Failure of Brittle Materials Under Shock and Impact

Experiments and Modelling

Springer

Arunachalam M. Rajendran
Department of Mechanical Engineering
University of Mississippi
Oxford, MS, USA

Srinivasan Gopalakrishnan
Department of Aerospace Engineering
Indian Institute of Science
Bengaluru, Karnataka, India

Stephan J. Bless
Department of Civil and Urban Engineering
New York University Tandon School
of Engineering
Brooklyn, NY, USA

ISSN 2524-5384 ISSN 2524-5392 (electronic)
Materials Horizons: From Nature to Nanomaterials
ISBN 978-981-97-7025-0 ISBN 978-981-97-7026-7 (eBook)
https://doi.org/10.1007/978-981-97-7026-7

This Springer imprint is published by the registered company Springer Nature Singapore Pte Ltd.
The registered company address is: 152 Beach Road, #21-01/04 Gateway East, Singapore 189721, Singapore

Dr. A. M. Rajendran dedicates this book to his wife, Pankajam. Her unwavering support, sacrifices, encouragement, and boundless love have been a constant inspiration throughout his life's journey. He also dedicate this book to his precious daughters, Nithya and Prenya, whose presence added immeasurable joy and meaning to every chapter of his life.

Dr. Srinivasan Gopalakrishnan dedicates this book to his late mother, Saraswathi, his father Srinivasan, his wife Anu and his children Karthik and Keerthana.

Dr. Stephan Bless dedicates this book to Prof. Thomas Ahrens, from whom he learned the basics, profundities, beauty, and subtleties of shock physics.

Preface

This book aims to prepare engineers and research students engaged in the study of brittle material failures under high strain rate and impact loading conditions, guiding them toward performing design analyses of blast and ballistic resistance of structural systems. The fundamental concepts and modelling methodologies presented in the book are not only applicable to brittle materials but also extend to the dynamic loading response of metallic materials. The deformation and damage mechanisms in brittle materials under dynamic loading conditions are relatively complex when compared to homogeneous isotropic metallic materials. For example, millions of micro and mesoscopic flaws such as cracks and voids/pores are the principal sources of fracture in ceramics and cementitious materials. In addition, it is also possible to generate atomistic defects like dislocations in crystalline ceramics under extreme pressure and very high strain rates. The defects in general are produced by the material processing or fabrication including chemical reactions and surface conditions. The elevated local stresses at microscopic length scales cause nucleation, growth, and coalescence of voids or cracks, and crack propagation leading to failure/fracture and fragmentation of brittle materials.

Understanding and modelling the structural response to extreme pressures and temperatures under high-velocity impact loading conditions necessitates exposure to continuum mechanics, shock wave propagation, fracture mechanics, continuum damage mechanics, and plasticity theory. The book emphasizes the importance of modelling and simulating impact problems using finite element or other solvers to gain a deeper understanding of the impact response of brittle solids.

While numerous articles over the past five decades have explored deformation and damage mechanisms in brittle materials like ceramics, concrete, glass, and composites under high strain rate and shock loading conditions, there remains a gap in easily accessible information. This book seeks to address this gap by compiling salient information and data from various sources into a monograph. It is intended to benefit researchers, practicing engineers, and graduate students who may be new to the subject matter.

The primary objective of this book, therefore, is to provide research students and practicing engineers with a comprehensive understanding of brittle materials' response in terms of deformation and damage evolutions leading to failure under impact loading conditions. Each chapter begins with a prologue, offering heuristics and historical discussions to readers unfamiliar with the topics, given the multidisciplinary and complex nature of the subject matter. The book also reviews stress and strain tensors, and wave propagation for completeness, outlining the modelling and simulation approach or methodology to seek solutions involving fracture and fragmentation of brittle solids. While crash worthiness studies involving automobiles or projectile penetration processes in vehicular and body armour systems are not covered, the application of constitutive and damage models to predict the response of armours is included for model and computational methodology validations.

Dr. Rajendran expresses his heartfelt gratitude to his former boss, mentor, and friend, Dr. Joseph P. Gallagher, of the University of Dayton Research Institute (UDRI) in Dayton, Ohio. His steadfast support and guidance played a pivotal role in steering him towards success in his career, shaping his professional journey. He extends his gratitude and thanks to Dr. Theodore Nicholas at Wright-Patterson Air Force Base, Dayton, Ohio, a friend and mentor, for his continuous guidance and the opportunities that fuelled his passion for high strain-rate modelling and experiments, significantly impacting his career trajectory. Special appreciation goes to Dr. Stephan Bless, his esteemed colleague and friend at UDRI, for his mentorship in shock physics. He is also grateful to his exceptional colleague, Mr. David Grove, at the US Army Research Laboratory. His outstanding proficiency in mathematics and coding was instrumental in the creation of the Rajendran-Grove ceramic model. This book wouldn't have been possible without the unwavering support and encouragement of Prof. Jagadeesh Gopalan and Prof. Srinivasan Gopalakrishnan at the Indian Institute of Science, Bengaluru, India.

Dr. Gopalakrishnan would like to acknowledge his Ph.D. research supervisor Prof. James Doyle at Purdue University, West Lafayette, all his teachers and graduate students, through whom he learnt all the mechanics reported in this book.

Dr. Bless wishes to acknowledge his gratitude to his research colleagues and students, from whom he has learned so much, not only about behaviour of brittle materials, but more generally how to think about science. Foremost among these are his Israeli associates Yehuda Partom and Zvi Rosenberg, as well as Dattatraya Dandekar of the US Army Research Laboratory. He also treasurers the many years that he spent through mutual visits collaborating with Gennady Kanel from Chernogolovka. Together with Dr. Rajendran, he also remembers the profound pleasure of sharing a good portion of his professional life with Nachhattar Brar.

All the authors greatly acknowledge the hard work and dedication of Ms. Rohiniben Patel from University of Mississippi, Oxford, towards obtaining permissions for publishing figures and tables from external sources.

Oxford, MS, USA
May 2024

Arunachalam M. Rajendran
Srinivasan Gopalakrishnan
Stephan J. Bless

Contents

About the Authors

Dr. Arunachalam M. Rajendran obtained his master's degree from the IIT Kanpur and his Ph.D. from the University of Washington, Seattle, WA in 1981. Arunachalam M. Rajendran serves as Chair and University Distinguished Professor for the Mechanical Engineering Department at the University of Mississippi, Oxford, MS. Prior to joining the University of Mississippi, Dr. Rajendran served the US Army Research Office as Chief Scientist for the Engineering Sciences Directorate. Dr. Rajendran primarily works on failure of materials under shock and impact. He has published more than 150 articles, and edited 10 books, and chaired more than 50 international conferences and symposiums. He has given invited/keynote talks at more than 100 universities and symposiums throughout the world.

Dr. Srinivasan Gopalakrishnan received his B.E. degree from UVCE, Bengaluru, master's degree in engineering mechanics from Indian Institute of Technology Madras, Chennai and Ph.D. from School of Aeronautics and Astronautics from Purdue University, USA, in December 1992. After his Ph.D., he was Postdoctoral Fellow in the Department of Mechanical Engineering at Georgia Institute of Technology. In November 1997, he joined the Department of Aerospace Engineering at Indian Institute of Science Bangalore, where currently he works as Senior Professor. His main areas of interest are wave propagation in complex media, computational material science, computational mechanics, smart structures, structural health monitoring, mems, and nanocomposite structures. He has published a total of 234 international journal papers, eight graduate level textbooks, two undergraduate books, 13 book chapters, and 160 international conference papers. He has given invited/keynote talks at more than 100 universities and symposiums worldwide.

Stephan J. Bless is a world-renowned shock physicist, who has made significant contribution in understanding propagation of shock waves in structures and materials. He got his Sc.D. degree (Earth and Planetary Sciences, 1970), S.M. degree (Earth and Planetary Sciences, 1968), and B.Sc. degree (Physics, 1965) all from Massachusetts Institute of Technology. His areas of expertise are impact and shock wave physics, penetration of soils including sand and clay, dynamic behaviour of natural and engineering materials under extreme loading, and terminal ballistics. He is currently Research Professor at New York University in the Department of Civil and Environment Engineering. Dr. Bless has more than 100 referred publications and two patents.

Chapter 1
Introduction

With the advent of high-performance computing, complex numerical algorithms for handling fracture and fragmentation, an enhanced understanding of wave propagation in complex materials, and advanced material constitutive and failure modelling, computational modelling of a wide range of shock and impact problems has become a standard engineering practice. These applications span a wide range, from assessing car crash worthiness, designing drilling tools, evaluating machinability, and even planning building demolitions to creating windshields and developing new high-strength and reinforced cements. Geotechnical concerns, such as rock drilling, petroleum extraction technologies, and the effects of underground explosions, are also addressed. Additionally, military and civil defence applications involve protecting against terrorist threats, examining weapon effects on protective structures, and designing ceramic armour.

In all these design applications, a fundamental understanding and modelling of the dynamic response of brittle materials require a grasp of three major topics: shock wave propagation, constitutive theories, and computational modelling.

Rocks: Many of the fundamental principles that underlie our understanding of brittle structural elements are rooted in rock mechanics. Since 1969, the "*Rock Mechanics and Rock Engineering*" journal [1] has been publishing articles related to experimental and theoretical aspects of rock mechanics, encompassing laboratory and field testing, computational methods, design principles, and onsite applications. Studies have included the investigation of confinement pressure effects on strength through triaxial compression tests. The book by Chun'an Tang and John A. Hudson [2], "*Rock Failure Mechanisms: Explained and Illustrated*", published in 2010, is particularly valuable for understanding the processes of rock failure under different loading conditions. Among its insights is the recognition that tensile failure is often the dominant failure mode in rocks, making the tensile strength of rock a vital parameter for analysing mining and excavation processes. Even under axial compression,

A. M. Rajendran et al., *Failure of Brittle Materials Under Shock and Impact*, Materials Horizons: From Nature to Nanomaterials, https://doi.org/10.1007/978-981-97-7026-7_1

wing-crack-based mechanisms can induce tensile micro-cracks leading to splitting and faulting failure [3].

Concrete: Another bulk brittle material, concrete, has garnered intense interest from civil and military engineers for decades. The compressive response of various concretes to a wide range of loading conditions has been thoroughly documented. Given concrete's widespread use in construction, understanding its stress–strain behaviour under compressive loading takes precedence over its behaviour under tensile loading. In their book, Bazant and Planas [4] explain fracture and size effects in concrete and other quasi-brittle materials, using fracture mechanics. Chen and his Suzuki [5] propose constitutive models for concrete, employing plasticity theories to describe pre- and post-crush behaviour under compression and fracture mechanics to describe tensile failure. Most constitutive and failure descriptions of rocks, concrete, and soils developed during the 1950s to 1970s were intended for quasi-static loading conditions.

Ceramics: Ceramic materials play a crucial role in space vehicle and space structural systems, protecting against high-velocity space debris impacts. Military armour systems capitalize on the exceptionally high compressive strength of crystalline ceramics. A special volume [6] on "*Ceramic Armour Materials Design*" underscores the importance of micro-structural analysis, high strain rate experiments, shock wave theories, and constitutive and damage modelling. Extensive studies involving high-velocity impact on ceramic materials, such as aluminium oxide and boron carbide, have been conducted through modelling and simulations to comprehend projectile penetration into thick armour plates.

These brittle materials can exhibit a range of microstructures, being isotropic, anisotropic, heterogeneous, or amorphous. For instance, rocks are heterogeneous and anisotropic, glass is amorphous, most ceramics are homogeneous, and concrete or cementitious materials are isotropic and heterogeneous. Although the compressive strength of brittle materials like rocks, concrete, and ceramics is significantly higher than their tensile strength under quasi-static and high strain rate loading conditions, they are vulnerable to tensile failure, with their fracture toughness notably lower than metals and metallic alloys. While it's feasible to eliminate tensile stresses under static or quasi-static loading conditions, this is rarely achievable under impact loading conditions due to wave reflections from stress-free lateral boundaries. Surprisingly, the most common mode of impact failure in brittle systems is tensile. Erzar and Forquin [7] have introduced an experimental method to determine the tensile strength of concrete at high strain rates. The recent article by Forquin [8] on brittle failure at high strain rates outlines the challenges and issues involved in obtaining dynamic properties.

As the applied load increases, many brittle solids that initially obey Hooke's law begin to develop permanent deformation (inelastic deformation or strain). For instance, the inelastic strains in rocks are due to microscopic heterogeneities that are either random distributions of anisotropic particles, or preferred particle orientation, or both [9]. Under high pressures and very high strain rates, rocks can fail due to faulting along certain preferred orientations. Failure of rocks under compression has been extensively studied during 60s through 80s. There are numerous constitutive models for rock fractures, both empirical and theoretical; a good summary can be found in Jing [10].

The stress–strain (constitutive) response of these brittle materials can be obtained through a wide variety of quasi-static (Q-S) and shock and impact experiments. Unlike the "*plastic strain*" in metallic materials, the definitions of "*inelastic strain*" as well as the "*stress–strain curve*" for brittle materials are statistical and complex due to strain or stress softening. With the advent of split Hopkinson pressure bar (SHPB) developments during the 1980s and 1990s, researchers in both rock and concrete obtained high strain rate data, primarily under compression. Ross et al. [11] utilized a 2-inch bar to characterize soils under dynamic loading, and Forquin et al. [12] also employed SHPB techniques to assess the effect of strain rate on concrete compressive strength. In 2014, Zhang and Zhao [13] reviewed dynamic experimental techniques for determining the mechanical behaviours of rocks. Generally, the compressive strength of most brittle materials tends to be highly sensitive to strain rate.

Understanding and modelling brittle material systems under shock and impact often necessitate a commitment to wave propagation analysis, impact experiments for calibration or validation, constitutive equations for relating stress and strain, equations of state for computing volumetric changes, damage and failure processes for representing material degradation, and computational methods involving time-marching codes. A plethora of research and commercial codes is available for the design analysis of impact-resistant structures, but users must input realistic model constants and select appropriate contact and boundary conditions for obtaining meaningful solutions to impact problems. It is also vital to choose suitable computer codes and models to obtain accurate solutions for highly nonlinear problems involving large deformations and fracture. This book aims to provide the most salient aspects of each of these topics, enabling individuals to develop expertise in solving a wide range of impact problems.

The application of constitutive (stress–strain relationship) and failure models for brittle materials is generally intended for computational modelling of high strain rate, shock, and projectile penetration processes. One critical requirement is the use of realistic values for material model parameters. Often, these values are not readily available. If a code user has a basic understanding of dynamic experiments and can access data for similar materials, it may be possible to determine or assign appropriate values for model parameters through simulations of existing experiments. However, in some cases, it's necessary to conduct tests using new experimental configurations to evaluate parameters in a constitutive model. This is especially true when dealing with relatively new or uninvestigated materials or novel and unique scenarios. It's important to conduct multiple types of experiments since there is often uncertainty in both the form of the constitutive equations and the values of their parameters. Experiments aimed at calibrating constitutive parameters should be as straightforward as possible to ensure that errors are attributed to problems with parameter values rather than conceptual flaws in the model itself. For this purpose, a chapter on dynamic experimental configurations with corresponding diagnostics and interpretation is included in this book.

In general, designing impact-resistant systems, such as structural or appliqué armour, and infrastructure components that use brittle materials becomes more complex due to stress and shock wave reflections from interfaces and boundaries. Damage can initiate and grow due to the arrival of compressive or tensile stress waves. Therefore, a solid understanding of shock wave phenomena is essential to leverage the complex response of brittle materials under dynamic loading conditions.

1.1 Wave Propagation

Impulsive loading, whether through impact or explosion, generates shock waves. Understanding the shock response of brittle materials is a multidisciplinary field that requires knowledge of continuum mechanics, mechanics of materials, computational methods, shock physics, and constitutive modelling.

Waves are the means by which energy is transported from one location to another, and the propagation of mechanical waves depends on the properties of the medium. The speed of a wave is determined by the material properties of the medium. The physics of wave propagation can vary between different media and can be influenced by factors such as space, time, stress, and strain states. The wave equation (Eq. (1.1)), derived from the equation of motion and Hooke's law, is represented as

$$u_{,tt} = C_0^2 u_{,xx} \tag{1.1}$$

where u is the displacement in the x-direction and C_0 is the sound speed in the solid medium.

The solution to this equation is discussed in many elementary texts on mechanics and especially in the stress and shock wave propagation chapters in *Impact Dynamics* [14]. Under high-intensity shock wave propagation, the shock speed increases with increasing shock amplitude or impact velocity. In a great many applications, there is a linear relationship between shock speed, U_s and material particle velocity, U_p, as given in Eq. (1.2):

$$U_s = C_0 + \alpha U_p \tag{1.2}$$

where α is a material property. Under very high pressures, the above relationship will become nonlinear with higher order term in U_p. Therefore, the shock wave speed is always equal or higher than the sound speed, except in rare cases where sound speed decreases with pressure. In general, the sound speed is given by the square root of modulus over density, where the modulus is the slope of the stress–strain curve. For instance, the wave speed, C_0, under *uniaxial stress state* as in a slender rod is given by

$$C_0 = \sqrt{\frac{E}{\rho}} \tag{1.3}$$

where E is Young's modulus and ρ is the material density. However, under *one-dimensional strain* condition, such as in a plate impact configuration where the longitudinal (thickness) direction is much less than the lateral dimension, causing the strain to be uniaxial, the wave speed, C_L, is given by

$$C_L = \sqrt{\frac{\left(K + \frac{4}{3}G\right)}{\rho}} \tag{1.4}$$

where K is the bulk modulus and G is the shear modulus. Similarly, shear wave speed is given by the square root of the shear modulus divided by the density.

The literature on shock compression in condensed matter is very rich and extensive, with references such as Jerry Forbes [15], Lee Davison [16], and Chap. 3 of various resources. Engineers and research students conducting impact analyses of structural or material systems must grasp basic concepts of shock wave propagation. In this book, the conceptual foundation and governing equations for stress and shock waves are provided, offering sufficient knowledge for most engineering impact analyses or the interpretation of numerical predictions.

Understanding stress and shock wave propagation is crucial for designing dynamic damage-resistant structures. Multi-axial stress or strain states are challenging to study directly in shock wave experiments, so simpler stress and strain states, like uniaxial stress and one-dimensional strain, are often used in laboratory-based stress/shock wave experiments to improve understanding of wave reflection, transmission, and complex wave motion.

At low stresses, shock waves are elastic, similar to acoustic waves. At higher stress levels, shock waves typically carry inelastic deformation and are slower than acoustic waves. An acoustic wave arrives before the inelastic shock wave, with its amplitude representing the maximum elastic stress that can be sustained under that particular deformation. This is known as the Hugoniot Elastic Limit (HEL). The locus of states attainable in a plane shock wave is called the Hugoniot. Measurements of Hugoniot curves provide various material properties, including elastic moduli, failure strain, maximum shear stress at failure, and sound speed. Shock data can be directly analysed to evaluate these constitutive parameters, or they can be obtained by matching time-history signals computed through finite element modelling with measured signals. The equation of state (EOS), which relates pressure, temperature, and volume, can be constructed from one-dimensional strain (*plate impact*) experiments.

In problems involving structural response to low-velocity impact and elastic waves in slender rods and bars, the stress analysis does not require the use of EOS due to the relatively low pressures involved. However, for high-velocity impact or high-energy explosion scenarios with very high pressures, describing an appropriate EOS is essential. Engineers performing stress or failure analyses of protective structures or armour systems need to understand EOS descriptions like Mie–Gruneisen for various materials. Model parameters for EOS are often not readily available in commercial codes for many brittle solids, so this book provides a methodology to estimate EOS parameters from shock wave experiments.

Two main compendia of EOS parameters are available: one from Los Alamos researchers (Marsh 1980 [17]) and another from their Russian counterparts (Trunin et al. [18]). Additionally, readers can find data in specific research articles cited in Chap. 4. Generally, EOS parameters don't significantly differ between various alloys of the same material. For example, the EOS parameters for steel and its various alloys will be the same. Further details on shock wave propagation can be found in Chap. 3 of the book.

1.2 Constitutive and Equation of State Models

Constitutive models relate deformation to stress and are an essential component of engineering analysis of impact and explosive loading of structures. With the advent of computing capabilities, researchers have developed and implemented various physics- and fracture mechanics-based models within both commercial and in-house finite element and finite difference codes.

In continuum mechanics, the stress tensor can be expressed as the sum of deviatoric (associated with shape change or shear-induced distortion) and hydrostatic (related to volume change or bulk) components. The equations that describe the shear part represent the constitutive modelling, while the equation of state accounts for the bulk part or the pressure versus volume relationship. For most engineering material problems, these aspects are typically considered uncoupled. The stress tensor, σ_{ij}, is given by

$$\sigma_{ij} = S_{ij} + P\delta_{ij} \tag{1.5}$$

where S is the deviatoric stress tensor and P is the pressure or mean stress. δ_{ij} is the Kronecker delta that is *zero or one* according to $i \neq j$ or $i = j$, respectively. For the sake of completeness, stress and strain tensors and these indices' notations are reviewed in Chap. 2.

The mathematical basis of constitutive model formulations for brittle materials often follows the theory of plasticity that was developed for metallic materials. In the von Mises yield criterion-based constitutive modelling, the most important input parameter is the strength of the solid, Y. The von Mises criterion is given by

$$3J_2 = \frac{3}{2}S_{ij}S_{ij}, \quad \text{when} \quad \sqrt{3J_2} \geq Y, \quad \text{plastic flow occurs} \tag{1.6}$$

where J_2 is the second invariant of the deviatoric stress tensor **S.** The strength Y is a material property which has to be determined from both quasi-static and dynamic experiments for the material system of interest. It is also called the "*flow stress*" or "*yield stress*" as in stress–strain curve description.

Material strength, whether in tension or compression, depends on many parameters. It can be expressed as

$$Y = f(\varepsilon_p, \dot{\epsilon}_p, P, T, D, Z)$$

$$\epsilon_p = \sqrt{\frac{2}{3}\epsilon_{ij}^p \epsilon_{ij}^p}, \qquad \dot{\epsilon}_p = \sqrt{\frac{2}{3}\dot{\epsilon}_{ij}^p \dot{\epsilon}_{ij}^p} \tag{1.7}$$

where ϵ_p is the equivalent inelastic or plastic strain, $\dot{\epsilon}_p$ is the equivalent plastic strain rate, P is the pressure, T is the temperature, D is the scalar damage, and Z is a state variable. Z is the internal state variable that models the influence or effects of a certain micro-mechanism or loading history effects on flow stress.

As the material's strength changes with loading history and cycles, appropriate evolution laws or equations are necessary in the strength model. For example, one commonly used strength model for *metallic materials* is the Johnson–Cook model [19]:

$$Y = (A + B\epsilon_p^n)(1 + C \ln \dot{\epsilon}_p)(1 - T^{*m}), \qquad T^{*m} = \frac{T - T_R}{T_M - T_R} \tag{1.8}$$

where n is strain hardening exponent, m is thermal softening exponent, A is quasi-static yield strength, B and n describe strain hardening, and C is strain rate hardening parameter. Here, T is applied temperature, T_R is room temperature, and T_M is the melting temperature.

This empirical equation effectively captures the key effects on material strength for most materials. According to the Johnson–Cook strength model, the material strength reaches zero at the melting temperature. For practical purposes, metallic materials are often assumed to be homogeneous and isotropic, and therefore making the $J - C$ equation applicable for both tension and compression, especially under shock and high strain rate loading conditions. Later, Johnson and Holmquist [20] developed similar empirical model for *brittle solids*.

An alternative to the von Mises theory is Griffith's equations [21], which are rooted in fracture mechanics. This approach leads to a different relationship between the principal stresses at failure, including the well-known approximation that the compressive strength is eight times the tensile strength. Chapter 5 explores the implications of this approach on constitutive models.

Strength models for brittle materials are relatively more complex due to the differences in deformation mechanisms between tension and compression. Most models can be described as quasi-empirical, with equations based on postulated micro-level damage mechanisms. However, the macro-behaviour is described using a few empirically derived parameters. A wide range of such models are available to describe concrete and rock-type brittle materials under quasi-static loading conditions. Given that inelastic strains are primarily attributed to the sliding of broken elements, a Mohr–Coulomb friction model is commonly used to describe shear resistance.

On the other hand, strength models for crystalline ceramics are limited to a few empirical models. These materials exhibit inelastic behaviour primarily due to tensile stresses around heterogeneities, and shear strength is highly dependent on mean stress. In the late 1990s, Johnson and Holmquist [22] developed a widely used set of equations that capture these assumptions. The empiricism in this strength model for brittle materials follows the Johnson–Cook model for metallic solids. Chapter 5 provides an overview of various modelling methodologies available for ceramic materials.

Various expressions are available to describe the EoS. The Mie–Gruneisen is one of the most popular equations used in the literature for a wide variety of materials. It is often expressed by

$$
\begin{aligned}
P &= P_H(1 - 0.5\Gamma\eta) + \Gamma E_s(1 + \eta) \\
P_H &= b_1\eta + b_2\eta^2 + b_3\eta^3
\end{aligned}
\tag{1.9}
$$

where η is the elastic volumetric compressive strain ($\eta = \rho/\rho_0 - 1$), where ρ and ρ_0 are the current and initial densities, respectively; Γ is the Gruneisen coefficient; b_1, b_2, and b_3 are empirical constants used for a cubic fit to the Hugoniot curve for the virgin material; and E_s is the internal energy per initial volume.

In stress analysis for transient shock wave problems, the stresses, σ_{ij}, are determined by adding the deviatoric stresses obtained from the constitutive relationship or strength model to the pressure independently obtained from the EoS. For relatively low pressures, there is no need an EoS, and pressure is simply related to the volumetric strain by the elastic relationship, i.e. $P = Ke_v$, where e_v is the volumetric strain and K is the bulk modulus.

1.3 Damage and Failure Mechanisms

The stress and strain states under both tension and compression influence damage initiation and growth mechanisms at microscopic scales. There are a vast number of articles that provide physical evidence of damage mechanisms in a variety of brittle solids. One of the most recognized mechanisms under compression is damage initiation at pre-existing flaw tips due to local tensile stresses. Nemat-Nasser and his co-workers [23–25] through a combined experimental and analytical studies investigated the effects of axial compression and level of boundary-induced lateral confinements on the microscopic failure processes that lead to macroscopic splitting or faulting-type fracture. In general, the micro-mechanical failure processes could include existing micropores and micro-cracks, creation of new micro-cracks due to mismatch between adjacent grains, growth and coalescence of micro-cracks, dislocations, and twinning under compressive loading conditions.

The compressive strength of most brittle materials increases with higher strain rates. This increase is often attributed to greater resistance to defect nucleation and growth at microscopic length scales. Under extreme pressures and strain rates, irreversible inelastic deformation in crystalline ceramics like aluminium oxide is primarily due to atomistic defects such as dislocations and twinning. These subatomic defects often lead to extensive plastic deformation (Shockey et al. [26]). In addition to dislocation-based plasticity, inelastic strains in structural ceramic materials can also result from micro-crack-induced fracture and fragmentation. In porous ceramics, the mechanisms of microscopic pore collapse can also influence the values of elastic bulk and shear moduli. At very high shock pressures, phase transitions can occur, including the evolution of crystal structures and grain-level fracture mechanisms. These mechanisms manifest as features in time-resolved shock waves.

Given that structural ceramics are used in armour systems to resist projectile penetration, a significant body of work has been carried out to understand and evaluate energy absorption resulting from a range of fracture and fragmentation mechanisms. Post-impact analysis of ceramics recovered from targets provides insights into the failure processes. Based on fractographic and metallographic examinations of ceramic targets, Shockey et al. [26] reported circular and radial crack patterns leading to cone configurations that correspond to "*Hertzian*" cracks. This so-called "*destructive*" technique is often employed to examine and infer various failure mechanisms in brittle ceramic solids under impact loading conditions.

Some research efforts have included X-ray computed tomography (XCT) to non-destructively reveal the internal meso-scale damage morphology in carefully encapsulated ceramic target plates (Wells et al. [27]). This XCT technique avoids additional damage in ceramics due to cutting and sectioning of the recovered target for studying internal damage generated by the penetration loading conditions.

In rocks and concrete, *dislocation-like atomistic defects* are not expected to dominate, and instead deformation mechanisms are primarily due to brittle flaws that nucleate at grain and aggregate interfaces. When researchers working on concrete response to dynamic loading use the term "*plastic*", it does not imply dislocation-based plasticity but rather refers to "*inelastic*" strains in cementitious materials. The failure processes based on brittle flaws are dominated by mechanisms associated with the collapse and growth of micropores and micro-cracks distributed throughout the bulk material. The impedance mismatch between the various constituents in concrete and the faulting-type mechanisms along certain preferred planes in rocks generates inelastic strains. Under compressive pressure at high strain rates, the interactions between interfaces generate complex stress/strain states dominated by shear-induced dilatation and sliding. The compressive strength of concrete and rocks increases with strain rate due to micro-inertia, which delays or suppresses damage initiation and growth, providing additional time for the material to deform elastically, thus causing an increase in material strength.

According to Reinhart et al. [28], the micro-inertia effects in the fracture process zone are the primary cause of the rate dependency of concrete. In concrete, micro-cracks nucleate and grow along the weak aggregate-matrix interfaces to form macro-cracks. However, in cementitious materials without aggregates, the damage processes differ, and both compressive and tensile strengths substantially increase when compared to concrete. In general, all these micro/macro-damage processes result in energy dissipation in brittle materials under impact loading conditions.

Bischoff and Perry [29] stress the importance of obtaining the complete stress–strain response of concrete at varying strain rates and the need for extensive experimental results to understand the fundamental changes in brittle material behaviours. They suggest that the increase in strength and compressive strain at maximum stress under high strain rates can be explained through the analysis of strain rate-dependent tensile micro-crack growth. The implications of hypothesized micro-mechanisms on continuum-based constitutive models are discussed in Chap. 5.

1.4 Dynamic Experiments

Most of the important dynamic material properties can be determined directly from impact experiments that generate relevant stress/strain states. For instance, the split Hopkinson bar (SHB) experiment [30–32] provides uniaxial stress conditions, and from that we can determine the stress–strain curve, strain to failure, compressive yield strength (or flow stress), shear strength, tensile strength, strain rate sensitivity and strain hardening parameters. The following schematic in Fig. 1.1 is provided for readers who are not familiar with the SHB experimental setups.

Tension and torsion setups are often very challenging for brittle materials due to premature fracture. However, alternative test configuration such as the "*Brazilian Splitting*" test ([34, 40]) can be employed to obtain certain dynamic tensile properties for brittle materials. Recently Chen et al. [39] provided a detailed description of the Brazilian test to determine tensile strength of alumina ceramic, PMMA, and granite. In addition, the standard compression SHB test can be modified to obtain stress–strain data under lateral confinement of the specimen. Empirical model parameters that describe the effects of pressure and strain rate on compressive strength can be directly obtained from SHB setups.

Besides the SHB setup, a wide range of other experimental configurations are available for research and material model validation. These include

- **Plate impact**: A thin flyer plate impacting a thick stationary target plate.
- **Rod-on-rod impact**: A short, slender ceramic rod impacting a long, slender ceramic rod.:

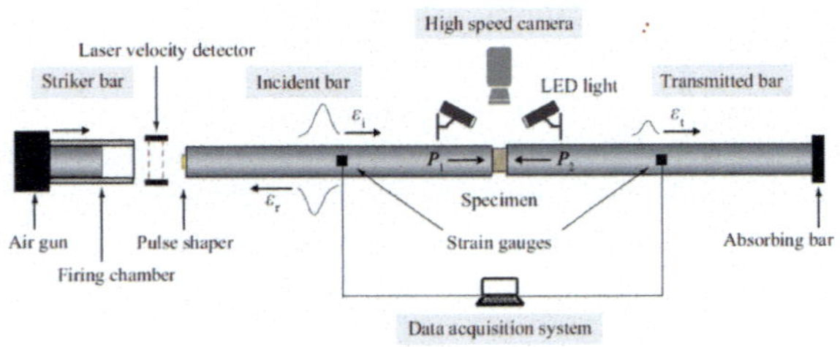

Fig. 1.1 Schematic of split Hopkinson pressure bar (SHPB) with diagnostics, including strain gauges, digital image correlation (DIC), and other measurement techniques [33]

- **Single-density plate-on-rod impact**: A thin flyer plate impacting a long ceramic rod.
- **Graded-density plate-on-rod impact**: A graded-density flyer plate impacting a slender ceramic rod with and without a confining steel sleeve.
- **Projectile penetration**: Also known as a DOP (depth of penetration) test, in which a long rod penetrator strikes a laterally confined ceramic tile backed by a semi-infinite block of steel or some other reference material. Most of these tests were primarily developed for characterizing the impact response of ceramic- and glass-type brittle materials. However, several other experimental configurations, such as the Brazilian disc test, three-point bend test, pressure-shear plate impact test, double flyer plate test, and various SHB test variations, are available for the determination of material properties and characterizations (see Chap. 4).

It is often desirable to test a brittle material using more than one experimental configuration. As diagnostic techniques have evolved, new experiments have emerged to study loading history effects on strength and damage. Experimentalists have developed newer test methods and advanced diagnostic tools, including high-speed cameras, photon-Doppler velocimetry (PDV), velocity interferometry system for any reflector (VISAR), stress gauges, cine X-ray photography, and synchrotron radiation. This diagnostic technological evolution is expected to continue.

In most modern plate impact and bar impact configurations, the goal is to obtain time-resolved stress pulse data, either with a VISAR, PDV, or an embedded piezoresistive stress gauge. In the penetration configuration, the measurement of interest is the projectile's residual DOP and/or residual length in the reference block. Constants for some model parameters, such as initial strength and moduli, can be directly evaluated from data. However, many others cannot be directly determined from experimental measurements; those constants must be indirectly calibrated by adjusting them until the model-generated time-resolved data matches the experimental records.

Numerical simulations can be used to help interpret the diagnostic measurements from complex impact experiments. Many models make predictions about the fracture evolution in brittle materials due to the very complex crack nucleation and propagation processes. It is often assumed that if the model correctly predicts observable features, such as velocity, stress, or DOP, then it can at least provide insight into the failure sequence, guiding the design of impact-resistant structures.

1.5 Computational Modelling

The analysis and utilization of high strain rate and shock wave experimental data often involve computational modelling and simulation of the experiments themselves. While certain material properties, such as bulk and shear modulus, fracture toughness, strength variations with strain rates, strain rate hardening, strain hardening, and thermal softening, can be directly determined from quasi-static and dynamic experiments, several other model parameters related to damage evolution and failure prediction require modelling and simulation. Those model parameters that cannot be determined directly from experimental measurements must be calibrated through a trial-and-error process using a model calibration scheme. It's important to note that it's often not possible to obtain a unique set of values for these model parameters.

The generality of model parameters is best verified through validation, which involves comparing multi-axial experimental measurements with results from computational modelling and simulation. For example, if model parameters were calibrated using data from uniaxial stress experiments (e.g. SHB) and one-dimensional strain experiments (plate impact), the accuracy of predictions for a multi-axial experiment (e.g. projectile penetration or the blast response of a panel) may be insufficient. Therefore, additional features may need to be added to the model, and the model constants may have to be adjusted using data from multi-axial experiments. If the constants for model parameters are determined from SHB tests performed at strain rates below $104/s$ and without using data from confinement tests, caution should be exercised when predicting structural responses to much higher strain rates (e.g. above $105/s$) and boundary conditions with lateral confinement, using the same constants. The deformation and fracture mechanisms in brittle material systems are sensitive to different stress and strain states, pressures, and strain rates.

Several solvers exist for modelling the impact response of brittle materials or structural systems. These solvers are based on various numerical methods and algorithms, including finite element, finite difference, element-to-particle conversion, discrete element method (DEM), material point method (MPM), smoothed particle hydrodynamics (SPH), and other meshless methods. Each method has its own benefits and advantages for obtaining realistic solutions for ballistic and blast applications. For example, to address the issue of extremely small time steps in computational modelling of projectile penetration into a thick target, special algorithms such as element-to-particle conversion are required. Instead of converting highly distorted elements into particles, it's also possible to erode away elements with aspect ratios approaching zero or extreme distortion. Although eroding elements can improve computational efficiency, it may violate the conservation of mass in principle.

Figure 1.2 illustrates the advancement of hydrocodes, such as the EPIC code [40], in simulating complex failure processes involving cavity formation, pulverization, fragmentation, and fracturing of a layered target. In this example, a long tungsten rod first perforates an oblique ceramic/elastomer/metal armour plate and then penetrates a standalone witness plate. The eroded mass from the rod and target was converted into particles, appearing as fragments with a distribution of sizes.

The element-to-particle algorithm in the EPIC code allowed the simulation to complete without crashing due to excessively small time steps. This was achieved by avoiding severe element distortion while preserving energy and mass conservation. The residual mass and shape of the tungsten long rod projectile matched exceptionally well with time-resolved X-ray radiographs from experiments.

Recently, particle-based methods like DEM (Discrete Element Method), MPM (Material Point Method), and XFEM (eXtended Finite Element Method) for modelling large deformation, fracture, and fragmentation problems have gained more attention due to the absence of element distortions and other advantages. Although

Fig. 1.2 EPIC finite element code simulation of a tungsten long rod penetration into a three-layer plate using element-to-particle conversion option. The behind armour debris (B.A.D) and the residual long rod after penetration of three-layer plate striking a witness plate in the simulation are also shown [41]

most particle-based codes are government-restricted, a few open-source codes with these useful features are available. Some commercial FEM codes do have options to use entirely particles (without any finite element) to discretize geometrically and materially nonlinear problems.

Incorporating fine-scale interfaces, such as laminations, fibre, or particle reinforcements, grain-level heterogeneity and faults in rocks into models may require the use of solvers like DEM and MPM, which can capture material behaviour at the micro/meso-level. These methods, particularly DEM and MPM, often prove more suitable than the FEM approach. It's important to note that rigorous computational modelling of meso-scale interfaces isn't commonly found in open literature; the existing research in this area is quite sporadic. However, cohesive zone models (CZM) have gained popularity over the last three decades for describing interfaces in FEM. CZM employs special constitutive equations to detail interfacial force versus displacement, making it a suitable choice for modelling interface failure, including delamination in layered composites.

When dealing with large deformation, fracture, and fragmentation, specialized algorithms come into play to address a range of phenomenological and numerical issues, such as handling smooth and non-smooth contacts, damage localization, non-local constitutive modelling, crack propagation, interfaces, and re-meshing. While this book doesn't delve into the numerical complexities of computational modelling and simulations, it does highlight their existence. Most advanced computer programs, or codes, offer options to tackle these challenges, but it's essential for code users to gain the necessary knowledge and experience to make the most of these options when simulating shock and high strain rate experiments.

While simulating shock and high strain rate experiments, it is important to perform a sensitivity study related to the mesh effects (such as the fineness of mesh or particle distribution) on solutions. It is possible that the results from two different meshes may substantially differ from each other. For instance, if the modelling involves localized failure processes or damage evolution within a band of few microns, realistic results could be obtained from a uniform grid with a micron-sized mesh. Perhaps damage evolution in a 10-micron interface on the delamination process can be realistically modelled with a 0.1 micron mesh size. These are only some thumb rules or somewhat a crude guideline to capture microscopic processes. In general, one should check that results are not sensitive to mesh refinement.

In computational simulations, the tensile failure of a brittle element can usually be assumed to occur instantaneously when the minimum principal stress reaches a critical value. However, under compressive loading, damage evolution and failure models due to various micro-structural defect mechanisms become complex and mathematically challenging. Continuum (macroscopic)-level models, based on assumed micro- and meso-scale processes, often involve a large number of parameters requiring determination from experiments and large-scale computer simulations.

Typically, due to mathematical and code-implementation complexities, micro-mechanical models are rarely included in commercial finite element or finite difference codes. These models describe defect nucleation, growth, and coalescence using internal state variables with appropriate evolution laws, demanding special numerical schemes for implementation. Determining model parameters is also challenging, and applications employing this approach often require collaboration among model and code developers. These models are typically confined to research codes developed by individual researchers for in-house projects. Occasionally, through government funding, computationally intensive micro-mechanical models are integrated into commercial codes like LS-DYNA, ABAQUS, and ANSYS. Additionally, codes such as CTH, DYNA3D, and EPIC, developed in US National and US Department of Defence laboratories with advanced strength, failure, and equation of state models, are often restricted for foreign use. For more information on these hydrocodes, please refer to Chap. 6.

The primary challenge facing the community using these codes is the lack of available model parameters for specific materials of interest. For example, while constants for the Johnson–Holmquist ceramic model may be available for materials like 99.8% alumina in open literature, constants for materials like titanium diboride (TiB_2) may not be readily accessible. Chapter 4 of this book provides information on the experimental data necessary for determining model parameters for brittle materials.

To utilize advanced high-resolution models effectively, it's often necessary to employ scalable parallel computing or multiprocessors. Enhanced computer capabilities now allow the use of hundreds, if not thousands, of processors to obtain accurate solutions for highly nonlinear geometrical and material systems under extreme loading conditions. It's important to note that not all design hydrocodes are scalable, and code users need expertise in selecting the appropriate code to take full advantage of scalable computing with a large number of processors. While some codes may not scale well beyond a certain number of processors, the use of cost-effective GPUs has gained popularity among researchers in universities and industries, offering an alternative to supercomputing.

1.6 Organization and Scope of the Book

In general, understanding the response of brittle materials to dynamic loading requires knowledge of shock wave propagation, constitutive modelling, high strain rate and shock experiments, and computational codes. These topics are covered in this book, along with a review of stress and strain concepts. Chapter 2 offers a concise review of continuum mechanics to aid in interpreting the subsequent chapters.

With the ongoing advancements in computing capabilities, modelling and simulating impulsive loading have become increasingly essential for solving various engineering problems. Realistic constitutive model constants are a crucial requirement for accurate simulations, making model constants' validation paramount for predicting the blast and ballistic responses of brittle materials.

Chapter 3 delves into shock wave propagation concepts, focussing on one-dimensional strain (1D-ϵ) states. It illustrates scenarios involving circular semi-infinite plates with finite thickness and infinite radius, impacting similar plates of the same material, leading to a 1D-ϵ state. This motion restricts lateral displacements and results in inertial confinement in the x- and y-directions. The chapter covers governing equations related to elastic and plastic shock wave propagation and explores elastic–plastic shock wave propagation when materials are plastically yielded.

Chapter 4 discusses high strain rate and shock experimental configurations used to determine fundamental dynamic properties of brittle materials. Various experiments and measurement techniques are presented, allowing researchers to leverage extensive literature on brittle material behaviour. This chapter equips researchers to design standard and new experiments for calibrating models under specific materials and loading conditions.

The complexity of deformation and damage mechanisms in brittle materials, often caused by porosity and micro-cracking, is addressed in Chap. 5. It also explores micro-plasticity in crystalline ceramics under lateral confinement at high velocities and "*ductility*" mechanisms in concrete-type cementitious materials. The chapter investigates the micro-mechanisms contributing to "*inelastic strain*" in brittle materials.

Chapter 5 emphasizes the importance of the compressive strength in military and civilian structures containing brittle materials subjected to impulsive loading. It discusses strength and damage model development and provides a review of ceramic strength and constitutive models. Often, model constants for specific brittle material systems are not readily available, and the chapter offers guidance on where to find such information. It also presents representative strength models for concrete.

In essence, computational modelling is crucial for understanding and designing complex failure-resistant brittle elements. Engineers can choose specialized finite element codes with appropriate algorithms to simulate impact-driven phenomena effectively. Chapter 6 provides insights into various computational methods for simulating large deformation, fracture, and fragmentation of brittle material systems, with applications also extendable to metallic materials.

In conclusion, a fundamental understanding of high strain rate material behaviour, shock wave propagation, and computational modelling is essential for solving impact problems. While there are numerous books on the dynamic response of metallic materials, this book focusses on the unique challenges posed by brittle materials. However, the experimental and modelling methodologies discussed here can also be applied to other brittle material systems, such as fibre-reinforced plastic composites and epoxy resins. The goal is to integrate these topics in one book to provide readers, whether practicing engineers or graduate students, with the knowledge needed to model the response of brittle material systems under extreme shock and impact loading conditions.

References

1. Rock mechanics and rock engineering - 56 Volumes; A journal published by Springer
2. Chun'an T, Hudson JA (2010) Rock failure mechanisms: explained and illustrated. CRC Press, ISBN-13: 978-0415498517/ISBN-10: 0415498511
3. Horii H, Nemat-Nasser S (1985) Compression induced micro-crack growth in brittle solids: axial splitting and shear failure. J Geophys Res 90(B4):3105–3125
4. Bazant ZP, Planas J (1998) Fracture and size effect in concrete and other quasi brittle materials (New directions in civil engineering), 1st edn
5. Chen WF, Suzuki H (1980) Constitutive models for concrete. Comput Struct 12(1):23–32
6. McCauley J, Crowson A, Gooch W, Rajendran AM, Bless SJ, Logan KV, Normandia M, (eds) Wax, Ceramic Trans 134:371–384
7. Erzar B, Forquin P (2010) An experimental method to determine the tensile strength of concrete at high rates of strain. Exp Mech 50(7):941–955
8. Forquin P (2016) Brittle materials at high-loading rates: an open area of research. Philos Trans R Soc A375. The Royal Society Publishing
9. Burdine NT (1963) Rock failure under dynamic loading conditions. Soc Petrol Eng 3(1):1–8. https://doi.org/10.2118/481-PA
10. Jing L (2003) A review of techniques, advances and outstanding issues in numerical modelling for rock mechanics and rock engineering. Int J Rock Mech Min Sci 40(3):283–353. https://doi.org/10.1016/S1365-1609(03)00013-3
11. Ross A, Nash P, Friesenhahn G (1986) Pressure waves in solis using a split-Hopkinson pressure bar, Technical Report ESL-TR-86-29. Tyndall Air Force Base, Florida
12. Forquin P, Gary G, Gatuingt F (2008) A testing technique for concrete under confinement at high rates of strain. Int J Impact Eng 35(6):425–446
13. Zhang QB, Zhao J (2014) A review of dynamic experimental techniques and mechanical behaviour of rock materials. Rock Mech Rock Eng 47:411–1478
14. Zukas JA, Nicholas T, Swift HF, Greszczuk LB, Curran DR (eds) Impact dynamics. Wiley
15. Forbes J (2012) Shock waves in condensed matters (a primer). Springer, Berlin, Heidelberg. ISBN: 978-3-642-32 534-2
16. Davison L (2008) Fundamentals of shock wave propagation in solids. Springer, Berlin, Heidelberg, ISBN: 978-3-540-74 568-6
17. Marsh SP (ed) (1980) LASL shock Huongiot data. University of California Press
18. Trunin RF, Gudarenko LF, Zhernokletov MV (2001) Experimental data on shock-wave compression and adiabatic expansion of condensed substances. VNIIEF, Sarnov
19. Johnson GR, Cook WH (1983) A constitutive model and data for metals subjected to large strains, high strain rates and high temperatures. In: Proceedings of seventh international symposium on ballistics, The Hague, The Netherlands

20. Johnson GR, Holmquist TJ (1992) A computational constitutive model for brittle materials subjected to large strains, high strain rates and high pressures. In: Meyers MA, Murr LE, Staudhammer KP (eds) Shock-wave and high-strain rate phenomena in materials. Marcel Dekker Inc, New York, pp 1075–1081
21. Griffith AA (1920) The phenomena of rupture and flow in solids. Philos Trans Roy Soc Lond Ser A 221:163–198
22. Johnson GR, Holmquist TJ (1994) An improved computational constitutive model for brittle materials. In: Schwartz SC, Shaner JW, Samara GA, Ros M (eds) High-pressure science and technology, American institute of physics conference proceedings, vol 309. AIP Press, New York, pp 981–984
23. Nemat-Nasser S, Horii H (1987) Compression-induced nonplanar crack extension with application to splitting, exfoliation, and rockburst. J Geophys Res 87(B8):6805–6821
24. Horii H, Nemat-Nasser S (1986) Brittle failure in compression: splitting, faulting and brittle-ductile transition. Philos Transa Roy Soc Lon A319:337–374
25. Sarva S, Namet-Nasser S (2001) Dynamic compressive strength of SiC under uniaxial compression.' Mater Sci Eng A317:140–144 1
26. Shockey DA, Marchand AH, Skaggs SR, Cort GE, Burkett MW, Parker R (1990) Failure phenomenology of confined ceramics targets and impacting rods. Int J Impact Eng 9(3):263–275
27. Wells JM, Rupert NL, Green WH (2002) Progress in the 3-D visualization of interior ballistic damage in Armour ceramics. In: McCauley J, Crowson A, Gooch W, Rajendran AM, Bless SJ, Logan KV, Normandia M, Wax S (eds) Ceramic transactions, vol 134, pp 441–448
28. Reinhardt HW, Weerheijm (1991) Tensile fracture of concrete at high loading rates taking account of inertia and crack velocity effects. Int J Fract 51(1):31–42
29. Bischoff PH, Perry SH (1991) Compressive behaviour of concrete at high strain rates. Mater Struct 24(6):425–450
30. Kolsky H (1949) An investigation of the mechanical properties of materials at very high rates of loading. Proc Phys Soc B 62:676–700
31. Harding J, Wood ED, Campbell JD (1960) Tensile testing of material at impact rates of strain. J Mech Eng Sci 2:88–96
32. Lindholm US (1971) High strain rate tests. In: Bunshah R (ed) Chapter 3, Appendix A, Techniques of metals research, vol 5 (1). Wiley
33. Xu Y, Dai F (2018) Dynamic response and failure mechanism of brittle rocks under combined compression-shear loading experiments. Rock Mech Rock Eng 51:747–764
34. International Society for Rock Mechanics (ISRM (1978), Suggested methods for determining tensile strength of rock materials. Int J Rock Mech Min Sci & Geomech Abstract 15:99–103
35. ASTM D 3967-08 (2008) Standard test method for splitting tensile strength of intact rock core specimens. ASTM International, West Conshohocken, USA
36. Li D, Ngai L, Wong Y (2012) The Brazilian disc test for rock mechanics applications: review and new insights. Rock Mech Rock Eng 46:269–287
37. John R, Antoun T, Rajendran AM (1991) Effect of strain rate and size on tensile strength of concrete. In: Proceedings of the American physical society conference on "Shock Compression of Condensed Matter" - 1991. Elsevier Science Publishers, pp 501–504
38. Chen P, Guo B, Chen J (2018) Dynamic Brazilian test using the Kolsky-Hopkinson bar machine. In: Othman R (ed) Chapter 4, in "The Kolsky-Hopkinson bar machine. Springer, Cham, pp 121–141
39. Wang QZ, Li W, Xie HP (2009) Dynamic split tensile test of flattened Brazilian disc of rock with SHPB setup. Mech Mater 41:252–260
40. Johnson GR, Stryk RA, Holmquist TJ, Souka OA (1990) Recent epic code developments for high velocity impact: 3D element arrangements and 2D fragment distributions. Int J Impact Eng 10(14):281–294
41. Johnson GR, Petersen EH, Stryk RA (1993) Incorporation of an SPH option into the epic code for a wide range of high velocity impact computations. Int J Impact Eng 14:35–394. Pergamon Press Ltd, UK

Chapter 2
Review of Continuum Mechanics

Prologue

A profound grasp of continuum mechanics is essential for comprehending the mechanics of brittle materials under multi-axial loading conditions. This chapter provides a concise overview of solid mechanics necessary to understand the material presented in subsequent chapters, avoiding an in-depth exploration of solid mechanics topics. To delve further into solid mechanics, readers are encouraged to refer to classic texts in the field [1–3].

Understanding continuum mechanics necessitates a clear knowledge of Tensor Analysis, a method to represent complex equations of deformation fields, stresses, their gradients, and governing equations through indicial notations. This chapter commences with a brief introduction to tensors and their operations, followed by a section on describing the deformation field. Here, two different approaches to representing motion: Lagrangian and Eulerian are presented based on the chosen frame of reference, and the corresponding strain–displacement relations are derived using these descriptions. Other topics covered in this chapter include the definition and derivation of different types of strain, stresses, and constitutive models for different material symmetries.

This chapter in addition to describing elastic stress–strain equations under different stress/strain states also introduces the reader to Hyper Elasticity, wherein various important hyperelastic constitutive models are reviewed. The hyperelastic models are useful for describing the unique stress–strain response of polymeric fibres in a composite laminate, such as the fibre-reinforced plastics.

It's important to note that the understanding of the behaviour of brittle materials under shock and impact loading requires a different set of non-linear constitutive models, which are deferred to Chap. 4. *While Chap.* 4 *discusses these models, a solid comprehension of the material presented in this chapter is crucial for understanding the content in Chap.* 4.

A. M. Rajendran et al., *Failure of Brittle Materials Under Shock and Impact*, Materials Horizons: From Nature to Nanomaterials, https://doi.org/10.1007/978-981-97-7026-7_2

It's essential to recognize that comprehending the behaviour of brittle materials under shock and impact loading necessitates a distinct set of non-linear constitutive models, which are addressed in Chap. 5. Although Chap. 5 discusses these models, a thorough understanding of the material presented in this chapter is crucial for grasping the content in Chap. 5.

2.1 Introduction of Tensor Analysis

In this chapter, we will deal only with tensors described in Cartesian coordinate system. As mentioned earlier, through the use of tensors, one can represent complicated equations in indicial notations. This requires understanding of *Summation Conventions*. Let us demonstrate the summation convention using a simple example.

The symbol x_i with a range $i = 1, ...n$ is normally used to denote any one of the variables in a set $\{x_1, x_2, ...x_n\}$, where the symbol i is called the *index* in one dimension. Similarly, we can represent a variable having two indices or in other words the variable in two dimensions such as t_{ij}, $i, j = 1, ...n$. That is the variable t_{ij} is represented through individual components in a set of n x n elements given by $\{t_{11}, t_{12}, ..., t_{mn}\}$.

Let us consider the following equation

$$p_1q_1 + p_2q_2 + p_3q_3 = R$$

We can clearly see that the above equation is a dot product of two vectors given by $\overrightarrow{p} = p_1i + p_2j + p_3k$ and $\overrightarrow{q} = q_1i + q_2j + q_3k$. The above equation can be written as

$$\sum_{i=1}^{3} p_iq_i = R$$

With the aid of summation convention, we can write the above equation. As

$$p_iq_i = R$$

According to the summation convention, the reputation of an index in a product term denotes a summation with respect to that index over its range. The repeated index is normally called *Dummy index* as opposed to one that is not summed, which is referred to as *Free index*. Since dummy index indicates summation operation, any index can be used without changing the result. For example, p_iq_i and p_jq_j have same meaning.

Following are some of the rules one can follow to avoid any mistakes while using summation:

- If a subscript occurs precisely twice in a term in an equation, then it must be summed over its full range. These are *repeated* or *dummy* indices. For example,

$$P_{ikk} = P_{i11} + P_{i22} + P_{i33}$$

- If a subscript occurs precisely once in a term, then it must occur precisely once in every other term in the equation. These are called *free* indices. To understand this rule, let us consider the following equation

$$\begin{aligned} F_1 &= ma_1 \\ F_i = ma_i \Rightarrow F_2 &= ma_2 \\ F_3 &= ma_3 \end{aligned}$$

- If a subscript occurs more than twice in a term, then it is a *mistake*. For example

$$A_{iij} B_{ij}$$

A proper example of a tensor notation is given below

$$\sigma_{ij} = S_{ij} + H_{ijkl} E_{kl}$$

In the above equations, i and j are the *free* indices, while k and l are *dummy* indices.

2.1.1 Contraction of Tensors

Contraction is a special summing operation performed over a tensor quantity. It is done by equating two indices and summing over the range of that index. For example, let us consider the following tensor

$$A_{ijkl} \qquad \text{(contraction over indices } i\text{)} \qquad A_{iikl}$$

After contraction, we see that a tensor with two free indices becomes a pair of dummy indices and the summation convention explained previously applies. The result of contraction of a tensor of order n is a new tensor of order $n - 2$.

2.1.2 *Special Symbols in Tensor Representation*

A number of special symbols are introduced to represent some complex equation in tensor notation. Two important symbols extensively used in tensor representation are $Kronecker - delta$ and $Permutation$ symbols, respectively. Some of their properties are summarized below:

Kronecker Delta

The Kronecker delta is represented as δ_{ij} and is defined as follows

$$\delta_{ij} = \begin{cases} 1 & \text{if} \quad i = j \\ 0 & \text{if} \quad i \neq j \end{cases} \tag{2.1}$$

In matrix form, the above relation can be written as

$$[\delta_{ij}] = \begin{bmatrix} 1 & 0 & 0 \\ 0 & 1 & 0 \\ 0 & 0 & 1 \end{bmatrix} \tag{2.2}$$

The matrix form of Kronecker delta is an identity matrix. Let us now discuss some of the properties of Kroneker delta, which are summarized below

- Kronecker delta is symmetric. That is : $\delta_{pq} = \delta_{qp}$.
- Trace of the Kronecker delta matrix: $\delta_{kk} = \delta_{11} + \delta_{22} + \delta_{33} = 3$.
- Contraction of tensor A_{ij} with δ_{ij}: $A_{ik}\delta_{ij} = A_{ik}\delta_{1j} + A_{2k}\delta_{2j} + A_{3k}\delta_{3j} = A_{jk}$..
- Dot Product of two unit vectors e_i and e_j is equal to δ_{ij}. That is $e_i.e_j = \delta_{ij}$

Note that a shared repeated index represents tensor $contraction$.

Permutation Symbol (PS)

The permutation symbol is defined as

$$\varepsilon_{ijk} = \begin{cases} 1 & ijk \text{ form an cyclic permutation 123, for example: 312} \\ -1 & ijk \text{ form an non-cyclic permutation 123, for example: 321} \\ 0 & \text{otherwise, that is for example 122} \end{cases} \tag{2.3}$$

Here are some of the properties of PS

- PS is Anti-symmetric. That is $\varepsilon_{ijk} = -\varepsilon_{ikj} = \varepsilon_{kij}$
- Trace of PS: $\varepsilon_{iik} = \varepsilon_{11k} + \varepsilon_{22k} + \varepsilon_{33k} = 0$
- Contraction: $\varepsilon_{ijs}\,\varepsilon_{pqs} = \delta_{ip}\delta_{jq} - \delta_{iq}\delta_{jp}$
- Determinant of a matrix A_{ij} is represented in terms of PS as $det[A_{ij}] = \varepsilon_{ijk}\, A_{1i}\, A_{2j} A_{3k}$

- Cross Product of two unit vectors can be expressed in terms of PS as $\hat{e}_i \times \hat{e}_j = \varepsilon_{ijk}\, \hat{e}_k$. For example, $\hat{e}_1 \times \hat{e}_2 = \varepsilon_{12k}\hat{e}_k$.

We will now perform vector *dot product* and *cross product* in terms of tensor notations. Let us consider two vectors **A** and **B**, which are represented in tensor notation as $A_i\hat{e}_i$ and $B_j\hat{e}_j$, respectively. The dot and cross product of these two vectors in terms of tensorial notation is given by

$$\begin{aligned}
&\hat{A} \bullet \hat{B} = (A_i\hat{e}_i) \bullet (B_j\hat{e}_j) = A_i B_j(\hat{e}_i \bullet \hat{e}_j) = A_i B_j\delta_{ij} = A_i B_i \\
&= A_1B_1 + A_2B_2 + A_3B_3 \\
&\hat{A} \times \hat{B} = (A_i\hat{e}_i) \times (B_j\hat{e}_j) = A_i B_j\hat{e}_i \times \hat{e}_j = A_i B_j\varepsilon_{ijk}\hat{e}_k \\
&= (A_2B_3 - A_3B_2)\hat{e}_1 + (A_3B_1 - A_1B_3)\hat{e}_2 + (A_1B_2 - A_2B_1)\hat{e}_3
\end{aligned}$$

2.1.3 *Coordinate Transformation*

Le us consider two Cartesian coordinate systems, namely, (x_1, x_2, x_3) and $(x_1^{'}, x_2^{'}, x_3^{'})$. The former has unit vectors as $\hat{e}_1, \hat{e}_2, \hat{e}_3$ while the latter has $\hat{e}_1^{'}, \hat{e}_2^{'}, \hat{e}_3^{'}$ as their unit vectors. These two coordinate systems are shown in Fig. 2.1.

Since these two coordinate systems are individually orthogonal to each other, we have

$$\hat{e}_i \bullet \hat{e}_j = \delta_{ij}, \qquad \hat{e}_i^{'} \bullet \hat{e}_j^{'} = \delta_{ij}$$

Any vector $\hat{x}$ can be projected onto these two coordinate systems and the result can be written as

$$\hat{x} = \hat{x}_j^{'}\hat{e}_j^{'} = x_j\hat{e}_j$$

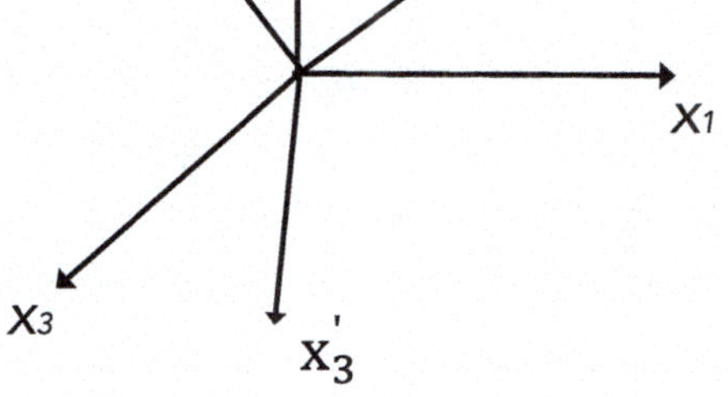

Fig. 2.1 Two different Cartesian coordinate systems with their orientation of unit vectors

Taking a dot product on the above equation with the unit vector $\hat{e}_i^{'}$, we get

$$\hat{x} \bullet \hat{e}_i^{'} = x_j^{'}\hat{e}_j^{'} \bullet \hat{e}_i^{'} = x_j\hat{e}_j \bullet \hat{e}_i^{'}$$

Using the properties of Kronecker delta, we can simplify the above equation as

$$x_j^{'}\delta_{ij} = x_j\hat{e}_j \bullet \hat{e}_i^{'} \qquad \text{or} \qquad x_i^{'} = x_j\hat{e}_j \bullet \hat{e}_i^{'}$$

From the above equation, we can infer that the dot product $\hat{e}_j \bullet \hat{e}_i^{'}$ is matrix of *direction cosines* between the two coordinate system normally designated as β_{ij}. That is

$$\beta_{ij} = \hat{e}_j \bullet \hat{e}_i^{'} \tag{2.4}$$

A similar procedure but taking the dot product with $\hat{e}_i$ leads to inverse relation. That is

$$x_i = \beta_{ji}x_j^{'} \tag{2.5}$$

From Eqs. (2.4) and (2.5), we can conclude that

$$\beta_{ij}\beta_{kj} = \delta_{ik}, \qquad \text{or} \qquad [\beta][\beta]^T = [I] \tag{2.6}$$

2.1.4 Properties of Second-Order Tensors

Second-order tensors are most prevalent in continuum/solid mechanics and they are used to represent stresses, and strains. Following paragraphs summarize some important properties of second-order tensors.

- **Symmetry and Anti-symmetry**:
  A tensor S_{ij} is said to be symmetric if $S_{ij} = S_{ji}$
  A tensor S_{ij} is said to be anti-symmetric if $S_{ij} = -S_{ji}$
- It follows that an antisymmetric tensor must have zeros on the diagonal.
- Any second-order tensor can be decomposed into sum of a *symmetric* and an *anti-symmetric* tensor as given below

$$\begin{aligned} B_{ij} &= \left(\frac{1}{2}B_{ij} + \frac{1}{2}B_{ij}\right) + \left(\frac{1}{2}B_{ij} - \frac{1}{2}B_{ij}\right) \\ &= \quad S_{ij} \quad + \quad A_{ij} \end{aligned}$$

- Contraction of a symmetric and an anti-symmetric tensor is zero. That is

$$S_{ij}A_{ij} = 0.$$

2.2 Description of Deformation

In this section, a brief description of how lengths, areas, and volumes of a solid changes during deformation. From this description of change in lengths, we derive the measure of strain. From the change in areas, we derive the measure of stress, and from the measure of change in volume, equations relating to conservation of mass is derived. While deformation gradient is shown here to be sufficient to describe the motion of a solid, description of strains is essential to derive the constitutive model of a solid. Hence, we provide different strain measures normally used in the analysis.

2.2.1 Description of Motion

Consider a body undergoing deformation to some applied loading (Fig. 2.2a). Let u_i^0 be its position at time $t = 0$ (undeformed) configuration and u_i^t is its position after some time t. In terms of unit vectors $\mathbf{e_i}$, they can be expressed as

$$\text{Undeformed position} \quad \mathbf{u^0} = \mathbf{u_i^0 e_i}$$
$$\text{Deformed position} \quad \mathbf{u} = \mathbf{u_i^t e_i}$$

Hence, the motion of the body can be expressed as

$$\mathbf{u} = \mathbf{u(x^0, y^0, z^0, t)} \quad \text{or} \quad \mathbf{u_i^0(x, y, z, t)} \tag{2.7}$$

The former represents the *Eulerian coordinates*, which is normally used to represent a fluid in motion. Here, the independent variables are the position vector $\mathbf{u_i}$ at a given instant and time. The latter represents *Lagrangian coordinates*, where quantities are expressed in terms of the initial position vector $\mathbf{u_i^0}$ and time. The difference between these two motion descriptions can be stated as:

- **Langragian**: If we put a rectangular grid on an undeformed body and visualize this grid after deformation, it will look like as shown in Fig. 2.3a
- **Eulerian**: If we put a rectangular grid on a deformed body and visualize in the undeformed state, it will look like as shown in Fig. 2.3b.

Here, the bold lowercase letters denote vectors, while bold capitals denote matrices.

Due to above definition of motion, the evaluation of the material derivative will differ in these two frames of reference. That is, in the Eulerian frame of reference, the derivative of $\mathbf{u(x^0, y^0, z^0, t)}$ is given by

$$\frac{d\mathbf{u}}{dt} = \frac{\partial \mathbf{u}}{\partial t} \tag{2.8}$$

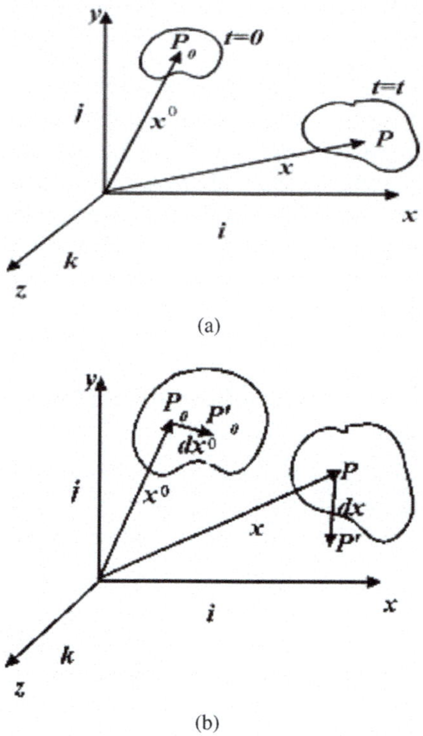

Fig. 2.2 **a** Undeformed and deformed configuration of a body **b** Deformation of neighbouring Points

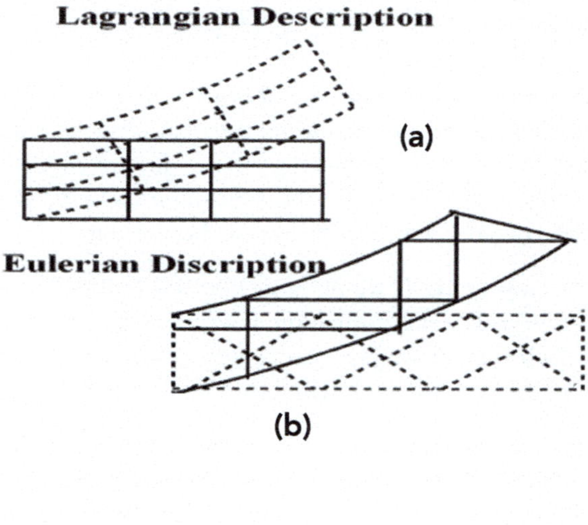

Fig. 2.3 Grids describing Lagrangian and Eulerian Motions

while the same in the Lagrangian frame, the derivative of $\mathbf{u^0(x, y, z, t)}$ is given by

$$\frac{d\mathbf{u}}{dt} = \frac{\partial \mathbf{u}}{\partial t} + \frac{\partial \mathbf{u}}{\partial x}\frac{dx}{dt} + \frac{\partial \mathbf{u}}{\partial y}\frac{dy}{dt} + \frac{\partial \mathbf{u}}{\partial z}\frac{dz}{dt} = \frac{\partial \mathbf{u}}{\partial t} + v_x\frac{\partial \mathbf{u}}{\partial x} + v_y\frac{\partial \mathbf{u}}{\partial y} + v_z\frac{\partial \mathbf{u}}{\partial z} \tag{2.9}$$

where v_x, v_y, and v_z are the convective velocity in the three material directions.

The motion of a particle is defined in terms of its coordinates attached to the particle. Displacement is for example is defined as the shortest distance travelled when a particle moves from one location to the other. If the position vectors of two points are $\mathbf{r_1}$ and $\mathbf{r_2}$, the displacement vector $\mathbf{u}$ is given by

$$\mathbf{u} = \mathbf{r_2} - \mathbf{r_1} = (\mathbf{x_2 i} + \mathbf{y_2 j} + \mathbf{z_2 k}) - (\mathbf{x_1 i} + \mathbf{y_1 j} + \mathbf{z_2 k})$$

In other words, the displacement vector can be written as

$$\mathbf{u} = (\mathbf{x_2} - \mathbf{x_1})\mathbf{i} + (\mathbf{y_2} - \mathbf{y_1})\mathbf{j} + (\mathbf{z_2} - \mathbf{z_1})\mathbf{k} \tag{2.10}$$

Deformation is defined as the comparison of two states, namely, the initial and the final configuration. The deformation gradient is a concept extensively used in the local theory of elasticity, which relates to the behaviour of the neighbouring particles. Consider points P_0 and $P_0^{'}$, shown in Fig. 2.2b, which at time $t = 0$ are at a distance $\mathbf{dr^0} = \mathbf{dx^0 i} + \mathbf{dy^0 j} + \mathbf{dz^0 k}$ (Fig. 2.2b). At time t, the two points move to locations P and $P^{'}$, so that the new distance is denoted $\mathbf{dr} = \mathbf{dxi} + \mathbf{dyj} + \mathbf{dzk}$. The location of $P^{'}$ with respect to P is given by $\mathbf{r} + \mathbf{dr} = (\mathbf{x} + \mathbf{dx})\mathbf{i} + (\mathbf{y} + \mathbf{dy})\mathbf{j} + (\mathbf{z} + \mathbf{dz})\mathbf{k}$.

Consider the x-direction vector, namely, $(x + dx)$. Expanding this term in Taylor series with respect to variables corresponding at time $t = 0$, gives

$$x + dx = x + \frac{\partial x}{\partial x^0}dx^0 + \frac{\partial x}{\partial y^0}dy^0 + \frac{\partial x}{\partial z^0}dz^0 \ldots\ldots\ldots \tag{2.11}$$

which gives the following relation:

$$dx = \frac{\partial x}{\partial x^0}dx^0 + \frac{\partial x}{\partial y^0}dy^0 + \frac{\partial x}{\partial z^0}dz^0$$

In similar manner, $y + dy$ and $z + dz$ can be expanded and simplified as

$$dy = \frac{\partial y}{\partial x^0}dx^0 + \frac{\partial y}{\partial y^0}dy^0 + \frac{\partial y}{\partial z^0}dz^0$$

$$dz = \frac{\partial z}{\partial x^0}dx^0 + \frac{\partial z}{\partial y^0}dy^0 + \frac{\partial z}{\partial z^0}dz^0$$

In tensorial notion, the above equations can be written as

$$\mathbf{dx_i^0} = \frac{\partial \mathbf{x_i^0}}{\partial \mathbf{x_j}} \mathbf{dx_j} \tag{2.12}$$

Similarly, the motion of particles at time $t = 0$ can be expressed in terms of the current time t as

$$\mathbf{dx_i} = \frac{\partial \mathbf{x_i}}{\partial \mathbf{x_j^0}} \mathbf{dx_j^0} \tag{2.13}$$

The quantities $\partial x_i/\partial x_j^0$ and $\partial x_i^0/\partial x_j$ are the *deformation gradients* and they form the basis of description of any deformation state. When expanded and written in the matrix form, Eqs. (2.12) and (2.13) become

$$\mathbf{dr} = \mathbf{F_0}\mathbf{dr^0}$$

$$\begin{Bmatrix} dx \\ dy \\ dz \end{Bmatrix} = \begin{bmatrix} \frac{\partial x}{\partial x^0} & \frac{\partial x}{\partial y^0} & \frac{\partial x}{\partial z^0} \\ \frac{\partial y}{\partial x^0} & \frac{\partial y}{\partial y^0} & \frac{\partial y}{\partial z^0} \\ \frac{\partial z}{\partial x^0} & \frac{\partial z}{\partial y^0} & \frac{\partial z}{\partial z^0} \end{bmatrix} \begin{Bmatrix} dx^0 \\ dy^0 \\ dz^0 \end{Bmatrix} \tag{2.14}$$

and

$$\mathbf{dr^0} = \mathbf{F}\mathbf{dr}$$

$$\begin{Bmatrix} dx^0 \\ dy^0 \\ dz^0 \end{Bmatrix} = \begin{bmatrix} \frac{\partial x^0}{\partial x} & \frac{\partial x^0}{\partial y} & \frac{\partial x^0}{\partial z} \\ \frac{\partial y^0}{\partial x} & \frac{\partial y^0}{\partial y} & \frac{\partial y^0}{\partial z} \\ \frac{\partial z^0}{\partial x} & \frac{\partial z^0}{\partial y} & \frac{\partial z^0}{\partial z} \end{bmatrix} \begin{Bmatrix} dx \\ dy \\ dz \end{Bmatrix} \tag{2.15}$$

The matrices $\mathbf{F}$ and $\mathbf{F_0}$ are the *Jacobian* matrices. Since the deformation is continuous, the determinant of the Jacobian matrices cannot be equal to zero. That is, a body with finite volume cannot be deformed to a body of zero volume or infinite volume, it is required that the following conditions are satisfied

$$0 < J_0 < \infty, \quad 0 < J < \infty \tag{2.16}$$

where $J = det[\mathbf{F}]$, $J_0 = det[\mathbf{F_0}]$. These conditions are very useful to verify that the deformation is physically possible and also allow derivation of the expression for the deformation of a *line*, an *area* or a *volume*.

2.2.2 Strain

Strain is a measure of relative displacement among particles within a body. There are three different measures of strain, which can be described on a specimen having initial and final length L_0 and L according to the following three definitions:

1. Engineering Strain:

$$\varepsilon = \frac{L - L_0}{L_0} = \frac{\Delta L}{L_0}$$

2. True Strain:

$$\varepsilon^t = \frac{\Delta L}{L} = \frac{\Delta L}{L_0 + \Delta L}$$

3. Logarithmic Strain:

$$\varepsilon^n = \int_{L_0}^{L} \frac{dl}{l} = \ln\left(\frac{L}{L_0}\right)$$

According to the above three definitions, the final Length L can be written in terms of all the strain measures, namely, the engineering strain:

$$L = L_0 + \Delta L = L_0 + L_0\varepsilon = L_0(1 + \varepsilon)$$

the true strain:

$$L = L_0 + \Delta L = L_0 + \frac{\varepsilon^t L_0}{(1 - \varepsilon^t)} = \frac{L_0}{(1 - \varepsilon^t)}$$

and the logarithmic strain:

$$L = L_0 \exp(\varepsilon^n)$$

The three strain measures are related by the following expressions

$$\varepsilon^t = \frac{\varepsilon}{1 + \varepsilon}, \qquad \varepsilon^n = \ln(1 + \varepsilon)$$

Strain measures are established by considering the distance change between two neighbouring material particles. Consider two material particles having coordinates (x^0, y^0, z^0) and $(x^0 + dx^0, y^0 + dy^0, z^0 + dz^0)$. After the motion, these particles will have the coordinates (x, y, z) and $(x + dx, y + dy, z + dz)$. The initial and final distances between these neighbouring particles are given by

$$(dr^0)^2 = (dx^0)^2 + (dy^0)^2 + (dz^0)^2 \tag{2.17}$$

$$dr^2 = (dx)^2 + (dy)^2 + (dz)^2 \tag{2.18}$$

Substituting Eq. (2.17) in Eq. (2.18) we get:

$$dr^2 = \mathbf{dr}^{\mathbf{t}}\mathbf{dr} = \mathbf{dr}^{\mathbf{0}^{\mathbf{T}}}\mathbf{F}_{\mathbf{0}}^{\mathbf{T}}\mathbf{F}_{\mathbf{0}}\mathbf{dr}^{\mathbf{0}} \tag{2.19}$$

After deformation, dr^2 is different from $(dr^0)^2$, hence we have

$$\begin{aligned} dr^2 - dr_0^2 &= \mathbf{dr}^{\mathbf{0}^{\mathbf{T}}}\mathbf{F}_{\mathbf{0}}^{\mathbf{T}}\mathbf{F}_{\mathbf{0}}\mathbf{dr}_{\mathbf{0}} - \mathbf{dr}^{\mathbf{0}^{\mathbf{T}}}\mathbf{dr}^{\mathbf{0}} \\ &= 2\mathbf{dr}^{\mathbf{0}^{\mathbf{T}}}\mathbf{E}_{\mathbf{0}}\mathbf{dr}^{\mathbf{0}} \end{aligned} \tag{2.20}$$

where $\mathbf{E_0} = \mathbf{F}_0^{\mathbf{T}}\mathbf{F_0} - \mathbf{I}$. The above measure gives the relative displacements between the two material particles, which are insensitive to rotations.

If the Eulerian frame of reference is used, then the relative displacement is given by

$$\begin{aligned} dr^2 - (dr^0)^2 &= \mathbf{dr}^{\mathbf{T}}\mathbf{dr} - \mathbf{dr}^{\mathbf{T}}\mathbf{F}^{\mathbf{T}}\mathbf{F}\mathbf{dr} \\ &= 2\mathbf{dr}^{\mathbf{T}}\mathbf{E}\mathbf{dr} \end{aligned} \tag{2.21}$$

where $\mathbf{E} = \mathbf{I} - \mathbf{F}^{\mathbf{T}}\mathbf{F}$. In Eqs. (2.18) and (2.19), the matrices $\mathbf{E_0}$ and $\mathbf{E}$ are the Lagrangian and Eulerian strain tensors. In tensorial form, they are given by

$$E_{0_{ij}} = \frac{1}{2}\left(\frac{\partial x_m}{\partial x_i^0}\frac{\partial x_m}{\partial x_j^0} - \delta_{ij}\right) \tag{2.22}$$

$$E_{ij} = \frac{1}{2}\left(\delta_{ij} - \frac{\partial x_m^0}{\partial x_i}\frac{\partial x_m^0}{\partial x_j}\right) \tag{2.23}$$

Let us now establish the physical significance of $E_{0_{ij}}$ and E_{ij}. For this we will consider a line element of length $dr^0 = dx^0$. The extension of the line element per unit length E_{0_1} is given by

$$E_{0_1} = \frac{dr - dr^0}{dr^0} \quad \text{or} \quad dr = (1 + E_{0_1})dr^0 \tag{2.24}$$

$$dr^2 - dr^{0^2} = 2E_{0_{11}}dr^{0^2} \tag{2.25}$$

Combining the above, we can establish the relationship between E_{0_1} and $E_{0_{11}}$ as

$$E_{0_{11}} = E_{0_1} + \frac{1}{2}E_{0_1}^2 \quad \text{or} \quad E_{0_1} = \sqrt{1 + 2E_{0_{11}}} - 1 \tag{2.26}$$

Expanding the right hand term by binomial expansion, we get

$$E = (1 + E_{0_{11}} - \frac{1}{2}E_{0_{11}}^2 + ...) - 1 \approx E_{0_{11}} - \frac{1}{2}E_{0_{11}}^2 \tag{2.27}$$

For small $E_{0_{11}}$, $E_{0_1} = E_{0_{11}}$, so that $E_{0_{11}}$ can be interpreted as an elongation per unit length only when the extension is very small. Similarly, we can write

$$E_{0_2} = \sqrt{1 + 2E_{0_{22}}} - 1, \qquad E_{0_3} = \sqrt{1 + 2E_{0_{33}}} - 1 \tag{2.28}$$

2.2.3 *Deformation in Terms of Displacement*

Next, we will establish the relations that deformation has with displacements. For establishing this, we use the following relations:

$$x = x^0 + u, \qquad y = y^0 + v, \qquad z = z^0 + w \tag{2.29}$$

Using these relations, we can write the derivatives as

$$\begin{aligned}
\frac{\partial x}{\partial x^0} &= 1 + \frac{\partial u}{\partial x^0}, & \frac{\partial y}{\partial x^0} &= \frac{\partial v}{\partial x^0}, & \frac{\partial z}{\partial x^0} &= \frac{\partial w}{\partial x^0} \\
\frac{\partial x}{\partial y^0} &= \frac{\partial u}{\partial y^0}, & \frac{\partial y}{\partial y^0} &= 1 + \frac{\partial v}{\partial y^0}, & \frac{\partial z}{\partial y^0} &= \frac{\partial w}{\partial y^0} \\
\frac{\partial x}{\partial z^0} &= \frac{\partial u}{\partial z^0}, & \frac{\partial y}{\partial z^0} &= \frac{\partial v}{\partial z^0}, & \frac{\partial z}{\partial z^0} &= 1 + \frac{\partial w}{\partial z^0}
\end{aligned}$$

In tensorial form, we can write the above equations as

$$\frac{\partial x_m}{\partial x_i^0} = \frac{\partial u_m}{\partial x_i^0} + \delta_{im} \tag{2.30}$$

Similarly, one can write

$$\frac{\partial x_m^0}{\partial x_i} = \delta_{im} - \frac{\partial u_m}{\partial x_i} \tag{2.31}$$

where, δ_{ij} is the kronecker delta. Substituting Eqs. (2.30) and (2.31) in the Lagrangian and Eulerian strain tensors (Eq. (2.23)), we get after some simplification

$$E_{0_{ij}} = \frac{1}{2}\left[\frac{\partial u_i}{\partial x_j^0} + \frac{\partial u_j}{\partial x_i^0} + \frac{\partial u_m}{\partial x_i^0}\frac{\partial u_m}{\partial x_j^0}\right]$$
$$E_{ij} = \frac{1}{2}\left[\frac{\partial u_i}{\partial x_j} + \frac{\partial u_j}{\partial x_i} + \frac{\partial u_m}{\partial x_i}\frac{\partial u_m}{\partial x_j}\right] \tag{2.32}$$

The first two terms in the above two equations represent the linear part of the strain tensors, while the last term represents the nonlinear part. Both these tensors are symmetric. When the displacement gradients are very small, we can neglect the nonlinear part of in above tensors. Thus, infinitesimal strain components have direct interpretations as extensions or change of angles. Further, the magnitudes of the strains are very small compared to unity, which means that deformations are very small. Hence, we can conclude for very small deformation,

$$E_{0_{ij}} = E_{ij} = \varepsilon_{ij}$$

where ε_{ij} represents the linear part of the Lagrangian or Eulerian strain tensors, which is obtained by expanding the first part of any one of Eq. (2.32) and is given by

$$\varepsilon_{11} = \frac{\partial u}{\partial x^0}, \quad \varepsilon_{12} = \frac{1}{2}\left[\frac{\partial u}{\partial y^0} + \frac{\partial v}{\partial x^0}\right], \quad \varepsilon_{13} = \frac{1}{2}\left[\frac{\partial u}{\partial z^0} + \frac{\partial w}{\partial x^0}\right]$$
$$\varepsilon_{22} = \frac{\partial v}{\partial y^0}, \quad \varepsilon_{23} = \frac{1}{2}\left[\frac{\partial v}{\partial z^0} + \frac{\partial w}{\partial y^0}\right], \quad \varepsilon_{33} = \frac{\partial w}{\partial z^0}$$

In addition, for small deformation, following condition is normally true. That is,

$$\frac{u_i}{L} << 1$$

where, L is the smallest dimension of the body under consideration. If the above condition is true, then we can conclude that $x_i^0 = x_i$. That is, we do not differentiate between Eulerian and Lagrangian coordinates. Hence, the functional form of displacement and their components become identical in these two frames of reference.

This book deals with behaviour of materials under high strain rate loading. Most of the analysis presented in this book deals only with highly non-linear and large deformation behaviour and hence we will be using extensively the equations given by Eq. (2.32).

2.3 Stress

When a body/continuum is subjected to forces, the body deforms and hence causing strains. This action takes place due to contact forces or moments, which are normally referred to as surface tractions as they occur on the surface of the body/continuum.

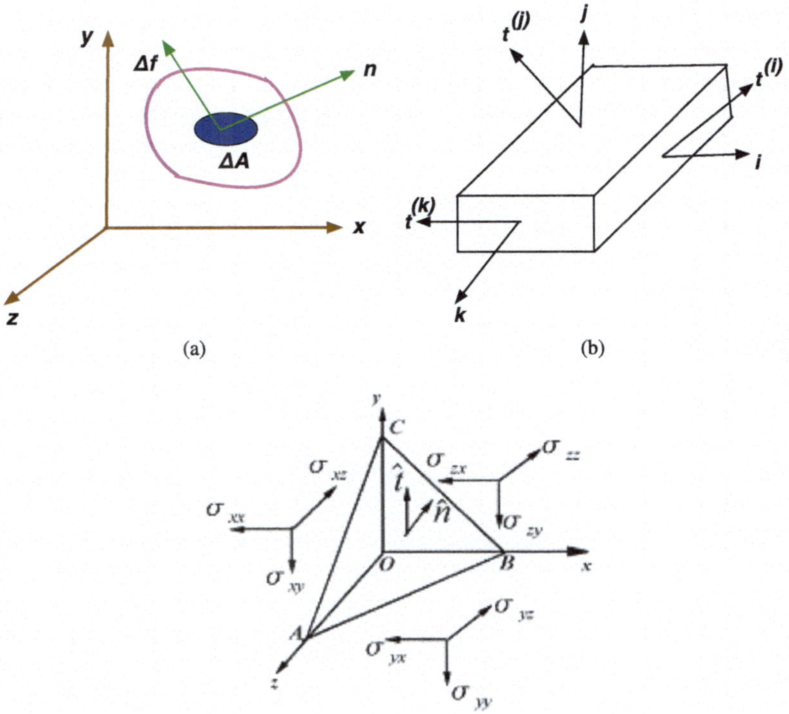

(a) (b)

(c)

Fig. 2.4 **a** A section of an arbitrary continuum **b** Stress Cube with stress vectors and outward unit normal **c** Stresses acting on an arbitrary surface of a tetrahedron

Depending upon the way these forces act, they can be classified as *extrinsic, mutual, or contact*. Extrinsic forces, such as gravity or electric or magnetic forces, always act outside the body. The most important type of force that is relevant to this book is the contact type force. These contact forces generate stress or pressure, which are the source of failure in a material.

To understand the concept of stress, let us consider a body shown in Fig. 2.4a. After the application of contact forces, a small surface element of area ΔA deforms and hence forces are generated on this small surface element such that they cancel each other or in other words, the small surface element is said to be in equilibrium. These generated contact forces act within the body. If we drop a unit vector perpendicular to this small surface element and designate it as $\mathbf{n}$ and if the resultant force acting on this element be $\Delta\mathbf{f}$, then in the limiting case of ΔA tending to zero, one can define the traction as follows:

$$\mathbf{t}^{(\mathbf{n})} = \frac{\mathbf{df}}{\mathbf{dA}} = \lim_{\Delta A \to 0} \frac{\Delta \mathbf{f}}{\Delta \mathbf{A}} \tag{2.33}$$

The above limit is valid only when the small element on which the forces are acting, is continuous. This traction vector is defined as force acting on a unit area on the surface. The traction vector having superscript (n) is dependent on the orientation of the area. To understand the meaning of traction vector, let us consider the elementary cuboid shown in Fig. 2.4b showing the components of the traction vector on the three faces of the cube.

Considering $\mathbf{n} = \mathbf{i}$, we have,

$$\mathbf{t^{(i)}} = \mathbf{t_x^{(i)}i} + \mathbf{t_y^{(i)}j} + \mathbf{t_z^{(i)}k}$$

Similarly for $\mathbf{n} = \mathbf{j}$ and $\mathbf{n} = \mathbf{k}$:

$$\mathbf{t^{(j)}} = \mathbf{t_x^{(j)}i} + \mathbf{t_y^{(j)}j} + \mathbf{t_z^{(j)}k}$$

$$\mathbf{t^{(k)}} = \mathbf{t_x^{(k)}i} + \mathbf{t_y^{(k)}j} + \mathbf{t_z^{(k)}k}$$

The above definition can be better understood by introducing a second-order tensor $\sigma_{ij} = t_j^{(i)}$, where the subscripts i and j describes the direction of traction and its plane of application. That is, in the $x - z$ plane, $\sigma_{xx} = t_x^{(i)}$, $\sigma_{xz} = t_z^{(i)}$, $\sigma_{zx} = t_x^{(k)}$. Hence, normal stresses σ_{xx}, σ_{yy} and σ_{zz}, are the projections of the traction vector $\mathbf{t^{(n)}}$ on the faces, while the projections perpendicular to outward normal $\mathbf{n}$ give the shear stress components $\sigma_{xy}, \sigma_{yz}, \sigma_{zx}, \sigma_{yx}, \sigma_{zy}$ and σ_{xz}. The σ_{ij} matrix containing all the 9 stress components is called the *Cauchy's Stress Tensor*, and this matrix is symmetric, that is. $\sigma_{ij} = \sigma_{ji}$.

It is next necessary to establish how traction vector $\mathbf{t}$ is related to outward normal $\mathbf{n}$. To derive this relation, we consider one of the surfaces of a tetrahedron shown in Fig. 2.4c. On the face perpendicular to the considered surface, the components of the three stress vectors are denoted by the corresponding stress components σ_{ij} on the plane on which they are acting. For example, the stresses acting on the face normal to x-axis are denoted as σxx, σxy and σxz.

The equilibrium of the tetrahedron requires that the resultant force acting on it vanishes. Imposing equilibrium along the x-direction gives

$$t_x dA - \sigma_{xx} dA_x - \sigma_{yx} dA_y - \sigma_{zx} dA_z + b_x \rho dV = 0 \tag{2.34}$$

where b_x is the x-component of the body force vector $\mathbf{b}$. Here t_x is the x-component of the traction vector, dA_x, dA_y, and dA_z are the areas of the face perpendicular to the coordinate axes x, y, and z axes, while dA is the area of the inclined surface ABC. Also, $dV = (1/3)hdA$ is the volume of the tetrahedron, where h is the smallest distance from any point to the inclined surface ABC. The outward normal vector can be written in terms of unit vectors along the three reference directions as $\mathbf{n} = \mathbf{n_x i} + \mathbf{n_y j} + \mathbf{n_z k}$. The elemental areas dA_i can now be written in terms of components of unit normal vector as

$$dA_x = n_x dA, \ dA_y = n_y dA, \ dA_z = n_z dA \tag{2.35}$$

Substituting Eq. (2.34) in Eq. (2.35) and letting $\mathbf{dA} \rightarrow \mathbf{0}$, we get the traction–stress relation in the x-direction as

$$t_x = \sigma_{xx} n_x + \sigma_{yx} n_y + \sigma_{zx} n_z \tag{2.36}$$

The above equations can be generalized and written in tensorial form as

$$t_i = \sigma_{ij} n_j \tag{2.37}$$

which is valid for any outward normal vector and in any coordinate system. Hence, it can be concluded that the state of stress in a body is completely known if the stress tensor σ_{ij} is given. In other words, given any surface and associated unit normal vector **n**, it is possible to determine a traction vector acting on that surface if the stress tensor is known.

2.3.1 *Properties of Stress Tensor*

Cauchy's stress tensor is a tensor of order two and it is symmetric. It has some unique properties, which are outlined below.

2.3.1.1 Principal Stresses

It is established in the previous section that the stress vector acting on a surface depends on the direction **n** and in general they are not parallel to **n** . Now we will attempt to find **n** such that the stress vector is acting in the direction of **n**. That is,

$$\mathbf{t_i} = \lambda \mathbf{n_i} \tag{2.38}$$

where λ is a scalar representing the magnitude to the stress vector. The direction $\mathbf{n_i}$ that satisfies the above condition is the principal direction and λ is called the principal stress. Substituting Eq. (2.37) in Eq. (2.38), we get,

$$\sigma_{\mathbf{ij}} \mathbf{n_j} = \lambda \mathbf{n_i}$$

The equation above can be written in matrix form as

$$[\sigma - \lambda \mathbf{I}]\,\mathbf{n} = \mathbf{0} \tag{2.39}$$

In other words, the above problem reduces to a standard eigenvalue problem of size 3×3 with eigenvectors giving the principal directions n_1, n_2, n_3 and the eigenvalues giving the principal stresses $\lambda_1, \lambda_2, \lambda_3$. This problem can be solved by equating the determinant of the matrix $[\sigma - \lambda \mathbf{I}]$ to zero. When the determinant of the above matrix is expanded, we get the following cubic polynomial equation in λ

$$\lambda^3 - I_1\lambda^2 + I_2\lambda - I_3 = 0 \tag{2.40}$$

where, I_1, I_2, I_3, are called *Stress Invariants*. They are called invariants as their numerical value does not change irrespective of the type of coordinate system adopted to solve the problem. They will find its use also in developing constitutive models for hyperelastic materials to be presented later in this chapter. They can be written both in terms of components of stress tensor σ_{ij} or the principal stress components λ_i as

$$\begin{aligned} I_1 &= \sigma_{xx} + \sigma_{yy} + \sigma_{zz} = \sigma_{ii} = \lambda_1 + \lambda_2 + \lambda_3 \\ I_2 &= \frac{1}{2}\left(I_1^2 - \sigma_{ij}\sigma_{ij}\right) = \lambda_1\lambda_2 + \lambda_2\lambda_3 + \lambda_3\lambda_1 \\ I_3 &= \frac{1}{3}\left(3I_1I_2 - I_1^3 + \sigma_{ij}\sigma_{jk}\sigma_{kl}\right) = \lambda_1\lambda_2\lambda_3 \end{aligned} \tag{2.41}$$

Also, the principal directions will have to satisfy the relation

$$n_1^2 + n_2^2 + n_3^2 = 1 \tag{2.42}$$

From the theorem relating to eigenvalue analysis, it can be easily shown that the eigenvalues are always real since the stress tensor is symmetric and hence the principal directions are mutually orthogonal.

2.3.1.2 Normal Stress

To obtain normal stresses from the Cauchy's stress tensor, we define the normal stress σ_n as the component of the stress vector in the direction of unit normal $\mathbf{n}$ to the surface of interest. This is obtained by taking the dot product of the stress vector and the normal vector. That is,

$$\sigma_n = t_i n_i = \sigma_{ij} n_i n_j$$

If the coordinate axes are chosen such that they are aligned in the direction of principal stresses, then the Cauchy's stress tensor becomes diagonal with

$$\sigma_{xx} = \lambda_1, \qquad \sigma yy = \lambda_2, \qquad \sigma_{zz} = \lambda_3$$

Hence, the normal stress can be evaluated from the expression

$$\sigma_n = \lambda_1 n_1^2 + \lambda_2 n_2^2 + \lambda_3 n_3^2 \tag{2.43}$$

If $\lambda_1 > \lambda_2 > \lambda_3$, then by using Eq. (2.41) in Eq. (2.72), we can conclude that, from which we get that λ_1 and λ_3 are the maximum and minimum normal stresses, respectively, at the point under consideration.

2.3.1.3 Shear Stress

The shear stress denoted by τ is the projection of the stress vector onto a surface. That is

$$\mathbf{t} = \sigma_n \mathbf{n} + \tau \qquad \text{or} \qquad \tau = \mathbf{t} - \sigma_\mathbf{n} \mathbf{n}$$

We will denote the magnitude of shear stress as τ_{mag} and it is given by

$$\tau_{mag}^2 = |\mathbf{t}|^2 - \sigma_n^2 = t_i t_i - \sigma_n^2 \tag{2.44}$$

where $|\mathbf{t}|$ is the magnitude of the stress vector. Let the principal directions be oriented in the 1, 2, and 3 directions, then we can write the tractions as

$$t_1 = \sigma_1 n_j = \sigma_1 n_1, \;\; t_2 = \sigma_2 n_j = \sigma_2 n_2, \;\; t_3 = \sigma_3 n_j = \sigma_3 n_3$$

Expanding $|\mathbf{t}|^2$, we get

$$|\mathbf{t}|^2 = t_1^2 + t_2^2 + t_3^2 = (\sigma_1 n_1)^2 + (\sigma_2 n_2)^2 + (\sigma_3 n_3)^2$$

Using the above relation, we can write the shear stress magnitude as

$$\begin{aligned} \tau_{mag}^2 &= (\sigma_1 n_1)^2 + (\sigma_2 n_2)^2 + (\sigma_3 n_3)^2 - (\sigma_1 n_1^2 + \sigma_2 n_2^2 + \sigma_3 n_3^2)^2 \\ &= n_1^2(1 - n_1^2)\sigma_1^2 + n_2^2(1 - n_2^2)\sigma_2^2 + n_3^2(1 - n_3^2)\sigma_1^3 - 2\sigma_1\sigma_2 n_1^2 n_2^2 \\ &= -2\sigma_2\sigma_3 n_2^2 n_3^2 - 2\sigma_1\sigma_3 n_1^2 n_3^2 \end{aligned} \tag{2.45}$$

Using Eq. (2.71) in Eq. (2.74) and simplifying, we can write shear stress magnitude as

$$\tau_{mag}^2 = n_1^2 n_2^2 (\sigma_1 - \sigma_2)^2 + n_2^2 n_3^2 (\sigma_2 - \sigma_3)^2 + n_3^2 n_1^2 (\sigma_3 - \sigma_1)^2 \tag{2.46}$$

Equation (2.75) suggests that $\tau_{mag} = 0$ on surfaces with

- $n_1 = 1$ with $n_2 = n_3 = 0$
- $n_2 = 1$ with $n_3 = n_1 = 0$
- $n_3 = 1$ with $n_1 = n_2 = 0$.

On these surfaces, the normal stress σ_n reaches maximum value or the minimum value.

Let us now consider x_2 surface where $n_2 = 0$ with $n_1 = n_3 \neq 0$. Shear stress on this surface using Eq. (2.75) becomes

$$\tau_{mag}^2 = n_3^2 n_1^2 (\sigma_3 - \sigma_1)^2 = (1 - n_1^2) n_1^2 (\sigma_3 - \sigma_1)^2 \tag{2.47}$$

On this x_2 surface, we have $n_1^2 + n_3^2 = 1$, and the extreme value of τ_{mag} occurs at

$$\frac{\partial(\tau_{mag}^2)}{\partial n_1} = 0 = (2n_1 - 4n_1^3)(\sigma_3 - \sigma_1)^2 \tag{2.48}$$

Solving the above equation, we get

$$n_1 = \pm\frac{1}{\sqrt{2}} \quad \text{and} \quad n_3 = \pm\frac{1}{\sqrt{2}}$$

That is, the maximum shear stress occurs on the surfaces bisecting the angle between x_1 and x_3 axes. The corresponding value of shear stress is

$$\tau_{max}^2 = \frac{1}{4}(\sigma_3 - \sigma_1)^2$$

2.3.2 *Lagrangian and Kirchhoff Stress Tensors*

In analysing a continuum undergoing finite deformation, it is always advantageous to use Lagrangian variables. In reality, true loading exists only in the deformed state and hence the corresponding loading and stresses in the body at the undeformed state are considered fictitious and are introduced for mathematical convenience. To understand the motivation in introducing the new definition of stress, it is necessary to keep in mind the following:

- The traction vector is defined in terms of force divided by the area
- The stress vector will be defined according to a transformation relation for the traction and the area normal.

Consider a surface of a body shown in Fig. 2.5, which is in undeformed configuration.

On this surface, the stress vector **t** acting on an area **dA** must be defined. The introduction of such a stress vector is arbitrary and two usual definitions will be discussed. To appreciate these two definitions, we will consider the Cauchy's stress first.

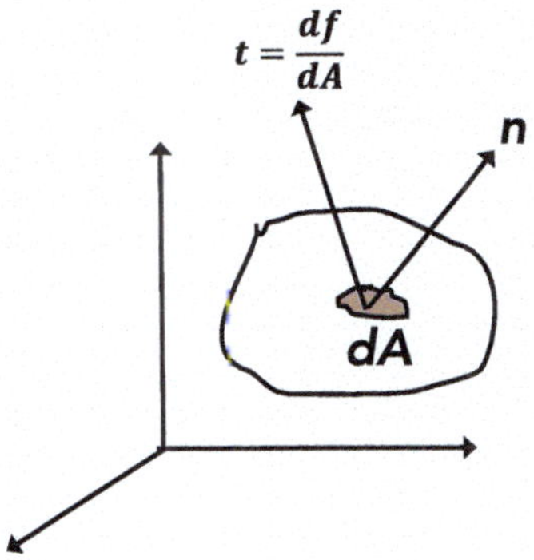

Fig. 2.5 Stress vector definition of a body in undeformed state

2.3.3 *Cauchy Stress Tensor*

In this stress measure, in the deformed state on every plane passing through a point, there is a stress vector t_i defined in terms of deformed surface area. That is, if **t** is the stress vector, and the total resultant force acting on a surface area **dA** be **df**, then

$$t_i = \frac{df_i}{\mathbf{dA}}$$

Here, the stress vectors and unit normals in the deformed body form two respective vector spaces. Then the Cauchy stress tensor σ_{ij} can be defined as the transformation between these two vector spaces. That is,

$$t_i = \sigma_{ij} n_j \tag{2.49}$$

Cauchy stress is purely a mathematical quantity. However, on special plane surfaces wherein the unit normals are in the direction of respective unit vectors, namely, $\mathbf{e}_1, \mathbf{e}_2, \mathbf{e}_3$, the nine components of σ_{ij} can be related to the stress vectors and hence can have physical meaning. Hence, the meaning of σ_{ij} are the components of stress derived from the force vector df_i divided by the deformed area. That is, the stress components are referred to the deformed area. This in elementary terms is called the *true stress*.

2.3.4 *Lagrangian Stress Tensor*

In this stress measure, unlike in the Cauchy stress measure, we will develop a stress measure that uses both undeformed and deformed state in developing the stress tensor. Here we consider a body whose undeformed state with area dA^0 and deformed state area dA is shown in the Fig. 2.6.

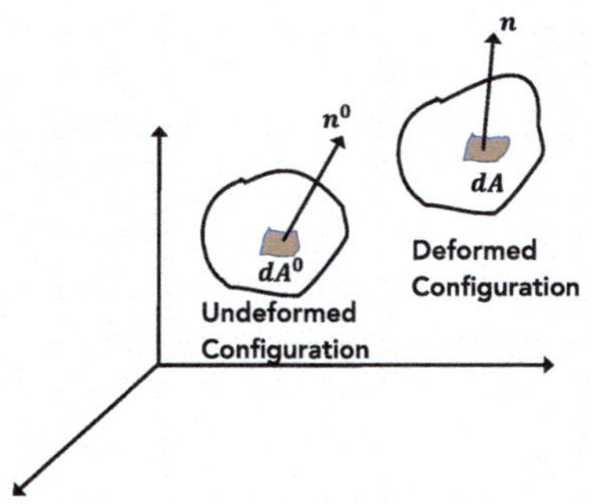

Fig. 2.6 The undeformed state with area dA^0 and Deformed state with area dA of a body

Let $\mathbf{df^0}$ be the resultant force corresponding to undeformed configuration and let this force be identical $\mathbf{df}$, the resultant force in the deformed configuration, then the Lagrangian stress vector t_i^0 is defined as

$$t_i^0 \equiv \frac{df_i^0}{dA^0} \tag{2.50}$$

The Lagrangian stress tensor σ_{ij}^L is defined as a the second order tensor, which relates the two vector spaces t_i^0 and n_i^0 in a manner as given below

$$t_i^0 = \sigma_{ij}^L n_j^0$$

The meaning of the stress tensor σ_{ij}^L are the components of stress derived from the resultant force df_i divided by the original area dA_0. In elementary terms, this definition of stress is called the *Engineering Stress*.

2.3.5 *Kirchhoff Stress Tensor*

As before let $\mathbf{df^0}$ be the resultant force in the undeformed configuration and it is related to the resultant force in the deformed configuration using a transformation that was used to transform line segments and it is given by

$$df_i^0 = \frac{\partial x_i^0}{\partial x_j} df_j \tag{2.51}$$

The Kirchhoff stress vector is defined as

$$t_i^0 \equiv \frac{df_i^0}{dA_0}$$

This leads to the definition of Kirchhoff stress σ_{ij}^K as

$$t_i^0 = \sigma_{ji}^K n_j^0$$

This relation looks very similar to the definition of the Lagrangian stress. However, in this case, the components of force are quite different.

Next, the meaning of the components of the Kirchhoff stress tensor σ_{ij}^K needs to be established. It is nothing but the stress components derived from the transformed components of the force vector, divided by the original area. All components of the stresses are referred to the original area.

2.3.6 *Relations Among the Stress Tensors*

From the definition of Lagrangian stress vector, we have

$$df_i = t_i^0 dA_0 = t_i dA$$

Using the traction–stress tensor relations for Lagrangian and Cauchy stress (Eqs. (2.49) and (2.50)), we can write the above equation as

$$\sigma_{ji}^L n_j^0 dA_0 = \sigma_{pi} n_p dA$$

Area transforms similar to lines discussed earlier in this chapter. Hence, we can write

$$n_p dA = J^0 \frac{\partial x_j^0}{\partial x_p} n_j^0 dA_0$$

Substituting this in Eqs. (2.49) and (2.50), we get

$$\sigma_{ji}^L n_j^0 dA_0 = J^0 \sigma_{pi} \frac{\partial x_j^0}{\partial x_p} n_j^0 dA_0$$

Hence, we can conclude

$$\sigma_{ji}^L = J^0 \frac{\partial x_j^0}{\partial x_p} \sigma_{pi} \tag{2.52}$$

We have $J^0 = \rho_0/\rho$, and hence the relation between Lagrangian and Cauchy stress tensors can be written as

$$\sigma_{ji}^L = \frac{\rho_0}{\rho} \frac{\partial x_j^0}{\partial x_p} \sigma_{pi} = \frac{\rho_0}{\rho} \left(\delta_{pj} - \frac{\partial u_j}{\partial x_p} \right) \sigma_{pi} \tag{2.53}$$

It is to be noted that *Lagrangian stress tensor* σ_{ij}^{L} *is not a symmetric tensor*. The inverse relation that relates Cauchy stress and Lagrangian stress can be written as

$$\sigma_{ji} = \frac{\rho}{\rho_0} = \frac{\partial x_j}{\partial x_p^0}\sigma_{pi}^{L} = \frac{\rho}{\rho_0}\left(\delta_{pj} + \frac{\partial u_j}{\partial x_p^0}\right)\sigma_{pi}^{L} \quad (2.54)$$

The relation between Kirchhoff stress tensor and Cauchy stress tensor can be derived in a similar way. The forward and inverse relation between these two stress tensor is given by

$$\sigma_{ji}^{K} = \frac{\rho_0}{\rho}\frac{\partial x_i^0}{\partial x_m}\frac{\partial x_j^0}{\partial x_n}\sigma_{mn}, \quad \sigma_{ji} = \frac{\rho}{\rho_0}\frac{\partial x_i}{\partial x_m^0}\frac{\partial x_j}{\partial x_n^0}\sigma_{mn}^{K} \quad (2.55)$$

It is to be noted that Kirchhoff stress tensor is a symmetric tensor. Since strain tensors are always symmetric, it is more convenient to use Kirchhoff stress tensor in formulating constitutive laws.

2.4 Governing Equations of Motion

The form of governing equations of motion of a 3D solid depends on the type of stress tensor (Cauchy, Lagrangian or Kirchhoff) used in the formulation. We will first derive the governing equation assuming that the stress tensor is Cauchy. Once this form is obtained, we will use the relations between Cauchy and Lagrangian stress tensor or Cauchy and Kirchhoff stress tensor and obtain the governing equations based on other two stress tensors. Here, we will use Principle of Virtual Work to derive the equations of motion.

Assuming the stress system on a 3D sold is defined by Cauchy stress tensor, we will use Newton's second law of motion for not only deriving the governing equation but also establishing the symmetry of the Cauchy's stress tensor.

In order to derive the equations of equilibrium, we will go back to Fig. 2.7, where a body of volume V, having density ρ and are subjected to body force $b_i = b_x\mathbf{i} + b_y\mathbf{j} + b_z\mathbf{k}$ and surface tractions $\mathbf{t} = t_x\mathbf{i} + t_y\mathbf{j} + t_z\mathbf{k}$. Here, $i,\ j,\ k$ are the unit vector in the three coordinate directions. Applying Newton's second law of motion to the elemental volume shown in Fig. 2.7 gives:

$$\int_A \mathbf{tdA} + \int_V \rho\,\mathbf{bdV} = \int_V \rho\ddot{u}dV \quad (2.56)$$

$$\int_A \mathbf{x} \times \mathbf{tdA} + \int_V \mathbf{x} \times \rho\mathbf{bdV} = \int_V \mathbf{x} \times \rho\ddot{u}\mathbf{dV} \quad (2.57)$$

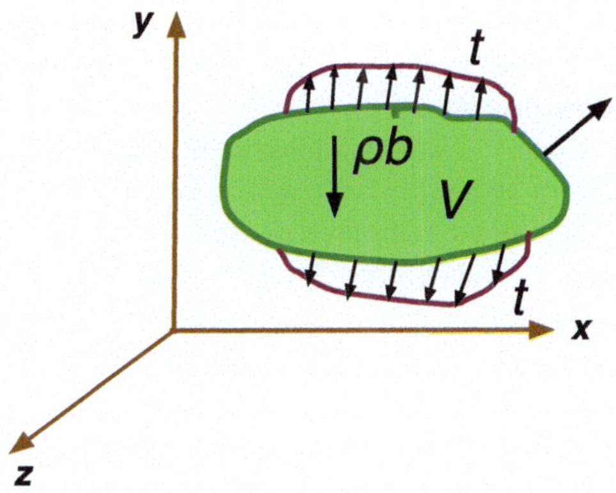

Fig. 2.7 Arbitrary small volume under the action of forces

These two equations in tensorial notation can be written as

$$\begin{aligned} &\int_A t_i dA + \int_V \rho b_i dV = \int_V \rho \ddot{u}_i dV \\ &\int_A \epsilon_{ijk} x_j t_k dA + \int_V \epsilon_{ijk} x_j b_k \rho dV = \int_V \epsilon_{ijk} x_j \rho \ddot{u}_k dV \end{aligned} \tag{2.58}$$

Using Eq. (2.49) in Eq. (2.58), and using the Divergence theorem [4], we get

$$\int_A t_i dA = \int_A \sigma_{pi} n_p dA = \int_V \frac{\partial \sigma_{pi}}{\partial x_p} dV$$

which can be simplified further as

$$\begin{aligned} &\int_V \left[\tfrac{\partial \sigma_{pi}}{\partial x_p} + \rho b_i - \rho \ddot{u}_i \right] dV \\ &\int_V \left[\tfrac{\partial}{\partial x_k} (\epsilon_{ijk} x_j \sigma_{pk}) + \rho \epsilon_{ijk} x_j b_k - \rho \epsilon_{ijk} x_j \ddot{u}_k \right] dV \end{aligned} \tag{2.59}$$

The first term in the second equation can be written as

$$\frac{\partial}{\partial x_p}(x_j \sigma_{pk}) = \sigma_{jk} + x_j \frac{\partial \sigma_{pk}}{\partial x_p}$$

Using the above equation in Eq. (2.59), we can get the governing equilibrium equation as

$$\frac{\partial \sigma_{pi}}{\partial x_p} + \rho b_i = \rho \ddot{u}, \qquad with \qquad \varepsilon_{ijk} \sigma_{jk} = 0 \tag{2.60}$$

While the first equation gives the governing differential equation of a continuum in terms of Cauchy stresses, the second equation states that Cauchy's stress tensor σ_{ij} is symmetric, that is $\sigma_{ij} = \sigma_{ji}$.

In the next few paragraphs, we will see how the governing equation given in terms of Cauchy stress (Eq. 2.60) will change if we use Lagrange or Kirchhoff stress tensors. In doing so, we will have to express the body force b, the area A, density ρ, and volume V in undeformed and deformed state. Hence, all quantities with superscript "0" represent quantities in the undeformed state. Let us first define body force per unit mass in the undeformed state as b^0 such that

$$b_i \rho_0 dV_0 = b_i \rho dV$$

From the mass conservation law, we have $\rho dV = \rho_0 dV_0$ Using this in the above equation, we have

$$b_i^0 = b_i$$

The body–force relation also refers to the inertial forces. We will now consider the governing equation based on Cauchy stress tensor derived earlier and given in Eq. (2.60). We can write the first term of the governing equation as

$$\frac{\partial \sigma_{ji}}{\partial x_j} = \frac{\partial \sigma_{ji}}{\partial x_p^0}\frac{\partial x_p^0}{\partial x_j} = \frac{\partial}{\partial x_p^0}\left(\sigma_{ji}\frac{\partial x_p^0}{\partial x_j}\right) - \sigma_{ji}\frac{\partial^2 x_p^0}{\partial x_p \partial x_j} = \frac{\partial}{\partial x_p^0}\left(\sigma_{ji}\frac{\partial x_p^0}{\partial x_j}\right) \tag{2.61}$$

Using Eq. (2.61) in Eq. (2.60), the governing equation becomes

$$\frac{\partial}{\partial x_p^0}\left(\sigma_{ji}\frac{\partial x_p^0}{\partial x_j}\right) + \rho b_i = \rho \ddot{u}_i \tag{2.62}$$

We know that the Lagrangian stress tensor is defined as

$$\sigma_{ji}\frac{\partial x_p^0}{\partial x_j} = \frac{\rho}{\rho_0}\sigma_{pi}^L$$

Using the above relation in Eq. (2.62), we can write the *Governing equation in terms of Lagrangian stress tensor*, which is given by

$$\frac{\partial}{\partial x_p^0}\left(\frac{\rho}{\rho_0}\sigma_{pi}^L\right) + \rho b_i = \rho \ddot{u}_i \tag{2.63}$$

Previously we have shown in this chapter that $\rho/\rho_0 = J$, the Jacobian and we also know that $\partial J/\partial x_p^0 = 0$, and hence, we can simplify the above equation as

$$\frac{\partial \sigma^{L}_{pi}}{\partial x^0_p} + \rho_0 b^0_i = \rho_0 \ddot{u}_i \tag{2.64}$$

We know that Cauchy stress tensor is symmetric. That is $\sigma_{ji} = \sigma_{ij}$. Using these relations, the symmetry condition yields

$$\sigma^{L}_{pi} \frac{\partial x_j}{\partial x^0_p} = \sigma^{L}_{pj} \frac{\partial x_i}{\partial x^0_p} \tag{2.65}$$

Following similar steps, we can derive the *Governing equation in terms of Kirchhoff stress tensor*, which is given by

$$\begin{aligned} &\frac{\partial}{\partial x^0_k}\left(\frac{\partial x_i}{\partial x^0_j}\sigma^K_{kj}\right) + \rho_0 b^0_i = \rho_0 \ddot{u}_I \\ &\sigma^K_{ij} = \sigma^K_{ji} \end{aligned} \tag{2.66}$$

2.5 Constitutive Modelling

Constitutive relations relate the stresses developed with the strains through a material matrix. There are many constitutive models available for different kinds of material. Elastic materials follow Hook's law. Some of the well-known non-linear material models include elastic-strain hardening, elastic-perfectly plastic models and Hyper-elastic models as shown in Fig. 2.8.

The constitutive relations can be developed by two different methods, one is based on the approach called Hookean Elastic Solid, and the second method is by the use of Principle of Virtual work . The constitutive law obtained based on the above theorem is called the Green's Elastic Solid . Both these methods essentially will give the same material matrix for a linear elastic material. We will use the latter approach to obtain the constitutive relation for hyperelastic materials, which will be discussed in the next section.

2.5.1 *Hookean Elastic Solids*

The constitutive model derived by this procedure outlined below is normally referred to as *Hooke's Law*. It is based on the assumption that for an elastic body, the stress depends only on deformation and not on the history of deformation. This can be mathematically expressed as

$$\sigma_{ij} = f_{ij}(\varepsilon_{ij}) \tag{2.67}$$

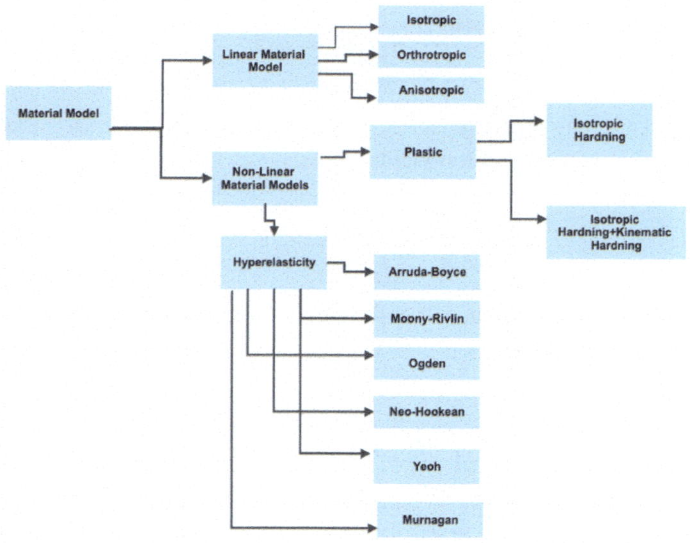

Fig. 2.8 Different material constitutive models

Expanding the above term by Taylor series about the initial configuration ($t = 0$), we get

$$\sigma_{ij} = f_{ij}(0) + \left[\frac{\partial f_{ij}(0)}{\partial \varepsilon_{kl}}\right]\varepsilon_{kl} + \frac{1}{2}\left[\frac{\partial^2 f_{ij}(0)}{\partial \varepsilon_{kl}\varepsilon_{mn}}\right]\varepsilon_{kl}\varepsilon_{mn} + \ldots\ldots.. \quad (2.68)$$

If the assumption of zero initial stress is true, then we require $\sigma_{ij} = 0$ when $\varepsilon_{ij} = 0$. This condition leads to $f_{ij} = 0$. The second term in Eq. (2.68) is the linear term and all other terms in the expression are nonlinear. Retaining only the linear term due to small strain assumption, we can write Eq. (2.68) as

$$\sigma_{ij} = C_{ijkl}\varepsilon_{kl} \quad \text{with} \quad C_{ijkl} = \left[\frac{\partial f_{ij}(0)}{\partial \varepsilon_{kl}}\right] \quad (2.69)$$

Equation (2.69) is called the *Hooke's Law*, which states that the stress tensor is linearly related to the strain tensor through a fourth-order tensor called the *Constitutive Matrix*.

The term in Eq. (2.69) C_{ijkl} is a fourth-order tensor of elastic constants, which are independent of either stress or strain. The tensorial quality of the constants C_{ijkl} follows the quotient rule, according to which for a fourth-order tensor, it should have $3^4 = 81$ elements. Due to symmetry of the stress tensor ($\sigma_{ij} = \sigma_{ji}$), we should have $C_{ijkl} = C_{jikl}$. Furthermore, since the strain tensor is also symmetric ($\varepsilon_{kl} = \varepsilon_{lk}$), we have $C_{ijkl} = C_{ijlk}$. Under these conditions, the fourth-order tensor C_{ijkl} will have only 36 independent constants. Hence, the total number of elastic constants cannot exceed 36, since the maximum independent elements in the stress and strain tensors are only 6 each. With these reductions, the generalized Hook's law can be written in the matrix form as

$$\begin{Bmatrix} \sigma_{xx} \\ \sigma_{yy} \\ \sigma_{zz} \\ \tau_{yz} \\ \tau_{xz} \\ \tau_{xy} \end{Bmatrix} = \begin{bmatrix} C_{11} & C_{12} & C_{13} & C_{14} & C_{15} & C_{16} \\ C_{21} & C_{22} & C_{23} & C_{24} & C_{25} & C_{26} \\ C_{31} & C_{32} & C_{33} & C_{34} & C_{35} & C_{36} \\ C_{41} & C_{42} & C_{43} & C_{44} & C_{45} & C_{46} \\ C_{51} & C_{52} & C_{53} & C_{54} & C_{55} & C_{56} \\ C_{61} & C_{62} & C_{63} & C_{64} & C_{65} & C_{66} \end{bmatrix} \begin{Bmatrix} \varepsilon_{xx} \\ \varepsilon_{yy} \\ \varepsilon_{zz} \\ \gamma_{yz} \\ \gamma_{xz} \\ \gamma_{xy} \end{Bmatrix} \tag{2.70}$$

where all the τ's represent the shear stresses in their respective planes, while all the γ's are the corresponding shear strains. For most elastic solids, the number of elastic constants can further be reduced by exploiting the material symmetry about different reference planes.

2.5.2 *Green's Elastic Solid*

An alternate method of deriving the constitutive relationship is by using the work and energy principles. This method is normally referred to as *Green's Elastic Solid*. For elastic materials, it will give the same material matrix as that of Hookean Solid. This method is based on the assumption that the work done by the elastic forces is completely transformed into potential energy, and further the potential energy is entirely due to the deformation a body undergoes due to applied tractions (forces). The details of the Principle of Virtual Work on which this method is based is explained in Sect. 2.4.

Consider a body of volume V, density ρ and subjected to surface tractions $\mathbf{t} = t_x\mathbf{i} + t_y\mathbf{j} + t_z\mathbf{k}$ on the surface S and the body force $\mathbf{b} = b_x\mathbf{i} + b_y\mathbf{j} + b_z\mathbf{k}$ as shown in Fig. 2.7. Here, we will again use the statement of Principle of Virtual work, which is given by Eqs. (2.56) and (2.57).

Here, as before, $\mathbf{t}$ is the traction vector on the boundary surface of area A, V is the total volume of the body with $\ddot{\boldsymbol{u}}$ being the acceleration vector. We begin by considering the total forces and moments acting on a body, which is given by Eqs. (2.56) and (2.57). The total incremental (virtual) work dW_e done by the forces or tractions t_i (acting on a surface S) of a body of volume V in displacing by an incremental (virtual) displacement du_i is given by

$$dW_e = \int_S t_i du_i dA + \int_V \rho b_i du_i dV \tag{2.71}$$

Using the Divergence theorem [4], the surface integral can be converted to the volume integral and the above equation becomes

$$dW_e = \int_V \sigma_{ij} d\left(\frac{\partial u_i}{\partial x_j}\right) dV = \int_V \sigma_{ij} d\varepsilon_{ij} \tag{2.72}$$

The change in the potential energy (also called Strain Energy) is given by

$$dU = \int_V dU dV \tag{2.73}$$

where U is the potential energy per unit volume (which is also called strain energy density function). Assuming that U is a function of only deformations (strains), which is the basic hypothesis on which this material model is based, we can write

$$dU = dU(\varepsilon_{ij}) = \frac{\partial U}{\partial \varepsilon_{ij}} d\varepsilon_{ij}$$

Using this in Eq. (2.73), we get

$$dU = \int_V \frac{\partial U}{\partial \varepsilon_{ij}} d\varepsilon_{ij} \tag{2.74}$$

Comparing Eqs. (2.74) and (2.72), we can say that

$$dU = dW_e, \qquad \text{or} \qquad \int_V \sigma_{ij} d\varepsilon_{ij} = \int_V \frac{\partial U}{\partial \varepsilon_{ij}} d\varepsilon_{ij} \tag{2.75}$$

Since the change in strains $d\varepsilon_{ij}$ is arbitrary, we can equate the integrands and in doing so we get,

$$\sigma_{ij} = \frac{\partial U}{\partial \varepsilon_{ij}} \tag{2.76}$$

Equation (2.75) is the famed Principle of Virtual Work (PVW), which is the heart of many numerical methods such as Finite Element Method (FEM). Returning to the constitutive modelling, Eq. (2.76) can be expanded using Taylor series about its initial state as

$$\sigma_{ij} = \frac{\partial U}{\partial \varepsilon_{ij}} = \left[\frac{\partial U(0)}{\partial \varepsilon_{ij}}\right] + \left[\frac{\partial^2 U(0)}{\partial \varepsilon_{ij} \varepsilon_{kl}}\right] \varepsilon_{kl} + \ldots\ldots\ldots\ldots\ldots = \sigma_{ij}^0 + C_{ijkl}\varepsilon_{kl} \tag{2.77}$$

Assuming that stress is zero when the strains are zero, the above equation reduces to

$$\sigma_{ij} = C_{ijkl}\varepsilon_{kl}$$

which is same as what we derived earlier. Because of the symmetry of both stress and strain tensors, the number of independent constants in C_{ijkl} tensor is 36. Since

$$C_{ijkl} = \left[\frac{\partial^2 U(0)}{\partial \varepsilon_{ij} \varepsilon_{kl}}\right] = \left[\frac{\partial^2 U(0)}{\partial \varepsilon_{kl} \varepsilon_{ij}}\right] = C_{klij}$$

This additional symmetry reduces the number of independent constants to 21.

2.5.3 *Elastic Stress–Strain Relationships*

The elastic stress vs. strain relationship for an isotropic homogeneous solid is given by,

$$\sigma_{ij} = \lambda \varepsilon_{kk} \delta_{ij} + 2G\varepsilon_{ij} \tag{2.78}$$

where $\varepsilon_{kk} = \varepsilon_{11} + \varepsilon_{22} + \varepsilon_{33}$. Lame's constants λ and G are related to bulk modulus K, shear modulus G, Young's modulus E and the Poisson's rail ν as

$$K = \lambda - \frac{2}{3}G, \quad E = 2G(1+\nu), \quad \text{and} \quad E = 3K(1-2\nu) \tag{2.79}$$

Under **uniaxial stress** conditions, only the axial stress (σ_{11} or σ_x) is non-zero, while all other stresses are zero. Therefore, the simplified stress–strain relation is: written as $\sigma_{11} = \lambda\varepsilon_{11} + 2G\varepsilon_{11}$.

Using Eq. (2.78), we can show $\sigma_{11} = E\varepsilon_{11}$ or simply $\sigma_x = E\varepsilon_x$ (Hooke's Law). The slope of the uniaxial stress–strain plot is E. The sound wave speed, C_0, in rods and bars under uniaxial stress is given by, $C_0 = \sqrt{E/\rho}$.

This condition is typical in uniaxial stress tests conducted on **metallic materials** using MTS or Instron machines, whether under compression or tension. The material yield stress or strength (Y) is the stress level at which the elastic deformation becomes plastic; we can also call this level as stress at the initial yield. However, for **brittle materials** like concrete or rocks, Y corresponds to the peak stress at fracture initiation.

Under **uniaxial strain** conditions, only the axial strain (ε_x) in the shock direction is non-zero and all other strains are zero. In plate impact configurations, material particles move in a single direction, resulting in a one-dimensional strain state. Further details will be provided in Chap. 3 and 4.

The simplified elastic stress–strain relationships (Eq. (2.78)) under uniaxial strain conditions can be expressed as,

$$\sigma_{11} = \lambda\varepsilon_{11} + 2G\varepsilon_{11}, \quad \sigma_{22} = \lambda\varepsilon_{11}, \quad \sigma_{33} = \lambda\varepsilon_{11} \tag{2.80}$$

It is important to note that the lateral stresses in Eq. (2.80) are equal to each other ($\sigma_{22} = \sigma_{33}$). The von Misses criterion for uniaxial strain conditions simplifies to a relationship between the lateral stresses and material strength:

$$Y = \sigma_{11} - \sigma_{22}, \qquad \text{or} \qquad Y = \sigma_x - \sigma_y \tag{2.81}$$

By combining Eqs. (2.79), (2.80), and (2.81) and replacing λ with K and G, a relationship between material strength (Y), and shock stress at the Hugoniot Elastic Limit (HEL) is obtained as,

$$Y = \frac{\sigma_x}{(K/2G + 2/3)} \text{ or } Y = \frac{\sigma_{HEL}}{(K/2G + 2/3)}, \tag{2.82}$$

It is worth noting that at the Hugoniot Elastic Limit, the shock stress is denoted as σ_{HEL}. Chapters 3 and 4 discuss the stress–strain curve under uniaxial strain or one-dimensional motion Eq. (2.82) in-depth. The slope of the linear portion of this curve is given by, "$K + 4G/3$" and the elastic shock speed, C_L is given by, $C_L = \sqrt{(K + 4G/3)/\rho}$.

Materials of interest can be broadly classified into those exhibiting almost purely elastic responses (possibly with some energy dissipation during rapid loading, like elastomers such as rubber or solid propellants); those that yield and exhibit considerable ductility beyond yield (such as mild steel, other commonly used metals, ice at low strain rates, and clay); those that flow through particle rearrangement with interactions dominated by some frictional mechanism (such as sand); and brittle materials (rocks, concrete, ceramics). Each of these situations requires different constitutive models, and some commercial finite element software, like LS-DYNA, ABAQUS, ANSYS, etc., have material libraries with models suitable for many situations. However, these libraries are not comprehensive, and the range of physical material behaviour is too broad for full coverage.

For instance, this book focuses on the behaviour of brittle materials under high strain rate loading, a topic not fully addressed in current material libraries. Although the John–Cook Material model is commonly employed for high strain loading in ductile materials, its direct application to brittle materials poses challenges. A comprehensive discussion of various constitutive models and failure modes specific to brittle materials is provided in Chap. 5.

Another category of materials, such as glass or polymeric fibre-reinforced plastics (FRP or GRP), where the simple linear Hooke's law is not applicable, necessitates nonlinear elastic constitutive models. Computational solutions for shock and impact problems involving composite laminates often require the use of hyperelastic material models. The following section presents a few salient hyperelastic models available in textbooks and in the open literature.

2.6 Hyperelastic Material Model

Constitutive relations are derived for hyperelastic material using strain energy density function. They are also normally referred to as Green's elastic material since their material model is derived using Green's solid approach . The constitutive law for these materials can be derived straight from linear Hooke's law by including the effects of material and geometric nonlinearities. In the literature, several hyperelastic laws have been well developed with special constitutive laws for a wide class of materials. For example, the reader may refer to [5]. Strain energy function is the bedrock for defining the feature of hyperplasticity of a material and many of these material laws have both non-dissipative and conservative behaviour. Such an approach of deriving the constitutive law enables to derive the stress–strain behaviour analytically and hence comparison of different hyperelastic models becomes simple and straightforward.

Most hyperelastic models developed till now are simple and are easily amenable for the development of numerical models. All of these models, in the limiting case of insignificant nonlinearities, reduce to standard linear Hook's law. This property can be effectively used to test the numerical validity of any new numerical algorithm proposed, especially in case of small strains. Most hyperelastic laws have conservation properties, and in order to exploit these properties, rigorous understanding of mathematics behind these laws is necessary. Several of these hyperelastic constitutive models are available in references such as [5–9], etc. This section summarizes the various different hyperelatic models presented by various researchers.

To derive the constitutive model for hyperelastic material, we go back to Fig. 2.2. A particle at the initial state x_i moves to a new state. The deformation gradient tensor F_{ij} is defined as before as .

$$F_{ij} = \delta_{ij} + \frac{\partial u_i}{\partial x_j} \tag{2.83}$$

where δ_{ij} is the Kronecker delta tensor, which takes the value 1 when the indices are equal and 0 otherwise. The Cauchy's stress tensor and Lagrangian stress tensor obtained through Green's Solid approach are normally referred to as Cauchy–Green tensor and Lagrange–Green tensor, respectively. .. We will now go back to Sect. 2.2.3, where in Eq. 2.32, we derived Lagrangian strain tensor E_{ij} and we also showed that it represents relative change in the length of a line segment. The Cauchy–Green tensor C_{ij} and the Lagrange strain tensor E_{ij} can be written as before as

$$C_{ij} = F_{ki} F_{kj} \tag{2.84}$$

$$E_{ij} = \frac{1}{2}(C_{ij} - \delta_{ij}) = \frac{1}{2}\left(\frac{\partial u_i}{\partial x_j} + \frac{\partial u_j}{\partial x_j} + \frac{\partial u_k}{\partial x_i}\frac{\partial u_k}{\partial x_j}\right) \tag{2.85}$$

The definition of the Cauchy–Green tensor Eq. (2.84) implies that it is symmetric an positive definite. This is also true of the Green–Lagrange strain tensor defined in

Eq. ((2.85)). Thus the strain invariants associated with the Green–Lagrange strain tensor, for instance, can be expressed as [5]

$$I_1 = E_{kk} \qquad I_2 = \frac{1}{2}(I_1^2 - E_{ij}E_{ji}) \qquad I_3 = det(E) \tag{2.86}$$

The invariants I_1, I_2, and I_3 of the strain measures provide an elegant means of handling the dependence of the strain energy function on the strain tensor, as will be illustrated in the subsequent development of hyperelasticity.

As mentioned before, hyperelastic materials are characterized through strain energy function W, which is then used in either Cauchy-Green tensor C_{ij} or the Green–Lagrange strain tensor E_{ij} to derive the constitutive relations. Note that, unlike normal elastic solids, here, we will not use deformation gradient tensor F_{ij} to derive these constitutive relations. This is done in order to preserve the frame invariance of the developed constitutive relations for these materials [10]. This procedure will ensure that the derived laws are not affected by rigid body transformations. According to [10], the hyperelastic constitutive laws, in addition to having frame indifference, should preserve material symmetry. In this chapter, only isotropic materials are considered and hence the stress or strain invariants derived from either C_{ij} or E_{ij} are scalars that are not affected by any rotation of the coordinate system. This dependence of the constitutive models on the stress or strain invariants can be justified mathematically and this can be found in [5, 10]. Some of the well-known hyperelasticity models reported in the literature are Neo-Hookean model[10], Mooney–Rivlin model [10], Ogden model [5], and Murnaghan model [11]. Some of these models are reviewed in this section.

We will now outline briefly the method to derive the constitutive model for hyperelastic materials. The isotropic elastic properties of a hyperelastic material are derived from the strain energy density function W, which defines the strain energy stored in the hyperelastic material per initial unit volume. The strain energy density comprises of two components, one due to deformation denoted by W_d and the second due to volume change denoted by W_v. That is, we can write the strain energy density function as

$$W = W_d(I_1, I_2, I_3) + W_v J \tag{2.87}$$

where I_1, I_2, and I_3 are the stress or strain invariants and J is the Jacobian or the determinant of deformation sensor. The stress or strain invariants are obtained from the Eigenvalues of the stress or strain tensor and they are given by λ_1, λ_2, and λ_3, respectively. The invariants are related to these eigenvalues as

$$\begin{aligned} I_1 &= \lambda_1 + \lambda_2 + \lambda_3 \\ I_2 &= \lambda_1\lambda_2 + \lambda_2\lambda_3 + \lambda_1 \\ I_3 &= \lambda_1\lambda_2\lambda_3 \end{aligned} \tag{2.88}$$

In the case of incompressible material, $I_3 = 1$ and hence, I_1 and I_2 reduces to

$$\begin{aligned} I_1 &= \lambda_1^2 + \lambda_2^2 + \lambda_1^{-2}\lambda_2^{-2} \\ I_2 &= \lambda_1^2\lambda_2^2 + \lambda_1^{-2} + \lambda_2^{-2} \end{aligned} \tag{2.89}$$

Once the invariants are determined, the strain energy density function W is constructed based on certain assumptions. Using the constructed strain energy density function, the constitutive relation of the hyperelastic material is obtained using Green's elastic solid approach . Many researchers have have developed different hyperelastic material models depending upon the way the strain energy density function is constructed. These are reviewed in the next few subsections without going into much detail. Only a brief summary is provided.

2.6.1 *Generalized Rivlin Model*

This is one of the first hyperelastic models to be formulated and it is also called *Polynomial Hyperelastic Model.* The details can be found in Refs. [12–14]. Here, the strain energy function W is expressed only in terms of two invariants, namely, I_1 and I_2 as

$$W = \sum_i^n \sum_j^m C_{ij}(I_1 - 3)^i (I_2 - 3)^j + \sum_i^n \frac{1}{D_i}(J_{el} - 1)^{2i} \tag{2.90}$$

where C_{ij} and D_i are the temperature dependent parameters and J_{el} is the elastic volume ratio. One of the more popular version of this model is the one where the strain energy density function is expressed as linear combination of the invariants as

$$W = C_{10}(I_1 - 3) + C_{01}(I_2 - 3) + \frac{1}{D_i}(J_{el} - 1)^2 \tag{2.91}$$

For an incompressible material, the above equation simplifies to

$$W = C_1(I_1 - 3) + C_2(I_2 - 3) \tag{2.92}$$

Equation (2.92) is the famous Mooney–Rivlin Hyperelasticity model. This model was developed by Mooney [12] and Rivlin [13] few years earlier. This model can be deduced from the generalized Rivlin model by setting $n = 1$, $C_{10} = C_1$, $C_{02} = C_2$ and $C_{11} = 0$. This model can be further simplified by setting $C_{01} = 0$ and the resulting equation becomes

$$W = C_{10}(I_1 - 3) + \frac{1}{D_i}(J_{el} - 1)^2 \tag{2.93}$$

Equation (2.93) represents the Neo-Hookean hyperelastic model . This model was first proposed by Treloar [15]. Similarly, we write 3 and 5 parameter Rivlin model as

$$
\begin{aligned}
W =& C_{10}(I_1 - 3) + C_{01}(I_2 - 3) + C_{11}(I_1 - 3)(I_2 - 3) + \frac{1}{D_1}(J_{el} - 1)^2 \\
W =& C_{10}(I_1 - 3) + C_{01}(I_2 - 3) + C_{20}(I_1 - 3)^2 + C_{11}(I_1 - 3)(I_2 - 3) \\
& + C_{02}(I_2 - 3)^2 + \frac{1}{D_1}(J_{el} - 1)^2
\end{aligned}
\tag{2.94}
$$

From these potentials, we can conclude the following:

- Higher number of terms can model more complex stress–strain relations but may require higher computational efforts and large amount of experimental data to determine the material constants. When the non-linearity becomes strong, literature on this model says that there may be convergence issues.
- As mentioned earlier, the two parameter Mooney–Rivilin model reduces to Neo-Hookean model when $C_{01} = 0$. In this case, the shear modulus $\mu = 2 \times C_{10}$. The non-zero C_{01} term makes the model under uniaxial tension more accurate, but it still cannot accurately simulate the multi-axial stress states. The parameter obtained from one type of deformation experiment cannot predict another type of deformation accurately.
- Three- or more-parameter Mooney–Rivlin models can describe unsteady shear modulus. However, solutions need to be carefully evaluated when the high-order terms are introduced, as it may produce strain energy values that can be unstable and yield non-physical solutions.

Some of the limitations of the Mooney–Rivlin model are as follows

- This model is not suitable in situations when deformation exceeds 150%.
- Since the higher order Mooney–Rivlin model has many constants to be experimentally obtained, it requires curve-fitting of experimental data.
- The model is not suitable for obtaining constitutive model of a compressible hyperelastic material such as foam.
- The predicted deformation error could be large when the strain/stress is beyond the range of input experimental data.

Once the strain energy density function is written, one can obtain the constitutive model using Green's solid approach.

2.6.2 Ogden Hyperelastic Material Model

The ogden model was introduced by Ogden [16] in the year 1972. More details of the model are also found by a book written by Ogden [5]. In this model, the strain energy density function is expressed as

$$W = \sum_{i=1}^{n} \frac{2\mu_i}{\alpha_i} \left(\lambda_1^{\alpha_i} + \lambda_2^{\alpha_i} + \lambda_3^{\alpha_i} - 3\right) + \sum_{i=1}^{n} \frac{1}{D_i}(J_{el} - 1)^{2i} \tag{2.95}$$

For incompressible materials, we have $\lambda_1\lambda_2\lambda_3 = 1$ and hence the above equation reduces to

$$W = \sum_{i=1}^{n} \frac{2\mu_i}{\alpha_i} \left(\lambda_1^{\alpha_i} + \lambda_2^{\alpha_i} + \lambda_1^{-\alpha_i}\lambda_2^{-\alpha_i} - 3\right) \tag{2.96}$$

where λ_i are the stretch ratio obtained from the strain tensors, μ_I and α_I are material constants that require determination.

Looking at the strain energy function, we can make following observations:

- The strain energy function uses three stretch ratio λ_1, λ_2, and λ_3, obtained using strain tensors
- When $n = 1$ and $\alpha_1 = 2$ then the model reduces to Neo-Hookean model.
- When $n = 2$, $\alpha_1 = 2$, and $\alpha_2 = -2$, then this model reduces to 2-parameter Mooney–Rivlin model .
- For ensuring stability, it is necessary that the material constant should satisfy the condition $\mu_i\alpha_i > 0$.

Some of the merits and demerits of Ogden material model are summarized below

- Ogden model is a very versatile model since it can be widely used to construct different hyperelastic constitutive models. This model is reportedly very accurate for the entire strain range of rubber materials.
- This model is ideal to handing problems with large strain levels. For higher order approximation, the model is very effective even for strain levels as high as 700%.
- Suitable for describing non-constant shear modulus and slightly compressed material behaviour.
- This model can describe the rapid stiffness rise in the late stage of deformation.
- In this model, the material parameters determined by one type of experiment cannot be used to predict another type of deformation. It is not recommended to use in the case of insufficient experimental data, such as only uniaxial tensile test data.

2.6.3 *Yeoh Hyperelastic Material Model*

This model was introduced by Yeoh [17] in the year 1990. In this model, the strain energy density function is expressed only in terms of first invariant I_1. It is a reduced polynomial representation of generalized Rivilin model with $n = 3$ and $m = 0$ (see Eq. (2.90)). That is, we write the strain energy density function as

$$W = \sum_{i=1}^{i=3} C_i(I_1 - 3)^I + \sum_{i=1}^{i=3} \frac{1}{D_i}(J_{el} - 1)^{2i} \tag{2.97}$$

For incompressible materials, the above equation reduces to

$$W = \sum_{i=1}^{i=3} C_i(I_1 - 3)^I = C_1(I_1 - 3) + C_2(I_1 - 3)^2 + C_3(I_1 - 3)^3 \tag{2.98}$$

Using the above strain energy function, Cauchy stress tensor is obtained through Green's elastic solid approach and is given by

$$\sigma = -p + 2\frac{\partial W}{\partial I_1} = -p + \sum_{i=1}^{i=3} iC_i(I_1 - 3)^{i-1} \tag{2.99}$$

where p is the hydrostatic stress.

From the strain energy function (Eq. (2.97)), we can conclude the following:

- Both Yeoh and Mooney–Rivlin are very similar, as both of them belong to the family of Polynomial models. In the order, both models are same, computer implementation of the Yeoh is simpler compared to Mooney–Rivlin model because it does not consider the role of the second invariant I_2. However, the volumetric terms of the Yeoh model are more complex compared to that of Mooney–Rivlin model..
- The initial shear modulus computed from Yeoh model is equal to $2 \times C_{10}$, and the initial modulus is equal to $2/D_1$. To ensure that the initial shear modulus is always positive, the parameter C_{10} must be greater than zero.
- Although the order parameter n can be large, in practical applications, we normally use $n = 3$. When $n = 1$, Yeoh model reduces to Neo-Hookean model. Therefore, it is generally recommended to use the third-order potential with $n = 3$, to take advantage of the superiority of the Yeoh model.
- When $n \geq 3$, a typical inverse S-shaped rubber stress–strain curve can be described using this model.

Some of the merits and demerits of Yeoh model are summarized below

- This Yeoh model is very simple and only a few parameters are required to describe the constitutive behaviour. The user only needs a small amount of experimental data to get reasonable numerical results, and the input parameters can be obtained only by the tensile experiment.
- Yeoh model can be described for a wide range of deformation. The reasonable results can be obtained under large uniaxial stretching and simple shear deformation.
- This model poses some problems in predicting the deformation of biaxial tension. It is shown to have some deviations

- For small strain deformation, the material parameters need to be carefully chosen. There may be a deviation between the values predicted by Yeoh model and the experimental data. However, in the finite element analysis, this deviation is not serious because the stress in the small strain region is small, and although the relative error may be large while the absolute error is small.

2.6.4 *Arruda–Boyce Hyperelastic Material Model*

This model was introduced by authors Arruda and Boyce [18] in the year 1993. It is based on an eight-chain network that considers eight orientations of chains in space with a non-Gaussian behaviour of the individual sub-chains in the proposed network. Unlike most other hyperelastic materials models, this model requires the determination of only two material parameters, an initial modulus and a limiting chain extensibility. The strain energy density function for this model is written as

$$W = C_1 \sum_{i=1}^{p} \alpha_i \beta^{i-1} \left(I_1^i - 3^i \right) + \frac{1}{D_1} \left(\frac{J^2 - 1}{2} - \log J \right) \tag{2.100}$$

where $C_1 = nkT$ is the material constant, n is subchain material density, k is the Boltzman constant, and T is the temperature. Here, $\beta = 1/N$ is the inverse of number of statistical segments between crosslinks N and J is the determinant of the strain gradient tensor. D_1 is another material constant dependent on the bulk modulus. Here α_i are coefficients of the series expansion of the inverse Langevin function, which are given by

$$\alpha_1 = 0.5,\ \alpha_2 = 1/20,\ \alpha_3 = 11/1050,\ \alpha_4 = 19/7000,\ \alpha_5 = 519/63750$$

Assuming that the material is incompressible, we can write the strain energy density function as

$$\begin{aligned} W =& C_1 \left[\frac{1}{2}(I_1 - 3) + \frac{1}{20N}(I_1^2 - 9) + \frac{11}{1050N^2}(I_1^3 - 27) \right] \\ &+ C_1 \left[\frac{19}{7000N^3}(I_1^4 - 81) + \frac{519}{673750N^4}(I_1^5 - 243) \right] \end{aligned} \tag{2.101}$$

The initial chain length in this model is assumed to be identical with that of a random walk of N steps of length l such that the fully extended chain length is equal to lN. For this case, the limiting extensibility (or chain locking stretch) is defined as the ratio of final length to initial length, and it is represented by $r_m = \sqrt{N}$. Using these, we can write the strain energy function for this model as

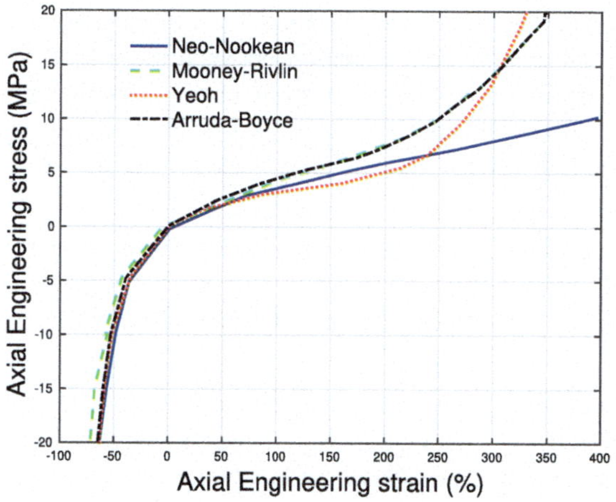

Fig. 2.9 Different hyperelastic material constitutive models

$$W = C_1\left[\frac{1}{2}(I_1-3)+\frac{1}{20r_m^2}(I_1^2-9)+\frac{11}{1050r_m4}(I_1^3-27)\right]$$
$$+C_1\left[\frac{19}{7000r_m^6}(I_1^4-81)+\frac{519}{673750r_m^8}(I_1^5-243)\right] \tag{2.102}$$

Some of the Arruda–Boyce model features are summarized below

- This model assumes that the material is incompressible by default. Compressible characteristics are introduced by adding the items containing volume change term J.
- When the limiting network stretch becomes infinite, then the Arruda–Boyce model becomes equivalent to the Neo-Hookean model.
- The limiting network stretch must always be greater than or equal to 1 for stability.
- Incompressible material parameter D_1 cannot be zero.
- This model is said to perform very well both in uni-axial and bi-axial stretching.
- In general, this model is easy-of-use and requires only three material parameters. It has accuracy advantages in high stretch conditions.
- The material parameters can be easily determined experimentally and also can be determined by the curve fitting numerical method of the test data.

This section discussed four different hyperelastic material models that are typically used in estimating failure of brittle materials. Once the strain energy density functional is explicitly written after the determination of the material constants, stress tensor is estimated through Green's solid approach . The stress–strain curve for rubber material for these four different models is shown in Fig. 2.9. At low strain levels, we see, all the four hyperelastic models predict similar behaviour. At very large strain, their behaviour differs significantly.

There is yet another hyperelastic model mentioned in Fig. 2.8 and not discussed here. It is the Murnagan hyperelastic model [11]. This model is mostly used in acousto-elasticity and dynamic nonlinear elasticity problems. It is not discussed here in this section. A more detailed account of this model is available in Ref. [19].

2.7 Summary

In this chapter, we present the fundamental aspects of continuum mechanics, emphasizing the importance of grasping this material to fully appreciate the subsequent chapters of the book. The chapter commences with a concise introduction to Tensor Analysis, elucidating the techniques for representing intricate equations in tensorial notation. Following this, the section on deformation introduces the concept of motion based on initial and final deformed configurations, known as Eulerian and Lagrangian motions, respectively.

Using these deformation descriptions, we establish the Strain Tensor based on both undeformed and deformed configurations. Utilizing these tensors, we determine expressions for the deformation of lines, areas, and volumes. Subsequently, we derive expressions for the Stress Tensor in terms of the deformed configuration (referred to as the Cauchy stress tensor) and the undeformed state (Lagrangian and Kirchhoff stress tensors). Employing the derived stress tensors, we then derive the governing equations of motion for a 3D continuum, incorporating all three different stress tensors, namely, Cauchy, Lagrangian, and Kirchhoff tensors. Lastly, we presented relationships among the various stress tensors.

The latter part of this chapter addresses the different constitutive models available for various materials and loading behaviours. Here, we present two different methods for deriving the constitutive matrix: the Hookean solid approach and the Green's solid approach. It is demonstrated that these two approaches yield the same constitutive matrix for elastic materials. Subsequently, using Green's solid approach, we derive constitutive models for hyperelastic materials, including the Neo-Hookean, Mooney–Rivilin, Yeoh, and Arruda–Boyce models, respectively. An understanding of the fundamentals of continuum mechanics and constitutive models is crucial for comprehending the materials presented in the subsequent chapters.

References

1. Malvern LE (1977) Introduction to the mechanics of a continuous medium. Prince Hall
2. Reddy JN (2013) An introduction to continuum mechanics, 2nd edn. Cambridge University Press
3. Irgens Fridtjov (2007) Continuum mechanics. Springer, New York
4. Krezig E (1992) Advanced engineering mathematics, 9th edn. McGraw Hill, New York, USA
5. Ogden RW (1997) Nonlinear elastic deformations. Dover, New York
6. Chadwick P (1999) Continuum mechanics: concise theory and problems. Dover, New York

7. Marsden JE, Hughes TJR (1994) Mathematical foundations of elasticity. Dover, New York
8. Ciarlet PG (2005) An introduction to differential geometry with applications to elasticity. Springer, New York
9. Wriggers (2008) Nonlinear finite element methods. Springer
10. Basar Y, Weichert D (1999) Nonlinear continuum mechanics of solids: fundamental mathematical and physical concepts. Springer
11. Murnaghan FD (1951) Finite deformations of an elastic solid. Wiley, New York
12. Mooney M (1940) Theory of large deformation. J Appl Phys 11:582–592
13. Rivlin RS (1948) Large elastic deformations of isotropic materials. II. Some uniqueness theorems for pure, homogeneous deformation. Philos Trans Roy Soc A 240:491–508
14. Rivlin RS, Saunders DW (1951) Large elastic deformations of isotropic materials. VII. Experiments on the deformation of rubber. Philos Trans Roy Soc Lond Ser A Math Phys Sci 243(865): 251–288
15. Treloar LRG (1943) The elasticity of a network of long chain molecules. J Chem Soc Faraday Trans 39:36–91
16. Ogden RW (1972) Large deformation isotropic elasticity - on the correlation of theory and experiment for incompressible rubberlike solids. Proc Roy Soc-A, Lond 326:565–584
17. Yeoh OH (1990) Characterization of elastic properties of carbon-black-filled rubber vulcanizates. Rubber Chem Technol 63(5):792–805
18. Arruda EM, Boyce MC (1993) A three-dimensional constitutive model for the large stretch behavior of rubber elastic materials. J Mech Phys Solids 41(2):389–412
19. Gopalakrishnan S (2016) Wave propagation in structures and materials. Taylor & Francis/CRC Press

Chapter 3
Shock Compression

Prologue

There are a number of books related to Shock Wave Propagation in Solids. Over the past 4 or 5 decades, the articles in the bound proceedings of "Shock Compression of Condensed Matter" provide state-of-the-art concepts and fundamental understanding of the shock response of a wide range of materials. The John-Wiley books, Dynamic Behavior of Materials by Marc A. Meyers [1] *and High Velocity Impact Dynamics by John A. Zukas* [2] *describe experiments, theories, and computational models towards developing an appreciation of the complexities in the study of shock waves in solids and other related topics. The recent books, James D. Walker* [3], *and Gennady Kanel,* [4] *will also provide theoretical and experimental insights into the complex deformation and failure processes in materials under shock and penetration loading conditions. The book by Grady* [5] *is especially useful for understanding non-ideal shocks. Davidson* [6] *also provides an overview from a chemical physics point of view. The book by Gathers* [7] *is very useful for non-planar shocks and for behaviour at ultra-high pressures.*

In laboratory settings, shock waves are normally produced by planar impact of a "flyer plate" onto a target. The impact produces plane shock waves in both the impact and the target. Interpretation of measurements in ideal plane impact scenarios is one of the major tools for investigating high-rate properties of brittle materials. Thus, the physics of shock compression undergirds both engineering and laboratory practice. For those reasons, a discussion of shock waves in compressed matter is provided in this chapter.

A. M. Rajendran et al., *Failure of Brittle Materials Under Shock and Impact*, Materials Horizons: From Nature to Nanomaterials, https://doi.org/10.1007/978-981-97-7026-7_3

3.1 Shock Compression

In this chapter, we take stress and strain as positive in compression. Since we are concerned with plane waves, equations are written for axisymmetric stress states, e.g. one-dimensional strain (1D Compression). The principal stress, σ_1, is in the direction of wave propagation. The transverse stresses, σ_2 and σ_3, are equal. The wave propagates with a velocity, U, and the particle velocity behind the wave is u. The mean stress or pressure is

$$P = \frac{\sigma_1 + \sigma_2 + \sigma_3}{3} \tag{3.1}$$

Shear stress is expressed in terms of the deviator, e.g. the principal stress minus the mean stress. There are two common expressions for this. One is the equivalent stress, written as σ_{eq} in engineering mechanics, or q, in geotechnical literature, defined as

$$\sigma_{eq} = \sigma_1 - \sigma_2 \tag{3.2}$$

The difference $(\sigma_1 - \sigma_2)$ is also equal to the yield strength Y, where Y is defined as the limit of elastic behaviour under uniaxial stress loading conditions, such as a conventional compressive or tensile test. Shear strength as normally used is equal to $(\sigma_1 - \sigma_2)/2$. More discussions on the material strength can be found in Chaps. 4 and 5.

The other expression for shear stress is the second deviatoric stress invariant which for 1D compression is given by

$$\sqrt{J_2'} = (\sigma_1 - \sigma_2)/\sqrt{3} \tag{3.3}$$

As an invariant, its value does not change if a different set of principal stress axis is used. Several sets of units are convenient for working out shock compression problems. The two most widely used are given in Table 3.1.

Conversion factors to other frequently used stress units are 1 kbar = 14.5 ksi = 109 dyn/cm^2 = 0.1 GPa. In this chapter, we mostly use SI units. These days most papers are written in SI units.

There are two types of waves to consider. They are most easily visualized in distance–time (x, t) space. The first is simple waves. In a simple wave, wave speed

Table 3.1 Systems of units for shock compression test

Quantity	SI(MKS)	Shock
Density ρ	kg/m^3	g/cm^3
Stress σ	Pa	GPa
Velocity	m/s	km/s = mm/μs
Distance, time (x, t)	m, s	mm, μs

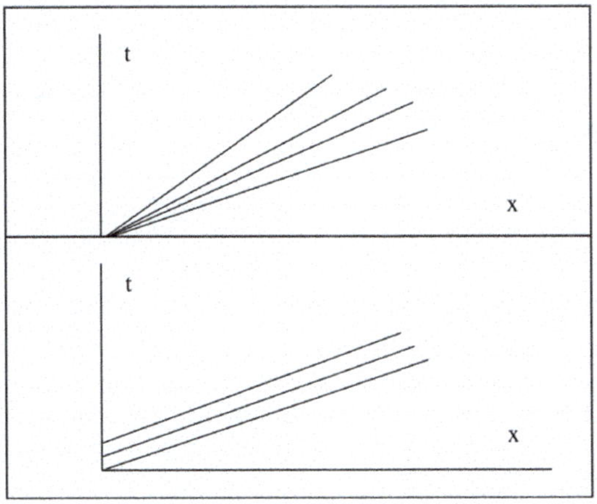

Fig. 3.1 Wavelets for simple waves (upper) and steady waves (lower)

is a function of stress, and each wavelet travels with a constant velocity. Thus the rise time of a wave will increase with propagation distance. When all the wavelets have a common origin, then it is a centred wave. The other type is a steady wave. In a steady wave, each wavelet has the same speed. Thus, the rise time of the wave does not change with propagation distance. Generally shock waves are examples of steady waves. Propagation of wavelets for these two types of waves is shown in Fig. 3.1. Shock waves are often described as a discontinuous jump between two states. But more precisely their defining characteristic is that they are steady waves whose rise time does not change as they propagate.

Planar acoustic and shock waves involve one-dimensional strain deformation in the direction of propagation. The only non-zero strain component is ε_1. Thus the volumetric strain is given by

$$\frac{\Delta\rho}{\rho} = \varepsilon_1 = -\frac{\Delta V}{V} \tag{3.4}$$

where V is the volume per unit mass, e.g. the reciprocal of density. If the stress deviators are zero, then

$$\sigma_1 = \sigma_2 = \sigma_3 = p \tag{3.5}$$

The slope of the stress strain curve for one-dimensional compression is known as the constrained modulus, M and it is given by

$$M = \frac{\partial\sigma_1}{\partial\varepsilon_1} \tag{3.6}$$

Many of the older publications in shock physics were concerned with such high pressures that P could be taken as equal to the principal stress. For that case of hydrodynamic compression, the constrained modulus is equal to the bulk modulus, K. In many shock problems, moduli must be considered to vary with stress. Although in most engineering problems involving ceramics or rocks, the change of moduli with compression will be negligible.

For isentropic compression, $K = K_S$, the isentropic bulk modulus can be written as

$$K_S = -V\left(\frac{\delta P}{\delta V}\right)_S = \rho\left(\frac{\delta P}{\delta \rho}\right)_S \tag{3.7}$$

where the derivatives are taken at constant entropy, S. Of course, there is also an isothermal bulk modulus, K_T, for which the derivatives in Eq. (3.7) are taken for fixed temperature. In this book, the bulk modulus will always refer to isentropic deformation, so we will simply use the symbol K without the subscript.

For elastic 1D strain deformation, M can be written as a function of two of the three commonly used elastic moduli, K, G (shear modulus), E (Young's modulus) and ν (Poisson's ratio). In particular, $M = K + 4G/3$, where G is the slope of a shear stress versus shear strain curve for simple shear (such as a rod in torsion). The relationship between the principal stresses for 1D compression is

$$\sigma_2 = \frac{\nu}{1-\nu}\sigma_1 \tag{3.8}$$

where

$$\sigma_1 = \left(\frac{4G}{3} + K\right)\varepsilon_1 \tag{3.9}$$

Note that the slope of stress–strain curve under uniaxial stress state as in a slender bar (or dog-bone specimen in a quasi-static test or split Hopkinson bar tensile or compressive stress tests as described in Chap. 4) is given by Young's modulus.

3.1.1 Jump Equations

Jump equations relate quantities in front of and behind steady waves. The speed of sound in solids can be derived from the wave equation, as explained in most books on continuum mechanics. However, there is a simple geometric derivation that is more in the spirit of shock analysis.

Consider a steady wave. Acoustic waves, being of small amplitude, are considered steady because the wave speed is not a function of stress. As shown in Fig. 3.2 in front of the wave the material is at rest, stress is zero, and the density is ρ. The wave advances at speed C. Behind the wave front material is moving with velocity u, the density has increased by $\Delta\rho$, and the stress is now σ_1.

Mass must be conserved across the wave front. If one considers how mass flows in and out of a tube over a time interval, mass in equals mass out translates to

$$\rho C = (C - u)(\sigma + \Delta\rho) \tag{3.10}$$

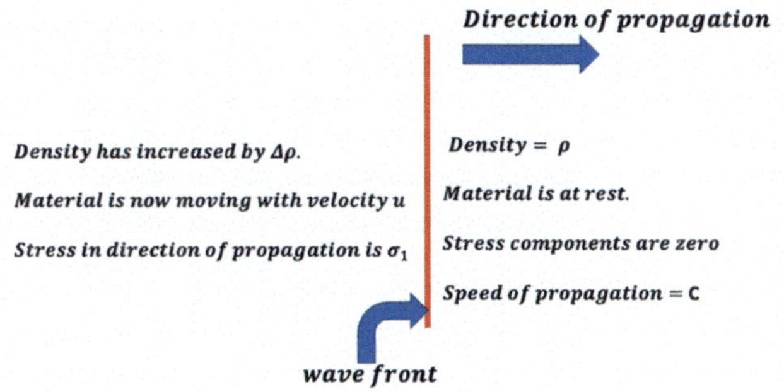

Fig. 3.2 Acoustic wave front

Likewise, if one considers that the force acting on material in the tube times the time that the force is applied (e.g. the time material remains in the tube) must equal the change of momentum of the material between entering and exiting the tube, one finds

$$\sigma = \rho u C \tag{3.11}$$

Eliminating u from Eqs. (3.10 and 3.11) gives for small strains:

$$C = \sqrt{\frac{\Delta\sigma}{\Delta\rho}} = \sqrt{\frac{M}{\rho}} \tag{3.12}$$

In words: the speed of an acoustic wave is the square root of the modulus divided by the density. If it is an elastic compression wave (i.e. a P wave) then M in Eq. (3.12) is the constrained modulus, $(K + 4G/3)$. If the material does not support shear, or shear stress is negligible, then $M = K$, since $G = 0$ for that case. The deformation in which shear stress is negligible is termed *"hydrodynamic"*. If the deformation is one-dimensional stress instead of one-dimension strain, then $M = E$ (Young's modulus). If the deformation is pure shear, as in a torsional wave in a bar, then $M = G$.

This discussion can be generalized to apply to shock waves. These are also steady waves, but they differ from acoustic waves in that in general the material ahead of the wave is not at rest, not at zero stress, the strains are not small, and the compression is not isentropic. Although the following conservation equations are strictly only true for steady waves, according to Grady [5], in almost all scenarios of interest, the steady wave assumption is adequate for relating conditions in front of and behind a wave front. Figure 3.3 illustrates this situation.

Equations for mass and momentum conservation are written the same way as in Eqs. (3.10 and 3.11), which are given by

$$\rho_1 (U - u_1) = \rho_2 (U - u_2) \tag{3.13}$$

$$\sigma_2 - \sigma_1 = \rho_1 (U - u_1)(u_2 - u_1) \tag{3.14}$$

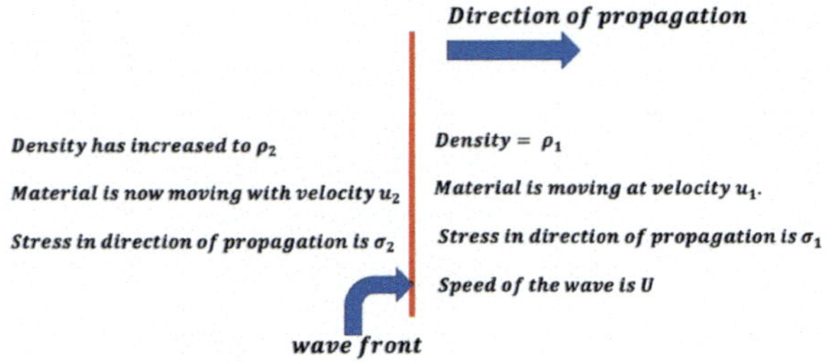

Fig. 3.3 A shock front. Variables ahead of the front are subscripted 1, behind are subscripted 2

There are four unknowns, namely, U, σ_2, ρ_2, and u_2. The two equations for mass and momentum conservation relate these four unknowns to the conditions in front of and behind the shock wave. Two of these unknowns must be measured and then the other two can be calculated. If the measured quantities are the particle velocity and wave speed, then Eqs. (3.13 and 3.14) are used to compute stress and particle velocity behind the shock. If the measured quantities are stress and density (density = $1/V$, where V is volume per unit mass) or shock speed and density, then Eq. (3.15) is very useful.

$$(U - u_1)^2 = V_1^2 \frac{\sigma_2 - \sigma_1}{V_1 - V_2} \tag{3.15}$$

Another useful equation is the relationship between strain and velocities, which is given by

$$\varepsilon = \frac{\rho}{\rho_0} - 1 = \frac{V_2 - V_1}{V_2} = \frac{u_1 - u_2}{U - u_1} \tag{3.16}$$

where V_1 is the volume of the material in form of the shock and V_2 is the volume of the material behind the shock. Note that, under compression, volume decreases while the density increases. Equations (3.13)–(3.16) are known as the *Hugoniot* equations.

Notice that the Hugoniot equations are simpler if, as they are often written, they are applied to a material initially at rest and unpressurized, e.g. $u_1 = 0$ and $\sigma_1 = 0$ and density is ρ_0. It is important to keep in mind that Eqs. (3.13)–(3.16) are written in terms of *Eulerian coordinates*, e.g. laboratory coordinates explained in Chap. 2.

An equation for energy conservation is often added to the *Hugoniot* equations. It is derived also from the framework of Fig. 3.3 by equating the mechanical work done on a volume of material, stress times distance, to the work done compressing that material, stress times volumetric strain given by

$$E_2 - E_1 = \frac{1}{2}(\sigma_1 + \sigma_2)(V_1 - V_2) \tag{3.17}$$

Equation (3.17) is not strictly correct if stress is not just a function of strain, but for problems of concern for this book, Eq. (3.17) should be adequate.

3.1.2 *Hugoniot Curve*

The experimentally determined relationships between the shock variables are collectively referred to as the Hugoniot curve. The Hugoniot defines the set of states that can be achieved through a single shock compression. It is a material property, like an isentrope. Usually the concern is the principal Hugoniot which is the stress states achievable by shocking a material that starts at zero stress and zero particle velocity. The principal Hugoniot curve thus expresses a relationship between any two of the shock compression variables: stress, particle velocity, wave velocity, or density. One can translate from one pair of variables to another by applying the Hugoniot Eqs. (3.13)–(3.16), written for u_1 and $\sigma_1 = 0$.

For many materials, the shock velocity is a linear function of the jump in particle velocity. That means that the principal Hugoniot is described by the equation:

$$U = C + su \tag{3.18}$$

(When applied to a material that is already in motion, the particle velocity of the material into which the wave moves should be subtracted from U and u.) Here C is the acoustic wave speed, given by Eq. (3.12). The slope of the $u - U$ curve is s. For a wave that only involves 1D elastic deformation, this speed is the longitudinal acoustic wave speed. For shocks provoking hydrodynamic behaviour, it is the bulk wave speed, given by $\sqrt{K/\rho}$. In other words, for a material that is first elastic and then plastic, there are two relevant relationships. For shocks in which the material remains elastic, wave speed is $\sqrt{K + \frac{4G}{3}/\rho}$. For shocks involving inelastic, or plastic deformation, the $U - u$ relationship is given by Eq. (3.18). Some values of initial density, K, G, C, and s that will be helpful in design of ceramics are given in Table 3.2. It includes SiC and WC computed from data in Grady [8], and alumina, taken from Reinhart et al. [9].

For a shock above the elastic regime, and for a given volumetric strain, the Hugoniot stress and the stress during hydrodynamic compression differ by the shear strength of the material, $\sigma - P = 2Y/3$, for a von Mises material. For example, Fig. 4.14 shows the offset between stress on the Hugoniot and hydrodynamic

Table 3.2 Reference Hugoniot data

Material	Density (kg/m^3)	K Bulk modulus, (GPa)	G Shear modulus, (GPa)	C (m/s)	s
Tungsten carbine	14930	365	259	5180	1.14
High purity alumina	3970	257	162	7027	1.675
Silicon carbide	3120	213	187	8556	1.01

compression for the common case of linear elastic compression with a modulus $K + 4G/3$ followed by inelastic compression with constant modulus K, followed eventually by reversible unloading. For historical reasons, the hydrodynamic compression curve is called a "*hydrostat*".

If Eq. (3.18) applies, then one can write the stress in terms of compression on the principal Hugoniot as

$$\sigma_1 = \frac{\rho_0 C^2 \eta}{1 - s\eta^2} \tag{3.19}$$

where $\eta = \frac{\rho}{\rho_0} = \varepsilon + 1$, and ρ_0 is the initial density.

3.2 Shock Stability

Consider the wave form in Fig. 3.4. Each point moves with the local sound speed with respect to the material in front of it. With respect to the lab it moves with the sound speed plus the particle velocity into which it is propagating.

In order for this wave to evolve into a shock, e.g. a discontinuity, the sound speed must increase with pressure, so that it becomes steeper over time. Consider P, V space, Fig. 3.5 (ignoring here the difference between P and σ_1)

The shock speed is $U = V_0[(P_H - P_0)/V_0 - V_H)]^{1/2}$. The sound speed is $c = V\sqrt{\partial P/\partial V}$. For shock stability, it is necessary that c increases with pressure, so that as a wave propagates the high-pressure part overtakes the low-pressure part. This means that the $P - V$ curve must be concave up, as shown for the isentrope in Fig. 3.5. The shock that jumps from state 0 to $state$ **H** travels with a slope given by the Rayleigh line in Fig. 3.5. The Rayleigh line is also the path taken by the shocked material. Said another way, the requirement for formation of a stable shock in a material is that $\frac{\partial^2 P}{\partial V^2} > 0$. Conversely, if the curve is convex, shocks are unstable.

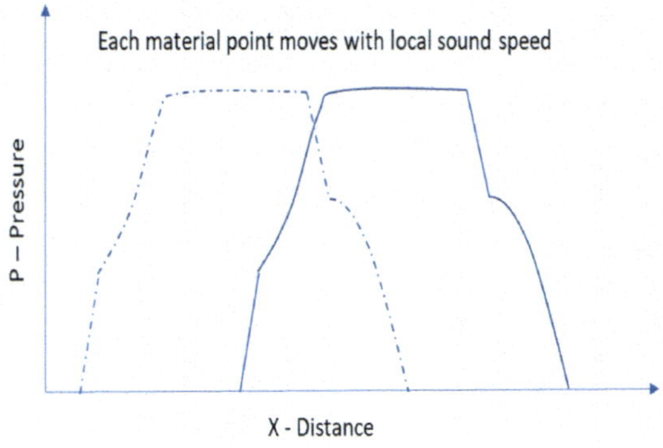

Fig. 3.4 A compression wave: loading and unloading, travelling left to right

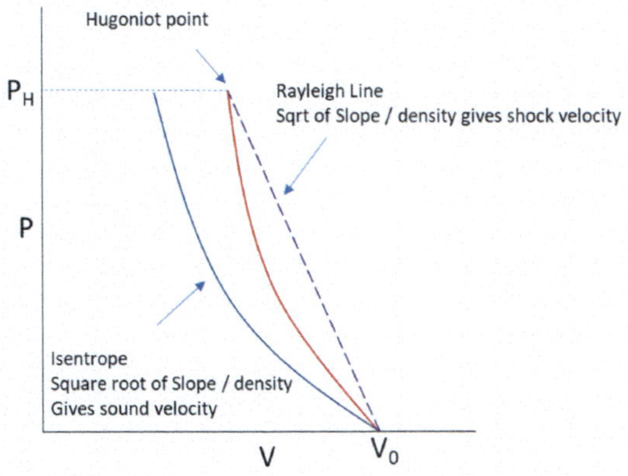

Fig. 3.5 Compression from initial specific volume V_0 and pressure P_0 to the Hugoniot point P_H, V_H

The reverse is true for release waves, which can propagate as shocks or dispersive waves according to the reverse criterion.

The energy in the shock is the area under the Rayleigh line. It is greater than if the material had reached P_H by a continuous compression along the Hugoniot or an isentrope. Hence, shock compression involves irreversibly heating, and a material that has been shocked and then isentropically unloaded is at a higher temperature than its initial state.

It is important to realize that the stability requirement often fails. There are three common reasons for concave down compression curves: collapse of void space, collapse of shear strength (e.g. elastic–plastic transitions), or collapse of crystal structure (e.g. phase changes).

3.2.1 *Elastic–Plastic Behaviour*

Elastic–plastic behaviour is the most common cause of shock instability. This is because the slope of the Hugoniot changes at the yield point. The limit of elastic behaviour is termed the Hugoniot Elastic Limit (HEL). Below the HEL the slope of the stress–strain curve is the constrained modulus, $K + 4G/3$. Above the HEL the slope is less than that. If deformation takes place with no additional shear strain, then the slope is K. The negative curvature of the Hugoniot around the HEL means that a high-pressure shock breaks up into an elastic wave, usually called the elastic precursor, and a plastic wave. In other words, the slope of the Hugoniot changes when the stress surpasses the HEL. That is,

$$\frac{d\sigma}{d\varepsilon} = K + \frac{4G}{3} \quad \text{for}\, \sigma < \text{HEL} \quad \text{and} \quad K \text{ for} \quad \sigma > \text{HEL} \tag{3.20}$$

This situation is pictured in Fig. 3.6. Above the HEL, the Hugoniot is normally concave up. This means that there is some pressure, point P in the figure, at which a

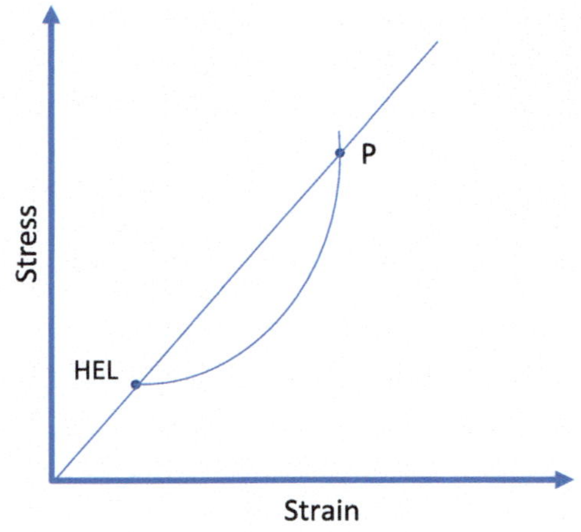

Fig. 3.6 Hugoniot for elastic–plastic behaviour

single shock will travel faster than the elastic shock. Thus at stress below H in Fig. 3.6 there is a single elastic shock. At pressures between HEL and P there is an elastic precursor. At pressures above P there is again only a single shock. However, for engineering applications involving ceramics, this situation is seldom encountered. The difference between the Hugoniot and the hydrostat is $2Y/3$, as discussed in Sect. 3.2.2. Thus, if both the hydrostat and Hugoniot are known, one can compute the strength at high pressure.

3.2.2 *Hugoniot in Stress-Particle Velocity Space*

The Hugoniot in (σ, u) space is very useful. The slope of this Hugoniot is the shock impedance. It is defined similarly to the acoustic impedance, $\rho_0 c$, namely, $\rho_0 U$. This follows from the Hugoniot equation (Eq. (3.14)). Shock impedance is designated by Z. It is often the case that it is effectively constant, as drawn in Fig. 3.7. It will be taken

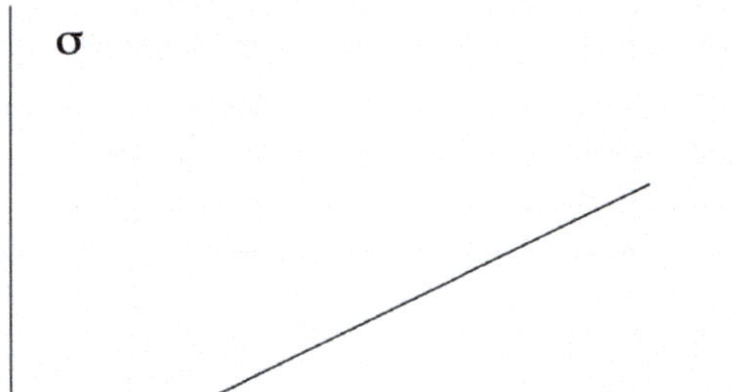

Fig. 3.7 Stress versus particle velocity for the case of constant shock impedance

as constant in this section, but solutions for non-constant Z proceed exactly as what follows, except they are carried out graphically or by solving quadratic equations. This is illustrated below in a worked-out example.

3.2.3 *Release Waves*

Consider a shock followed by a release wave shown in Fig. 3.8.

The incremental jump condition can be applied to integrate down the isentrope which gives the result

$$u = u_0 \int_{P_0}^{P} \frac{dP}{\rho c} \tag{3.21}$$

This is called the *Riemann integral*. It is very useful for determining the equation of state by integrating along a Hugoniot or isentrope.

3.2.4 *Impedance-Match Diagrams*

In Chap. 4, many specific examples of impact scenarios are discussed using layered targets or impactors. For example, Fig. 4.27 shows six different configurations that have proven useful for investigating dynamic response. The placement and thicknesses of layers can provide load/unloading, load/reshock, load/unload/reshock, etc., all of which can be checked against predictions of constitutive models.

The general method of analysing impacts involving materials in contact is by the use of impedance-match diagrams. In this method, loading paths are described in stress-particle velocity space. From Eq. (3.18), the equation of the Hugoniot with a linear $U - u$ relationship in $\sigma - u$ space is

$$\sigma = \rho_0 u(C + su). \tag{3.22}$$

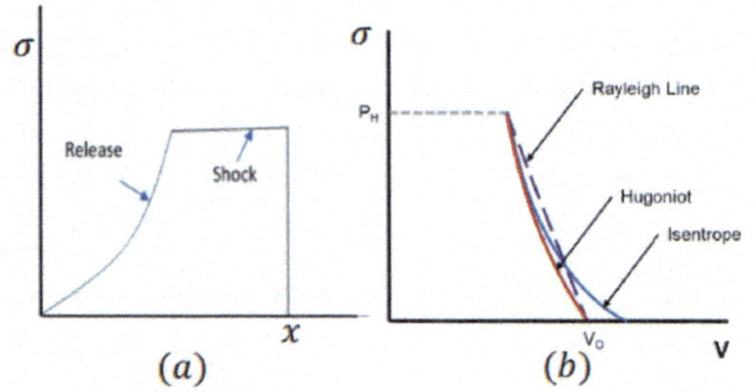

Fig. 3.8 A shock followed by a release wave along an isentrope

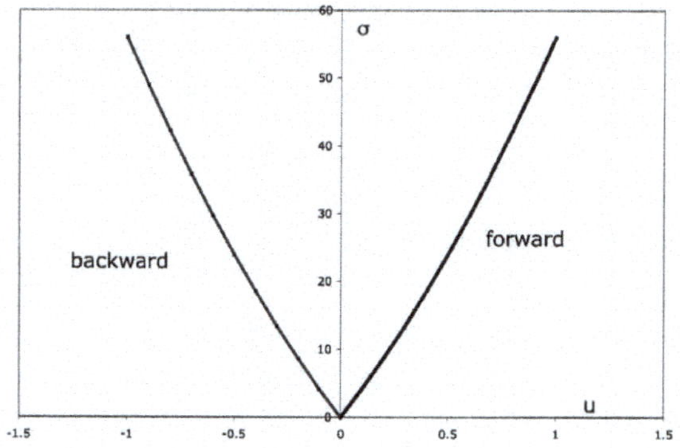

Fig. 3.9 Forward and backward running shock waves

Equation (3.22) applies to virtually all material of interest in this book. This is the principal of Hugoniot, footed at zero stress and zero velocity. When a material is shocked from that initial state, the stress and particle velocity will fall on this line. If the shock were running in the negative direction, this plot would be flipped around the vertical axis, so $u<0$ and $U<0$, but stress remains positive, as shown in Fig. 3.9.

If the material already has a particle velocity u_0 before impact, and the shock induces a velocity $\Delta u = u - u_0$ on top of that, then the jump in stress is

$$\sigma = \rho_0 \Delta u (C + s\Delta u) \tag{3.23}$$

In the often adequate approximation that the shock impedance $\rho_0 U$ is constant, Eq. (3.23) can simply be written as

$$\Delta\sigma = Z\Delta u \tag{3.24}$$

where $\Delta\sigma$ is the jump in stress and Δu is the jump in particle velocity.

The simplest multi-material consideration is an impact of a semi-infinite flyer plate of material B with velocity u_0 onto a semi-infinite target of material A. Impact is at $x = 0$, $t = 0$. In the σ, u-plane the impedance-match diagram looks like Fig. 3.10. Curve A is the Hugoniot for a right running wave in the target. It is footed at $u = 0$, the initial state of the target. Curve B is the Hugoniot for a left running wave in the flyer plate. It is footed at $u = u_0$, the initial state of the flyer plate.

Since the two materials are in contact after the impact, after impact the stress and particle velocity must be the same in both materials. The final state after impact must lie of both Hugoniots. The solution is that after impact both the flyer plate and the target must lie on state 1. Behind the right running shock, the material jumps to state 1. Behind the left running wave in the flyer it also jumps to state 1. For constant impedances, simple algebra gives state 1 as

$$\sigma = u_0 \frac{Z_A Z_B}{Z_A + Z_B} \tag{3.25}$$

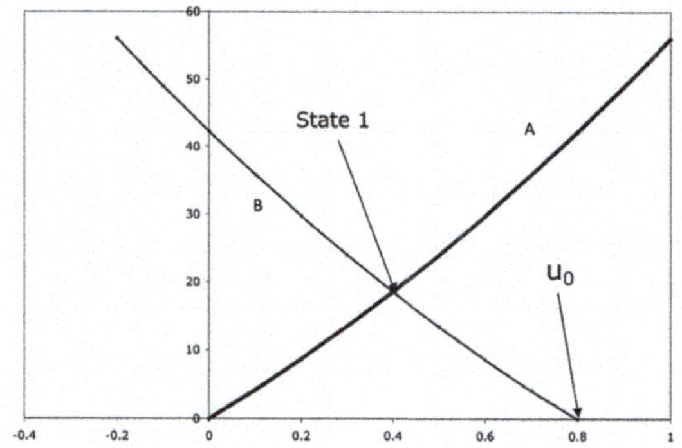

Fig. 3.10 Impedance-match diagram for impact on a target made of Material A struck by a flyer plate made of Material B, travelling at velocity u_0

Equation (3.25) is an adequate for a great many impacts involving stiff materials such as ceramics. If materials A and B are the same (e.g. a symmetric impact), then the resulting particle velocity is $u_1 = u_0/2$. This is a very useful insight for experiment design because it means that if we know the impact velocity, then we know the shock particle velocity, for like-on-like impacts.

The analysis of the more complex configurations discussed in Chap. 4 proceeds in exactly the same fashion. Every interaction between a shock and an interface produces left and right running waves. Also interactions between shocks can generate new waves as well. The Hugoniot equation (Eqs. (3.24)) always apply for each wave, and it is always true that after the shocks are generated the material behind the left and right running waves must be at the same stress and particle velocity. If there is no solution, then the interfaces separate and the stress is zero. That is the case of spall discussed in Chap. 4 (Sect. 4.2.3).

Some simple applications of this methodology reveal two important limiting cases. When compression waves arrive at a free surface, the reflected wave drops the stress to zero and the free surface velocity is twice the incoming particle velocity. This is the basis for many measurements, since it is much easier to measure free surface velocity than in situ velocity. For a rigid boundary, e.g. an interface with an infinite impedance material, the stress doubles and the particle velocity drops to zero.

In the case of release waves, the wave fronts in the impedance-match diagrams can disperse into fans of wavelets. But the above analysis still applies for both the jump across the fan and the stress path that follows from individual wavelets.

As an example, consider if the flyer plate is of finite thickness (Fig. 3.11). When the shock reaches the free surface behind the flyer plate, it reflects as release waves. Since these are unloading waves, they follow an isentrope, and a single shock is unstable. Depending on the experiments, the dispersion of the unloading waves may or may not be of interest. The stress behind the unloading waves is zero, because the stress at the free surface is zero. If materials A and B are the same, the particle velocity will be zero, and the flyer plate will be stationary behind the unloading wave. If the shock wave running in material A encounters a free surface, it will generate a fan of left-travelling release wavelets. Then the two release fans pass through each other,

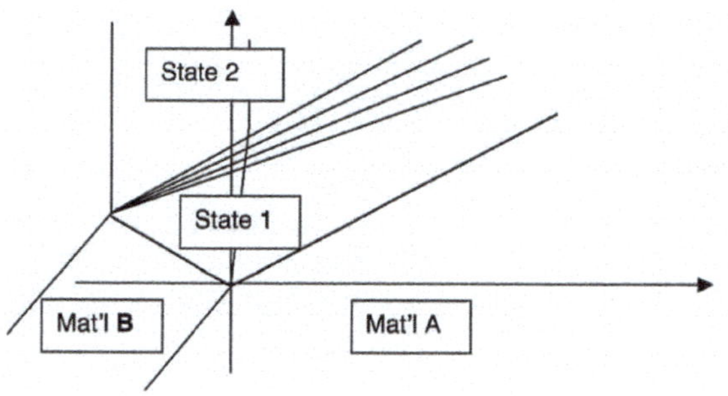

Fig. 3.11 Unloading from the rear surface of the flyer plate

the stress will fall to a tensile value equal in magnitude to the initial compressive stress. This can cause tensile failure, termed spall, as discussed in Chap. 4.

As explained above, a very important property of release waves is that they are faster than shock waves. So, for example, in the case of Fig. 3.11, the shock in Material A will eventually be "*erased*" by the release wave, and all that will remain is a right-travelling acoustic wave.

Release waves are also generated at lateral free surfaces. For a finite diameter flyer plate, this limits the extent of the shocked region in the target. To the extent to which the difference in shock and release speeds can be ignored, and the particle velocity is much less than the wave speeds, the region of the target that experience the full shock will be triangular, with a vertex angle of approximately 90°. This type of reasoning also applies to analysis in 1D stress scenarios, e.g. bar impact, as discussed in Chap. 4. Consideration of sample size the corresponding time during which shock analysis is valid is very important in design of experiments.

3.2.5 *Lagrangian Velocities*

The wave velocities discussed so far have been with respect to a laboratory observer. These, by definition, are Eulerian velocities. In most applications, Lagrangian velocities are more useful. These are the velocities with respect to the material points. There are three reasons why Lagrangian velocities are often used. The first is that in most applications this is the velocity that is measured. If a wave traverses a plate of material, the Lagrangian velocity is the original thickness of the plate divided by the transit time. The second reason is that Lagrangian x, t diagrams are simpler because there is no motion associated with material points. The third is that the Hugoniot equations are simpler when written in Lagrangian coordinates.

The Hugoniot equations are written with respect to Eulerian velocities. So in Eq. (3.12), for example, the wave front moves at speed U with respect to the material into which it travels, and velocity $U - u_1$ with respect to the material in front of the wave.

For example, for an elastic wave the wave speed with respect to the material into which the wave propagates is

$$C = \sqrt{\partial\sigma/\partial\rho} \tag{3.26}$$

If the propagation velocity is C, then with respect to the lab, the wave moves at

$$U = C \pm u \tag{3.27}$$

depending on which it is travelling upstream or downstream, where u is the particle velocity of the material into which the wave travels.

Lastly there is the Lagrangian velocity, which is with respect to material points. Material points are designated by their initial position, h. A Lagrangian velocity is $\Delta h/\Delta t$. This is the velocity that we measure when we divide an initial distance in the target by the travel time through it.

The Lagrangian framework is very useful when applied to impedance diagrams. The relationship between the jump in stress, $\Delta\sigma$, and the jump in particle velocity, Δu, across a shock front from Eq. (3.14) is

$$\frac{\Delta\sigma}{\Delta u} = \rho U \tag{3.28}$$

where ρ is the density in front of the wave and U of the velocity of the wave with respect to the material in front of the wave. Consider two points that are separated by Δx at some time. The two points were originally separated by Δh, but the material may be moving and it may have been compressed, so Δh is no longer equal to Δx. However, the mass between the points is the same in both Eulerian and Lagrangian coordinates, hence

$$\rho\Delta x = \rho_0\Delta h \tag{3.29}$$

Time is also the same in Eulerian and Lagrangian coordinates. The time it takes the wave to travel from the first point to the second is given by

$$\Delta t = \frac{\Delta x}{U} = \frac{\Delta h}{u_0} \tag{3.30}$$

where u_0 is particle velocity at the initial state. Hence, Eulerian and Lagrangian velocities (for a material initially at rest) are related by the density ratio as

$$\rho U = \rho_0 u_0 = Z \tag{3.31}$$

This is a considerable simplification when one needs to carry out successive applications of the jump conditions. One does not need to keep track of the current density or correct wave speeds for moving material points.

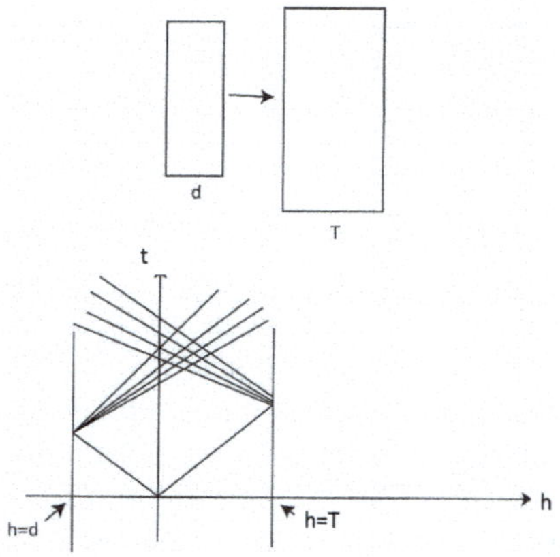

Fig. 3.12 Impact of a flyer and target in Lagrangian coordinates, showing shocks and release waves

If a wave can be broken down into a series of steps (e.g. it has a finite rise time), then Eq. 3.28 can be integrated. This means that integration through a record of C(t) and u(t) gives the stress–strain path followed by the material. Usually what is measured is stress versus time, so the integral is

$$u = u_0 \pm \int_{\sigma_0}^{\sigma} \frac{d\sigma}{\rho_0\, u_0} \tag{3.32}$$

As already mentioned, this is also known as a *Riemann integral*. In many experimental configurations, as discussed in Chap. 4, various techniques are used to avoid shocks, so that wave fronts have resolvable rise times, and then the integral in Eq. (3.32) can be evaluated. Strain, as usual, is given by Eq. (3.16).

The distance–time diagrams are much simpler in Lagrangian coordinates. For example, an impact that produces release fans from the backs of the target and flyer plate graphs like (h, t) the drawing in Fig. 3.12, which is much simpler than the corresponding (x, t) drawing in Fig. 3.11.

3.2.6 Example of a Lagrangian Impedance-Match Solution

Fluency in use of the Lagrangian framework for doing impedance-match solutions is essential for design and interpretation of experiments. In order to illustrate the procedures, we provide a worked-out example.

Consider the impact in Fig. 3.13. This is a buffer plate experiment. The purpose of such a configuration is usually to increase the effective rise time of compression shocks to investigate time-dependent behaviour. In this case, the target, material A, is a ceramic with generic ceramic properties. The buffer, aka driver plate, is a low-impedance metal, material B, for which we take properties of soft aluminium, for which strength effects can be ignored. Impact speed is 1000 m/s, the driver plate

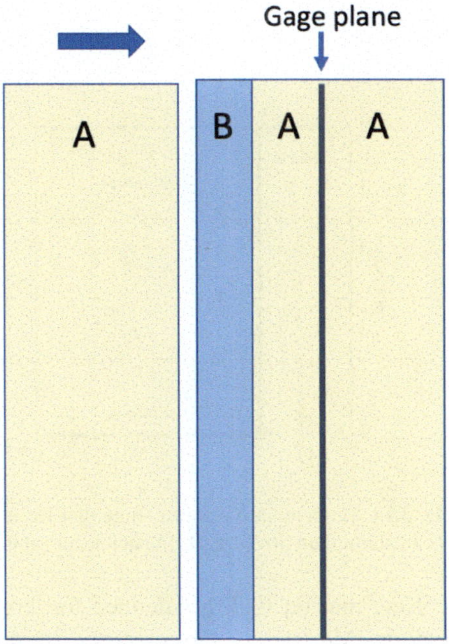

Fig. 3.13 Buffer experiment: **A** is Ceramic and **B** is Aluminium

thickness is 4 mm. There is a gauge plane in the ceramic where either stress or particle velocity measurements could be made, situated 5 mm behind the driver/ceramic interface. Impact occurs at $t = 0$. The flyer plate thickness and target thickness are taken as large enough that reflections from free surfaces will not affect the gauge plane during the times of interest. Likewise, the lateral dimensions are large enough that the strain can be considered one dimensional. That condition is approximately satisfied if the time for a wave to transit the flyer plate radius is significantly longer than the duration of the experiment.

The HEL of the ceramic, material A, is taken as $\sigma_H = 5\,\text{GPa}$. Its density is $\rho_{0A} = 3120\ \text{kg/m}^3$. Below the HEL the compression is elastic and the wave speed is the longitudinal sound speed, C_L which is taken as $12000\,\text{m/s}$. For plastic shocks, the shock speed will be described by

$$U = C_A + s_A u \tag{3.33}$$

where we will take $C_A = 8500\,\text{m/s}$ and $s_A = 1.1$. This is the principal Hugoniot, which applies to right-travelling shocks running into stationary material at zero pressure. Thus, using SI units, the principal stress, particle velocity Hugoniot of SiC is

$$\sigma = \rho_{oA} C_L u \quad \text{for} \quad \sigma < 5 \times 10^9 \tag{3.34}$$

$$\sigma = (3120)u(8500 + 1.1u) \quad \text{for} \quad \sigma > 5 \times 10^9 \tag{3.35}$$

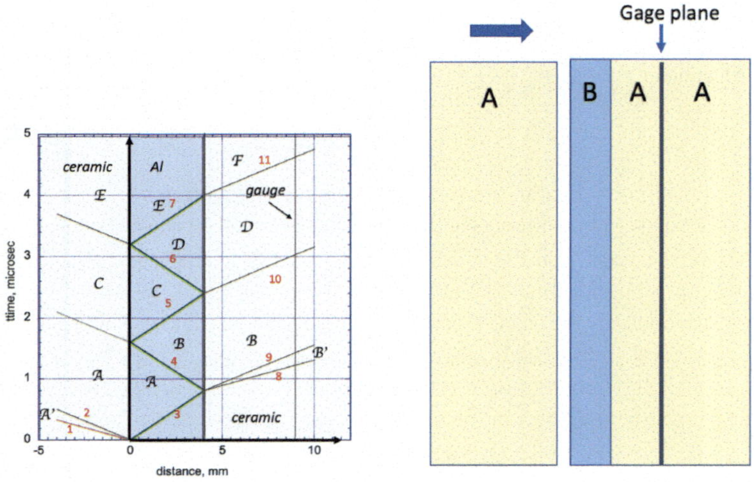

Fig. 3.14 (Lagrangian) distance–time plot for impact of ceramic flyer plate on a ceramic target, via an aluminium driver plate, Impact occurs at 0, 0. **A** is Ceramic and **B** is Aluminium

Note that Eq. (3.35) is defined for stress above HEL using Eq. (3.33). For aluminium, Material B, the reference density ρ_0 as 2700 kg/m^3 and the bulk wave speed is $C_B = 5000\,\text{m/s}$ so the principal Hugoniot is

$$\sigma = \rho_0 C_B u = (1.35 \times 10^7)u \tag{3.36}$$

We have solved this example graphically. That is often the easiest procedure for non-constant impedance, since precision better than three significant figures is seldom warranted in practical situations. This is the case in part because wave arrival times can normally be measured extremely precisely, and wave speeds are proportional to the square roots of the derivatives of the stress–strain relationships (moduli).

Next we will step through the Lagrangian distance–time (Fig. 3.14) diagram making use of the impedance-match diagram (Fig. 3.15).

Impact at $time = 0$ creates a left-travelling shock in the flyer plate and a right-travelling shock in the aluminium driver plate. The ceramic is elastic–plastic, so in the flyer there are two waves, front 1 is the elastic wave, and front 2 is the plastic wave. Line 3 is the shock front for a right-travelling shock in the driver plate. After the impact the flyer and the driver are in $state$ **A**. As discussed above, $state$ **A** is found from the impedance-match diagram, Fig. 3.15.

$State$ **A** is reached by a right-travelling wave into the Material B and a left-travelling wave into the flyer, Material A, and it must lie on the intersection of the Hugoniots of the two materials. In the flyer plate, the equation of the elastic part is line TA' for a left-travelling wave, that is, at zero pressure when the particle velocity is u_0, e.g.

$$\sigma = \rho_0 C_L (u_0 - u) = 3.46 \times 10^7 (1000 - u) \tag{3.37}$$

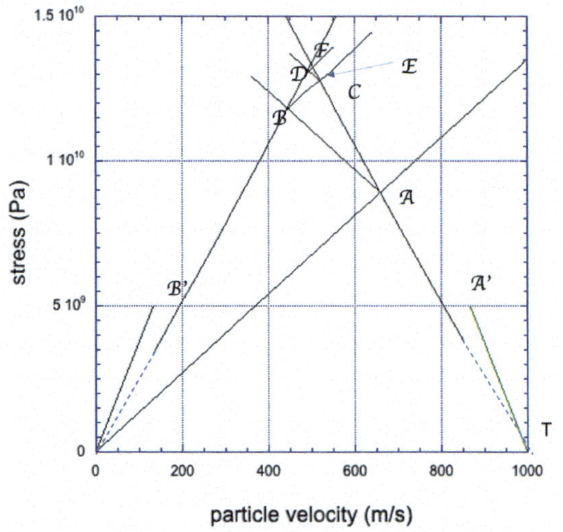

Fig. 3.15 Impedance-match diagram for Fig. 3.14

The limit of this line is at the HEL, where stress is $\sigma_{HEL} = 5\,\text{GPa}$ and $u_{HEL} = 856\,\text{m/s}$.

The plastic shock in the flyer must lie on line TA, whose equation is

$$\sigma = 3120(1000 - u)[(8500 - 1.1)(1000 - u)] \tag{3.38}$$

where u_0 is assumed as 1000 m/s. This comes from Eq. (3.36). These are states that can be reached by shocking SiC beyond its HEL. These are waves 1 and 2 in Fig. 3.14. These shocks in the flyer plate are not observed in this experiment.

The shock states in the driver plate (wave 3 in Fig. 3.14) must lie on the aluminium Hugoniot centred on the origin, that is, from Eq. (3.36), line OA in Fig. 3.15. The initial shock in the flyer and driver, *state* **A**, is determined by the intersection, and it is (659 m/s, 8.59 GPa).

When the wave 3 reaches the buffer/ceramic interface, a right-travelling shock is produced in the ceramic target and a left-travelling wave goes back into the buffer. Both waves start from *state* **A**. The elastic wave in the ceramic is determined from Eq. (3.34). This brings the ceramic target to *state* **A**$'$, where the stress is σ_{HEL}. The final shock state must lie on the plastic Hugoniot Eq. (3.35).

The reversed wave in the aluminium driver plate must lie on

$$\sigma = 1.37 \times 10^7(u - 659) + 8.59 \qquad \text{(units of m/s and GPa)} \tag{3.39}$$

which follows from Eq. (3.36). Their intersection produces *state* **B**, which occupies the material behind those waves, as shown in Fig. 3.14.

This construction continues for the waves that reverberate in the driver plate and transmit as shocks in the ceramic at the driver ceramic interface. Successive shock states in the ceramic target lie on line O, B', B, D, F. Note the limit state is from

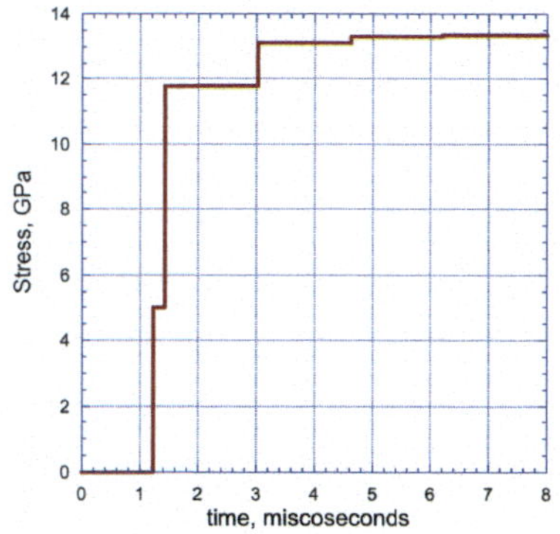

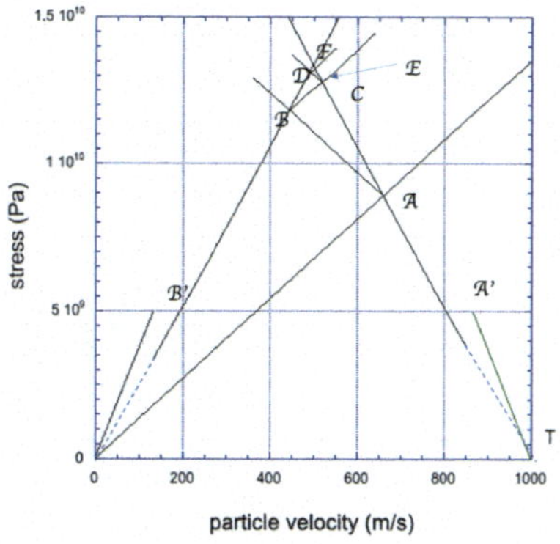

Fig. 3.16 Stress on gauge plane

the scenario in which the driver is not present. That is a like-on-like impact, and thus the particle velocity is approaching half the impact velocity, or 500 m/s. The stress is determined by the ceramic Hugoniot as 14.34 GPa.

The stress history on the gauge plane follows waves 8, 9, 10, 11 in Fig. 3.14. It would be as shown in Fig. 3.16.

This exercise shows how the gauge record is related to the assumed constitutive relationship. In experiments normally the task is to compute the latter from the former. The motivations for conducting such experiment might be to determine if the predicted final stress state is the same as that reached from a single shock, which could be affected by time-dependence, heating (from the altered Rayleigh line), or a more isentropic loading. Or it might check the assumption made in Fig. 3.15 that each stress step in the ceramic above the HEL is along the plastic Hugoniot, whereas, in some materials, additional loading follows an elastic stress path. This same configuration could also be used to study the driver plate material. It also is brought to 13.34 GPa but through steps A, B, C, D, E, F.

In this example, there was no unloading. If there were, then a judgement would have to be made as to whether unloading is elastic or plastic. Or more commonly, the point of the experiment would be to determine the unloading path from gauge measurements. In that case, the arrival times in the gauge record usually will determine the unloading modulus quite precisely.

3.2.7 *Entropy and Temperature*

Entropy increases along a Hugoniot but not much. To see this, consider a shock that starts from state 0. Its entropy can be expressed as a Taylor series:

$$S - S_0 = \left(\frac{dS}{dV}\right)_0 + \left(\frac{d^2S}{dV^2}\right)_0 + \left(\frac{d^3S}{dV^3}\right)_0 + \dots\dots\dots\dots \tag{3.40}$$

Using thermodynamic identities, it can be shown that the first two terms in this expansion are zero, and the change in entropy along a Hugoniot is

$$S - S_0 = \frac{1}{12T_0}(V_0 - V)^3\left(\frac{d^3P}{dV^3}\right) \tag{3.41}$$

In other words, it only depends on compression to the third order. It is very small, e.g. a Hugoniot is almost indistinguishable from an isentrope. Thermodynamically, along the isentrope the entropy remains constant; the process is reversible and adiabatic. This is especially true for relatively incompressible materials, such as ceramics.

This implies that the change in temperature from a shock wave is also very small for relatively incompressible materials. It depends on the Gruneisen parameter (γ), defined as

$$\gamma = V\left(\frac{dP}{dE}\right)_V \tag{3.42}$$

where E is the energy of the system. The rise in temperature along an isentrope is given by

$$T_S = T_0 e^{\left(\frac{\gamma \Delta V}{V}\right)} \tag{3.43}$$

where $\frac{\gamma}{V}$ can usually be taken as constant. Along the Hugoniot, we have

$$\frac{dT}{T} = \frac{\gamma dV}{V} + \frac{1}{2TC_v}(V_0 - V)dP + (P - P_0)dV \tag{3.44}$$

For typical metals this is only a few degrees/kbar, and less for ceramics.

3.3 Equation of State

An equation of state (EOS) is the relationship between the thermodynamic parameters of a material. There are usually three independent ones, for example, volume per unit mass, V, temperature, and pressure. Pressure, volume, and internal energy, E, are also often used as equation of state parameters. In shock physics, the main reason one needs on equation of state is to relate the Hugoniot to other thermodynamic paths, for example, the isentrope (Fig. 3.17).

To introduce the EOS concept, first consider a well-known relation for ideal gases. Its equation of state is given by

$$PV = nRT \tag{3.45}$$

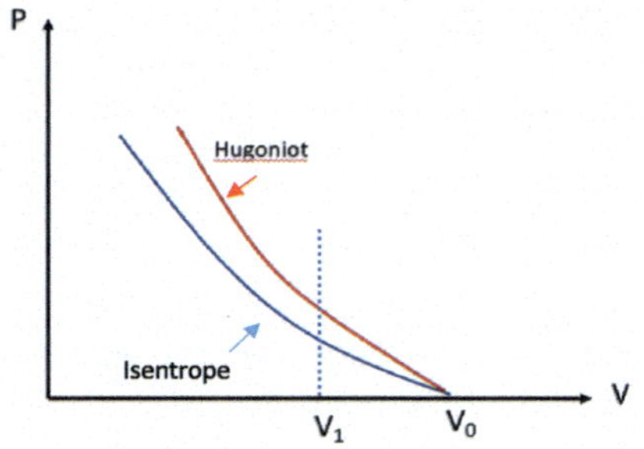

Fig. 3.17 Hugoniot pressure is higher than the isentropic pressure after compression from V_0 to V_1

where n is the number of moles, R is the gas constant, and T is temperature. For an ideal gas E is only a function of T, namely,

$$dE = C_V dT \tag{3.46}$$

Along an adiabat ($dS = 0$) the EOS can be found from the thermodynamic relationship given by

$$dQ = TdS = dE + PdV \tag{3.47}$$

$$dE = -pdV = C_V dT \tag{3.48}$$

$$T = -PdV/C_V \tag{3.49}$$

where C_V is the specific heat at constant volume. Also, we have

$$pdV + VdP = RdT = -PRdV/C_V \tag{3.50}$$

which can be manipulated to give the following equation:

$$(1 + R/C_V)(dV/V) + dP/P = 0 \tag{3.51}$$

There is a thermodynamic identity given by

$$C_V - C_P = TV\alpha^2/\beta \tag{3.52}$$

where

$$\alpha = -\left(\frac{1}{V}\frac{\delta V}{\delta T}\right)_P$$

and

$$\beta = -\left(\frac{1}{V}\frac{\delta V}{\delta P}\right)_T$$

Also for an ideal gas, we have

$$\left(\frac{\delta V}{\delta T}\right)_P = R/P \tag{3.53}$$

and

$$\left(\frac{\delta V}{\delta P}\right)_T = -\frac{RT}{V^2} = -\frac{P}{V} \tag{3.54}$$

Hence, $C_V - C_P = R$. Substituting into the dV, dP equation gives

$$\gamma\frac{dV}{V} + \frac{dP}{P} = 0 \tag{3.55}$$

where

$$\gamma = \frac{C_P}{C_V}$$

Is the ratio of specific heats. The result is the well-known equation for isentropic compression of an ideal gas:

$$PV^{\gamma} = constant \tag{3.56}$$

For solids, the most commonly used EOS is the Mie–Gruneisen, EOS. Here (and it what follows) the meaning of the symbol gamma is the Gruneisen parameter, defined in Eq. (3.42) as

$$\gamma = V\left(\frac{\delta P}{\delta E}\right)_V \tag{3.57}$$

With this, the Gruneisen EOS can be written as

$$E = E_0 + (P - P_0)\frac{V}{\gamma} \tag{3.58}$$

$$OR \tag{3.59}$$

$$P = P_{ref} + \frac{\gamma}{V}(E - E_{ref}) \tag{3.60}$$

This equation gives the change from a reference pressure (usually along the Hugoniot) when the internal energy changes at constant volume. In practice $\frac{\gamma}{V}$ is almost constant. It also can be shown from thermodynamic identifies that

$$\gamma = \frac{3KV\alpha}{C_V} \tag{3.61}$$

These coefficients are often known for materials of interest. For most materials, the Gruneisen parameter is between 1 and 3.

The offset between the Hugoniot and isentrope is given by the EOS Eq. (3.60) where the reference curve is the Hugoniot. E along the isentrope is found from $dE = TdS - PdV = -PdV$. The energy along the Hugoniot is

$$E_H = P_H(V_0 - V_H)/2 \tag{3.62}$$

The other common EOS used for shock analysis is based on the empirical linear relationship between u and U, e.g. Eq. (3.18). This can be put into the Mie–Gruneisen form using the definition of strain:

$$\varepsilon = \frac{V_0 - V}{V_0} \tag{3.63}$$

From which one obtains a very useful expression for the Hugoniot pressure:

$$P_H = \frac{P_0 C^2 \varepsilon^{\;\;2}}{(1 - S\epsilon)} \tag{3.64}$$

To put this in the Mie–Gruneisen form, we start with

$$P = \frac{\gamma}{V}E + f(V) \tag{3.65}$$

and solving for $f(V)$ using Eq. (3.62) to get the whole EOS.

$$P\,(V, E) = \frac{\gamma}{V}E + \frac{P_0 C^2 \varepsilon^{\;\;2}}{(1 - S\epsilon)}\left[1 - \frac{\gamma}{2V}(V_0 - V)\right] \tag{3.66}$$

This is also a very useful EOS.

There are other EOS formulas, but they are intended for extremely high pressures, much higher than are encountered in normal engineering practice. They should not be needed for the range of problems considered in Chap. 4.

3.4 Summary

Impacts produce much higher stresses than are observed in normal engineering practice. The field of shock physics provides the means to analyse impact effects.

The concept of a Hugoniot is central. When a material is shock loaded from normal conditions, the shocked state must lie on a curve called the Hugoniot. The Hugoniot is a material property. It is described by the Hugoniot equations which specify the values of two of the four variables that characterize a shock wave: density, particle velocity, shock wave speed, and stress.

The main analysis tool comes from impedance-match diagrams. This impedance analysis allows one to compute the shock stress that arises from an impact. It is based on the idea that when interfaces are in contact, which implies that stress and particle velocities are continuous, then the shock states must satisfy the Hugoniot conditions on both sides of the interface. Combined with a few other standard measurements, Hugoniot curves can be related to isentropes and pressure-dependent strength.

Plate impact experiments can be designed to determine constitutive relationships. Usually such experiments are designed using two analytic tools: distance–time graphs and impedance-match graphs. Chapter 4 provides several examples. There are subtleties involved in these techniques, and this chapter provides a worked-out example that shall be useful to new practitioners.

References

1. Mayers MA (1994) Dynamic behavior of materials. John Wiley & Sons Inc
2. Zukas JA (1990) High velocity impact dynamics. John Wiley & Sons Inc
3. Walker JD (2021) Modern impact and penetration mechanics. Cambridge University Press
4. Kanel GI (2019) Shock waves in solid state physics. CRC Press
5. Grady DE (1995) Dynamic properties of ceramic materials. Sandia Report SAND94-3266
6. Davidson L (2008) Fundamentals of shock wave propagation in solids. Springer Science & Business Media
7. Gathers GR (1994) Selected topics in shock wave physics and equation of state modelling. World Scientific
8. Grady DE (2017) Physics of shock and impact, vol 1. IOP Publishing
9. Reinhart WD, Grady DE, Chhabildas LC, Mashimo T (2001) Shock compression and release properties of Coors AD995 alumina. In: Ceramic armour materials by design. Ceramic transactions, vol 134. American Ceramic Society

Chapter 4
Impact and High Strain Rate Experiments

Prologue

A substantial body of research exists concerning the ductility, flow, and fracture of metallic materials when subjected to high-rate compressive, shear, and tensile loads. Ductility is a crucial property that generally varies with factors such as strain rate, temperature, and inertia. Additionally, tensile strength is influenced by strain rate, temperature, and loading history. To illustrate, in applications involving projectile penetration, penetration resistance—defined as the stress a material can withstand penetration—often correlates with compressive shear strength in most models. For brittle materials, compressive strength also holds significance for a slightly different reason. In effective ceramic armour applications, the resistance to penetration due to compressive stresses is so high that penetration only occurs after damage has developed. In many cases, tensile strength is so low that tensile damage becomes inevitable. Therefore, penetration takes place after damage has occurred, making the compressive strength of damaged ceramic material highly significant. Moreover, the degradation of strength resulting from damage is highly sensitive to confining pressure (or mean stress) in brittle materials.

However, quasi-static compressive strength tests using equipment like Instron™, UTS™, or MTS™ testing systems alone are insufficient to fully characterize the impact of strain rate and pressure on the mechanical properties of brittle materials. A range of diagnostic impact and strain rate experiments is necessary to assess the dynamic response of brittle materials. Specialized experiments are also required to recover specimens that have been shocked or deformed, helping to establish micro/meso/macro-damage mechanisms. The impact tests are based on various configurations, such as a normal or oblique plate impacting another normal or oblique plate, rod-on-rod tests, single-density plate-on-rod setups, and graded-density plate-on-rod configurations. High strain rate tests predominantly involve stress wave propagation experiments, including the use of the split Hopkinson pressure bar.

A. M. Rajendran et al., *Failure of Brittle Materials Under Shock and Impact*, Materials Horizons: From Nature to Nanomaterials, https://doi.org/10.1007/978-981-97-7026-7_4

4.1 Split Hopkinson Pressure Bar (SHPB)

Over the past four decades, the Split Hopkinson Pressure Bar (SHPB) experimental setup has garnered extensive attention in journal articles and books. Notably, a book by Chen and Song [1] delves into the design, testing, and applications of the Split Hopkinson (Kolsky) bar. More recently, Ramzi Othman [2] has edited a volume on SHPB-related topics, featuring a chapter by Stephen Walley [3] from the Cavendish Laboratory, Cambridge, UK, titled "*The Origins of the Hopkinson Bar Techniques*". The SHPB technique has seen significant evolution and advancement, thanks to contributions from researchers worldwide since its inception in 1918.

Initially, Kolsky [4] introduced a torsional setup to obtain shear stress–shear strain curves for metals. However, over time, the Kolsky bar testing concept was extended to encompass tension and compression loading conditions through the SHPB setup. With the advent of high-resolution digital amplifiers, oscilloscopes, highly sensitive strain gauges, and high-speed cameras, the application of compression SHPB techniques has expanded to testing brittle materials. Various SHPB configurations have been developed for testing brittle materials under high strain rates and confining pressure. In this book, our aim is not to provide exhaustive reviews of all available SHPB testing methodologies. Instead, we intend to offer a guide to the existing body of literature on SHPB techniques suitable for the compression testing of brittle materials.

This book primarily focuses on the characteristics of brittle substances and introduces modifications to the conventional bar setup necessary to obtain desired and valid measurements in compression testing of brittle materials. Unlike metals, where "*plastic*" or inelastic strain is observed under uniaxial compressive stress and high strain rate loading conditions, in high-strength ceramics, such strain is primarily a result of cracking and fragmentation. In most cases, it remains unclear whether softening under compressive load is an intrinsic property or an extrinsic behaviour associated with cracks originating from surface flaws, especially in the case of glasses, or stress concentrations at the anvils. Consequently, the "*ductility*" associated with 1D compression testing of ceramic materials under dynamic loading conditions holds little relevance in engineering applications. However, there are "*intrinsic*" strains associated with strength properties that can be measured through SHPB tests. The absence of "*plastic*" strain in ceramics presents significant challenges for measuring compressive strength because stress–strain relationships cannot be discerned during loading. Ravichandran and Subhash [5] have provided pertinent SHPB testing techniques for ceramics, cautioning that extreme care must be exercised when testing high-strength ceramics at high strain rates. They also discussed the inherent limitations in SHPB tests, test procedures for high-strength ceramics, design modifications, and the effectiveness of proposed adjustments. Chen et al. [6] offer further insights into the design of ceramic specimens for SHPB tests. It is worth noting that the Hopkinson bar loading mechanisms can be adapted for other geometries, such as a Brazil disc test (e.g. Li et al. [7]). The Split Hopkinson Pressure Bar (SHPB) experimental setup, as shown in Fig. 4.1, has evolved into a standard mechanical property

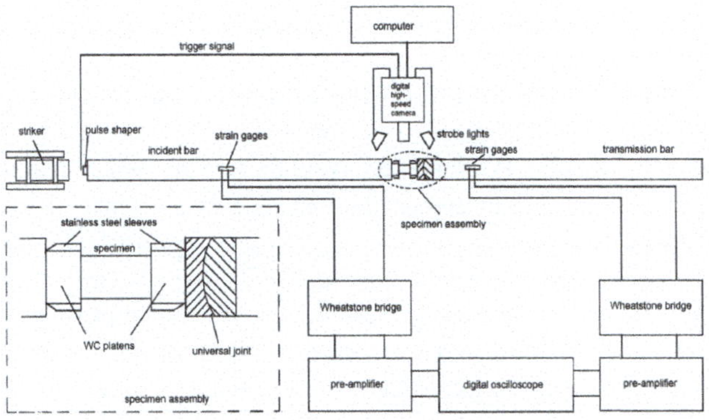

Fig. 4.1 A schematic of SHPB setup with mounted strain gauges on the incident and transmitted bars (Courtesy of Chen et al. [8])

testing technique for assessing the strain rate dependency of compressive strength in a wide range of materials, including metals, ceramics, rocks, and concrete. However, the schematic presented here features a unique specimen assembly designed specifically for testing high-strength ceramic materials. To prevent premature specimen edge failure, a pair of laterally confined tungsten carbide platens with mechanical impedance matching to the bars has been strategically placed between the specimen and the bars. Chen et al. [8] further improved this setup by introducing a pair of curved hardened tool steel discs (universal joints) between the tungsten carbide platen and the transmission bar to correct any minor misalignments. This reference is invaluable for gaining insight into the intricacies of modifying traditional SHPB setups.

In the conventional SHPB test, the specimen is positioned between two long bars: the incident bar, through which a stress pulse loads the specimen, and the transmission bar. A striker bar impacts the incident bar at a velocity that generates stress pulses in both bars with amplitudes below their respective yield strengths. Consequently, only elastic waves propagate through the bars. The SHPB essentially imposes a displacement boundary condition on the specimen as the gap between the bars closes upon the arrival of the stress incident wave. The stress in the specimen is transmitted into the incident and transmission bars, and their elastic response is monitored through strain gauges. The analysis of strain gauge signals is used to determine the engineering stress and strain in the sample.

The strain gauges are connected to a digital amplifier-oscilloscope recording system to measure voltage-versus-time histories. The strain gauge manufacturers provide calibration charts or equations for converting the recorded voltage into strains in the bars. SHPB wave propagation analysis yields the relationships between strain histories in the two bars and the load/displacement conditions on the sample, allowing for the straightforward calculation of true stress, true strain, and strain rate in the specimen. SHPB experiments are typically designed to maintain a constant strain rate, ensuring that the results can be directly related to input parameters in a constitutive model.

The bars are typically fabricated from high-strength steels for testing both brittle and ductile materials, except in cases involving relatively weak samples, where alternative materials may be used. For instance, Wayne Chen and his colleagues at Purdue University employed high-strength maraging steels with a density of 8.1 gF/cm^3 for the striker, incident, and transmission bars. Young's modulus (E) and the speed of sound (C_0) for these bars were 200 GPa and 4970 m/s (5 mm/μs), respectively. The incident and transmission bars had diameters of 19.05 mm (3/4") and lengths of 2130 mm (7 ft) and 915 mm (3 ft), respectively. The striker length is adjustable to control the loading duration as needed, but its diameter must match that of the incident bar. The choice of bar materials and dimensions may vary among different researchers and laboratories worldwide. It's not uncommon to encounter the use of low-impedance materials like aluminium, magnesium, and polymers for testing very soft materials such as biological tissues, rubbers, and soft polymers (see References Lee et al. [9], Gray [10], Chen et al. [11], Song and Chen [12], Chen and Song [13], and Chen [14]).

4.1.1 Testing Challenges and Issues

In a recent study, Chen [14] not only outlined the prerequisites for testing soft materials under high strain rate conditions but also offered insights into the general challenges and issues associated with the Split Hopkinson Pressure Bar (SHPB) testing methodology, along with suggested remedies for each challenge. We will delve into these challenges in the following sections. While the interpretation of strain gauge data has become increasingly precise and the data reduction and interpretation methods have been fine-tuned over the past two decades, measurements on brittle materials pose distinct challenges. These challenges revolve around:

(a) *Dynamic equilibrium (uniform stress) in the specimen*: Ensuring a uniform stress distribution within the specimen during dynamic loading.
(b) Specimen aspect ratio: Managing the specimen's aspect ratio towards determining realistic stress–strain data.
(c) *Correction for radial inertia of the bar*: Correcting for the radial inertia of the bars to achieve accurate measurements.
(d) *Homogeneous deformation in the specimen*: Ensuring homogeneous deformation in the specimen, which is essential for reliable results.
(e) *Pochhammer–Chree dispersion correction.* Compensating the incident bar strain gauge data due to wave dispersion issues.

Additionally, to maintain elasticity in the bars, very hard platens are often necessary. Moreover, imperfections in the interface geometry can lead to failure at stress concentrations rather than under average conditions. Sources of errors in data may include

1. *Inaccurate strain signal*: Arising from stress wave reflections from non-flat specimen surfaces.
2. *Imperfect Bar End Faces and Contacting Conditions*: These can lead to inaccuracies in data.
3. *Lubricating layers*: Such layers between the specimen, platens, and bar end faces can introduce errors. While these errors are often transient, they can obscure the intrinsic specimen response. To mitigate axial misalignment, many SHPB setups employ a laser beam.

Chen and Song [13] even discussed the use of a universal joint in conjunction with the laser beam to ensure nearly perfect alignment of all three bars. An insert in Fig. 4.1 illustrates a schematic of the universal joint between a hardened platen and the transmission bar used by Chen et al. [8].

4.1.2 Wave Shaper Effects

A "*wave shaper*" is often interposed between the striker and the incident bar to alter the shape of the loading pulse generated by the impact. In a review by Follansbee and Franz [15], and further expanded upon by Franz et al. [16], the concept of pulse shaping for SHPB (Split Hopkinson Pressure Bar) experiments involving metal samples was extensively discussed. These papers highlight the preference for a gradually rising incident pulse over a steeply rising one, aiming to minimize the impacts of dispersion and facilitate the achievement of dynamic stress equilibrium within the sample. They also discuss experimental techniques for pulse shaping, including a numerical procedure designed to rectify raw data affected by wave dispersion in the bars. To shape the incident pulse, these authors employed methods such as machining a large radius on the impact face of the striker bar or placing a ductile shim onto the incident bar's surface. These shims, acting as pulse shapers, typically comprised discs slightly larger than the bars and were 0.1 to 2.0 mm thick. Materials used for the shims ranged from paper to aluminium, brass, or stainless steel. Nemat-Nasser et al. [17] also proposed a "wave or pulse shaping technique" to mitigate several loading issues associated with SHPB experiments.

An example of how the shape (both amplitude and time duration) of the stress pulse can be modified by inserting thin discs of various soft materials like copper and rubber is shown in Fig. 4.2. In general, the wave or pulse shapers decrease the rise time, and may decrease the stress amplitudes. In this example, the wave forms with rubber probably correspond to the actual strength of the brittle specimen.

In particular, for brittle materials, achieving stress equilibrium is critical since failure might occur before reaching equilibrium. For instance, using an annealed copper disc as a pulse shaper for striking relatively low-strength ceramics led to generating a smoother stress pulse. Chen and Song in their book [1] employed an annealed C11000 copper disc with a diameter of 10.21 mm and a thickness of 0.79 mm as the pulse shaper for striking speed of 15 m/s in the Macor (a relatively low-strength

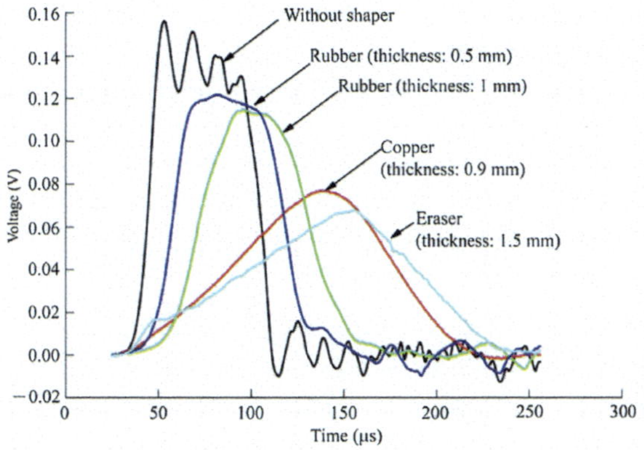

Fig. 4.2 Influence of wave shaper materials on the measured strain gauge signals in SHPB (Xia and Yao [18])

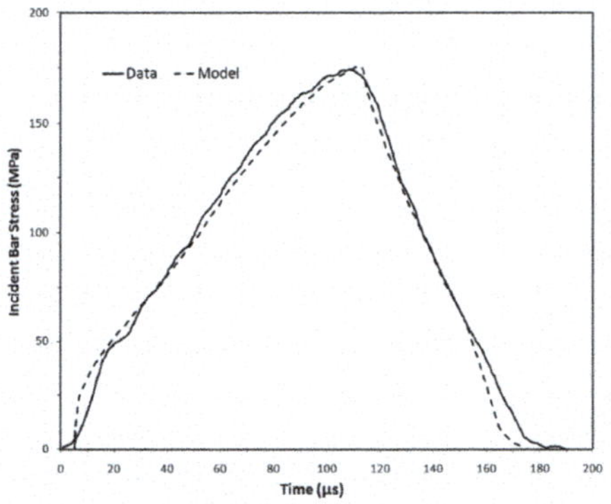

Fig. 4.3 Smooth incident pulse with increased rise time with the use of annealed copper pulse shaper for testing a relatively softer machinable glass ceramic material, Macor [19]

ceramics) characterization. The incident pulse with annealed copper wave shaper produced a smooth stress pulse as shown in Fig. 4.3.

In addition to the pulse shaping with copper inserts between the striker bar and incident bar, Forrestal and Warren [20] proposed a tapered (or a truncated conical) striker bar to obtain constant engineering strain rate for the specimen response. This modification to the striker bar enables shaping of the incident bar strains. Their analytical solution to this striker bar configuration showed the possibility of obtaining constant true strain rate for the specimen.

4.1.3 Specimen Length or Aspect Ratio Effects

Controlling the initial strain rate in a material can be achieved through tailoring the specimen length. The strain rate experienced by the specimen is inversely proportional to its length. Opting for a shorter specimen is ideal to achieve higher strain rates. The standard length-to-diameter ratio of 1.0 is commonly employed for ceramics. However, there's a limitation on strain rates for brittle materials due to their small failure strain, leading to premature fracturing. For brittle materials with lower wave speeds, like Macor, rocks, or porous cement, the specimen length is also limited by the requirement for dynamic equilibrium. A lengthy specimen tends to fail from the impact end before experiencing homogeneous deformation or before evenly distributing the load across its thickness. Xia and Yao [18] extensively reviewed dynamic rock tests using SHPB, covering various diagnostic systems and stress states. Given the low fracture strains ($< 1\%$) in brittle materials, utilizing a DIC (digital image correlation) system and strain gauges on the specimen provides more accurate stress–strain curves, as reported by Song and Chen [12] and Chen and Song [1].

The necessary specimen size also depends on the sample's heterogeneity. Larger specimens with an aspect ratio (L/D) of one and a diameter of 19.05 mm or above are typically used for concrete or rock specimens in bars ranging from 50.8 mm to 76.2 mm in diameter. Conversely, much smaller specimens (about 10 mm diameter or below) are employed for testing crystalline ceramic materials in smaller diameter SHPB. For instance, Grote et al. [21] used an aspect ratio between 0.3 and 1.55 for mortar materials. Smaller specimens prove beneficial in capturing stress–strain data at elevated strain rates using the SHPB setup, especially with ceramics known for their high compression strength, difficulty in deformation, and fracture resistance compared to cementitious materials.

In a recent study (2023), Zhou et al. [22] conducted SPHB tests on granite, varying the length-to-diameter ratios from 0.5 to 1. They observed that the compressive strength of granite increased with higher strain rates and increasing L/D ratio. The tests, performed with pulse shapers, revealed the significant influence of specimen size on the measured compressive strength of granite. Moreover, fracture strain and the number of fragments increased with the strain rate. Notably, a L/D ratio of 0.5 doubled the final strain compared to L/D of 1.

4.1.4 Pochhammer–Chree (P-C) Correction

The Pochhammer–Chree (P-C) correction is utilized to adjust the incident bar strain gauge data to account for wave dispersion between the gauge location and the specimen interface. It's crucial to note that SHPB data reduction relies on the assumption that the force applied to the specimen at the bar end equals the force calculated using the measured data at the gauge location. However, wave dispersion in the bar can introduce errors without P-C corrections. Forrestal et al. [23] modelled radial inertia effects on dispersion in SHPB and formulated expressions for three additional stress components induced by radial inertia in a valid SHPB test, applicable to incompressible and compressible, linear elastic material samples. Their research indicated that inertia-induced radial stresses are generally negligible in rocks and ceramics when compared with the axial stresses calculated for a state of uniaxial stress. However, in the case of large-diameter concrete specimens, lateral confinement from radial inertia could significantly impact test data. Closed-form solutions that encompass various SPHB attributes, such as the effects of radial inertia on longitudinal stress, are particularly useful in the initial stages of experimental designs involving high strain rates.

Gorham [24] demonstrated minimal or insignificant effects of dispersion in SHPB, suggesting that, for practical purposes, P-C correction might not be necessary for conventional high-strength metallic bars. However, when using plastic or other softer bars to derive stress–strain curves for softer materials, applying the P-C correction to the measured strain gauge signals becomes essential. P-C effects primarily relate to high-frequency components in the waveform resulting from rapid radial acceleration in the bar due to high-frequency motion. These effects can also manifest in the specimen, necessitating a smaller specimen diameter compared to the incidence pulse wavelength to reduce or eliminate radial oscillations.

The pulse shaping also ensures the desired testing conditions achieved on the sample. One of the benefits of pulse shaping is that the high-frequency components are physically filtered out. The fundamental frequencies are nearly non-dispersive. In this way, the sample is loaded by a well-defined and clean loading history (no oscillations). There is also no need to perform P-C corrections. Therefore, it is always better using a combination of metallic bars and pulse shapers for both strong and soft material specimens so that uncertainties associated with the wave dispersion can be minimized or eliminated. The books by Chen and Song [1] and Song and Chen [12] elaborate these aspects with underlying models and examples on SHPB testing of brittle as well as soft materials. There are several review articles, such as the one by Ramesh [25], to which readers may refer to learn more aspects of the SPHB testing methodology.

In addition to pulse shaping, the specimen length must be relatively short, ensuring about 3 to 10 wave reverberations within the specimen for stress uniformity or homogeneous deformation. Davies and Hunter [26] provided an analytical solution in 1963, including inertial and frictional terms in the governing equations. Their work suggested that frictional effects could be negligible if the compressive specimen's optimal thickness corresponds to an L/D ratio of $\vartheta\sqrt{3/2}$ where ϑ represents Poisson's ratio of the specimen material. To minimize or eliminate radial inertia and friction effects at specimen/bar interfaces, their analysis proposes using an L/D ratio between 0.3 and 0.5 for SHPB compressive specimens, especially applicable to metallic and polymer materials with high ϑ ($>$ 0.33).

4.1.5 SHPB Analysis

We present this analysis to ensure comprehensive coverage in our discussions about SHPB setups. The basic SHPB configuration allows for a wide range of modifications. The equations used to derive stress–strain curves from measured strain gauge records may slightly differ across various configurations. These differences stem from factors like the placement of strain gauges and the type of diagnostics employed, such as high-speed camera imagery in conjunction with strain gauges on the specimen. It's

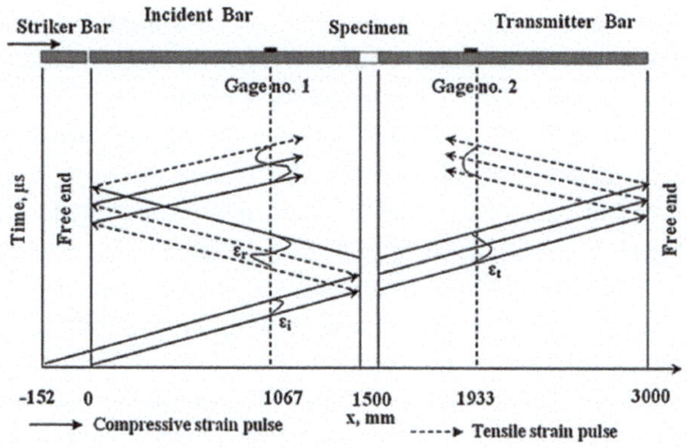

Fig. 4.4 A schematic of $X - t$ diagram for the compressive SHPB setup. The incident, transmitted, and reflected strain pulses are indicated by ϵ_I, ϵ_T, and ϵ_R, respectively

advisable for readers to acquaint themselves with stress wave propagation in the Hopkinson bar setup, illustrated through Lagrangian or $X - t$ (distance versus time) diagrams, as shown in Fig. 4.4.

Upon impact of the striker bar on the incident bar, compression waves travel through both bars. The wave in the striker bar rebounds from its free end as an unloading (release) wave, traversing back into the incident bar. Simultaneously, the incident wave travels through the specimen into the transmitted bar, with a portion of it reflecting back into the incident bar. The release wave from the striker bar and the partially reflected wave from the incident bar end/specimen interface reach the gauge locations, as shown in Fig. 4.4.

The derivation details for the three resulting equations can be found in various sources. Key references include Nicholas and Rajendran's work on material characterization at high strain rates in "*High Velocity Impact Dynamics*", edited by Zukas in 1990 [27]. Additionally, Michael Adam Kaiser's Master of Science Thesis on "*Advancements in the Split Hopkinson Bar Test*" at Virginia Polytechnic Institute [28] presents the essential theoretical developments associated with the SHPB. Ramesh in his review article on "*High Strain Rate and Impact Experiments*" [25] discusses the challenges of obtaining accurate data for ductile and brittle materials. Based on the wave propagation analysis, the following three main equations are obtained for the time histories of strain rate $\dot{\varepsilon_s}$, strain ε_s, and stress σ_s, in the specimen.

$$\frac{d\,\epsilon_s(t)}{dt} = \dot{\epsilon}_s = -\frac{2C_0}{L}\epsilon_R(t)$$
$$\epsilon_s(t) = -\frac{2C_0}{L}\int_0^t \epsilon_R(t)\,dt$$
$$\sigma_s(t) = E\frac{A_s}{A_b}\epsilon_T(t) \tag{4.1}$$

The recorded reflected pulse (ε_R) and transmitted pulse (ε_T) from the strain gauge are used to determine the stress and strain response of the specimen. Here, L represents the length of the specimen, C_0 is the sound speed in the bar, and As and Ab are the cross-sectional areas of the specimen and the bar, respectively. These equations are derived from the force equilibrium of the specimen and the application of Hooke's law for linear elastic behaviour in the bar.

Obtaining stress-strain data from wave propagation analysis (Eq. (4.1)), is straightforward for metallic systems. A stress vs. strain rate plot can be constructed using data from a series of SHPB experiments conducted at varying striker bar velocities. Notably, this analysis applies to both tensile and compressive loading conditions.

SHPB data for a wide range of ductile and brittle solids consistently indicate that stress (or strength) increases with higher strain rates. However, brittle materials, such as rocks, concrete, and ceramics, exhibit significant strain rate sensitivity in their maximum compressive strength. For example, in rock materials, average compressive strength increases from approximately 10 MPa at low strain rates to over 100 MPa at high strain rates, as shown in Fig. 4.5. This increase in flow stress (strength) with rising strain rates, evident in most brittle materials, is critical for developing constitutive (or strength) models suitable for high strain rate applications.

High strain rate stress–strain curves, especially for metals at elevated temperatures, can be obtained via Split Hopkinson Pressure Bar (SHPB) tests by heating the specimen to the desired temperature using methods such as an induction coil wrapped around the specimen [30] or an air-cooled infrared spot heater [31]. Yet, conducting high strain rate characterization of brittle materials at elevated temperatures remains relatively rare, potentially due to the limited scope of engineering applications. Nonetheless, there are instances of such experiments. Lankford [32], for instance, reported high-temperature compressive strength data for aluminium oxide.

Tension Tests: Direct tensile tests of brittle materials with an SHPB are inherently difficult because the strain-to-failure is usually so small that stress equilibrium is not achieved. Additionally, failure commonly arises due to stress concentrations at the specimen grip, posing another challenge. An alternative, as mentioned earlier, is using the compression setup but replacing the sample with a disc (whose face normal is orthogonal to the bars) as shown in Fig. 4.6. This is known as a Brazilian disc geometry, and it is a way to measure tensile strength even though the loading

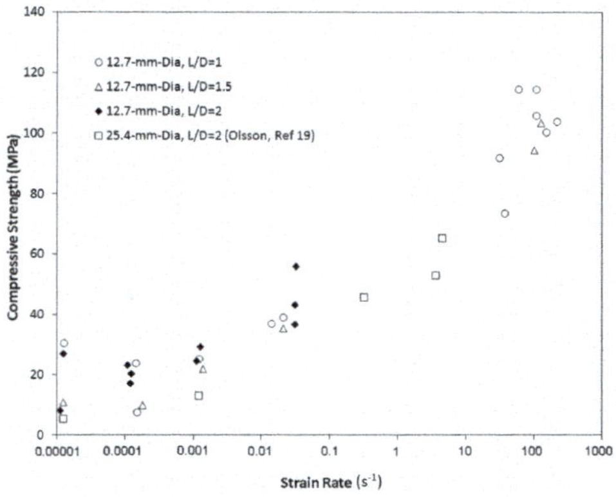

Fig. 4.5 Stress–log strain rate plots for rock materials [29]

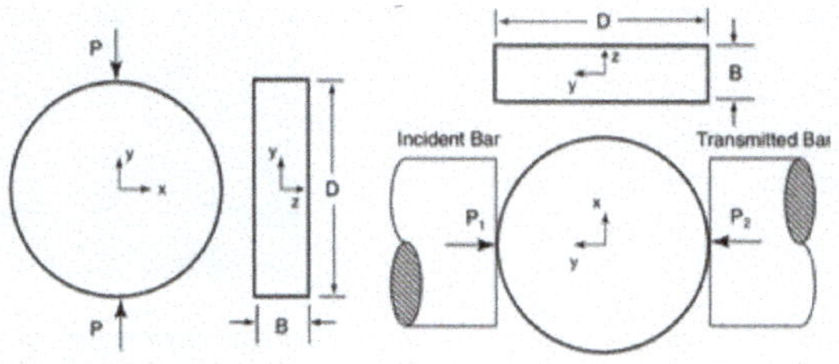

Fig. 4.6 A schematic of a Brazilian (splitting tensile) test in SHPB (Dai and Xia [39])

is compressive [33–38]. Opposing edges of the disc are loaded in compression, and the force is computed from the gauge records in a manner similar to conventional SHPB analysis. However, a brittle disc will usually fail due to the tensile stresses that develop in the centre of the disc. The validity of the stress–strain data is contingent upon crack nucleation occurring at the centre of the disc. Ensuring this validity necessitates the utilization of high-speed camera diagnostics during Brazilian tests. For example, recently, Chen et al. [33] provided a detailed description of how to use the Brazilian test to determine the tensile strength of alumina ceramic and other brittle solids.

The stress in the lateral (x) direction can be simply calculated from "$2P/\pi DB$", where P is the load, and B and D are the thickness and diameter of the disc, respectively. From the time histories of strain rate and stress determined from the strain gauges on the bars, the peak stress at fracture initiation and the corresponding strain rate can be determined. This test configuration is mostly used by the rock and concrete communities to determine the strain rate-dependent tensile strength.

Compression Tests: In brittle materials, compressive or shear failure often depends on mean stress, underscoring the need to acquire high strain rate data not only for unconfined compression (common in the SHPB configuration) but also for confined compression. Additionally, results obtained from unconfined samples frequently produce highly statistical outcomes. The research articles by Chen et al. [1, 8] provide examples of SHPB compression tests with confinement, emphasizing the importance of conducting precise, highly diagnostic experiments supported by computational analyses. Figure 4.7 shows high-speed camera images displaying the dynamic fracture and failure progression of a silicon carbide ceramic specimen under uniaxial compression. These images exhibit the gradual fracture of the specimen and can be temporally correlated with the engineering stress–strain data in Fig. 4.8. This figure corresponds to the stress history in the specimen, aligning with the images in Fig. 4.7. The images offer a detailed account of the evolving surface cracks and the specimen's response under stress, revealing a sequence of changes leading to eventual failure.

The first image was taken before the arrival of the loading pulse. The image 4 at 75 μs shows surface cracks near the edges just became visible. From Fig. 4.8, it can be seen that the overall compressive stress has reached about 83% of the eventual peak stress. Edge cracks have propagated along the specimen axial direction indicating splitting-type fracture. However, the number of cracks dramatically increases in images 6 and 7. The diameter of the specimen correspondingly increases indi-

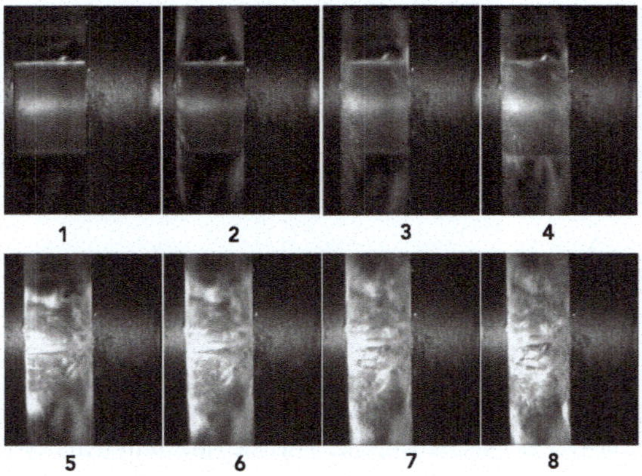

Fig. 4.7 Dynamic fracture and failure process of an SiC-N specimen under uniaxial compression (The 8 images are taken at: 0, 40, 70, 75, 80, 85, 90, 95, 100 μsec)] [1]

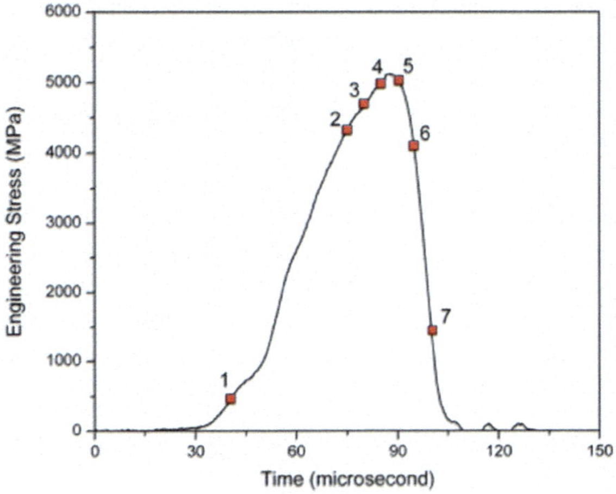

Fig. 4.8 The stress history in the specimen corresponding to the images in Fig. 4.7 ([1])

cating crushing of the specimen. The damage-induced softening continues until the specimen cannot support any load at all.

When the specimen is laterally constrained, as schematically shown in Fig. 4.9, isolated small cracks occur throughout the specimen. At some threshold load, these cracks coalesce into faults. The stress after failure relaxes to a value corresponding to the confined compressive strength of the broken material as shown by Nemat-Nasser and Horii [40].

The SHPB data for brittle materials rarely leads to an unambiguous "*stress–strain*" curve as in metals, especially for high-strength ceramics, due to micro-cracking and fragmentation, which causes heterogeneity. In many constitutive models for brittle materials, the behaviour is classified into two regimes: pre-failure and post-failure. The peak stress, Fig. 4.9c, is taken as the confined compressive strength (e.g. 1D strain) of the intact material. The post-failure strength, Fig. 4.9d, is considered to be the strength of the damaged or comminuted material. Both may be functions of strain rate. However, it should be cautioned that since the specimen may have

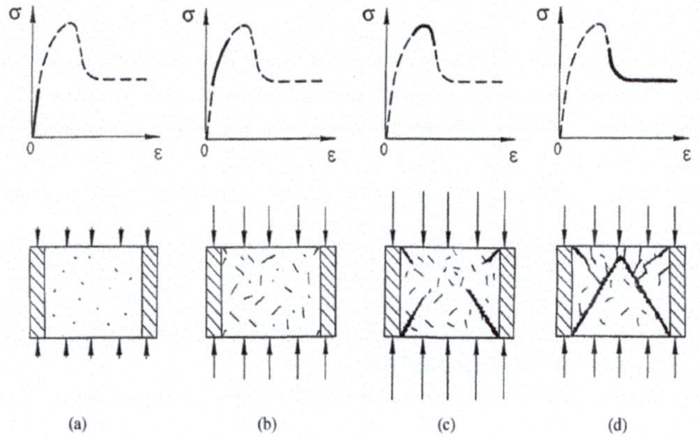

Fig. 4.9 Schematic of the dynamic process of the fault formation under compression (Courtesy by Chen et al. [1])

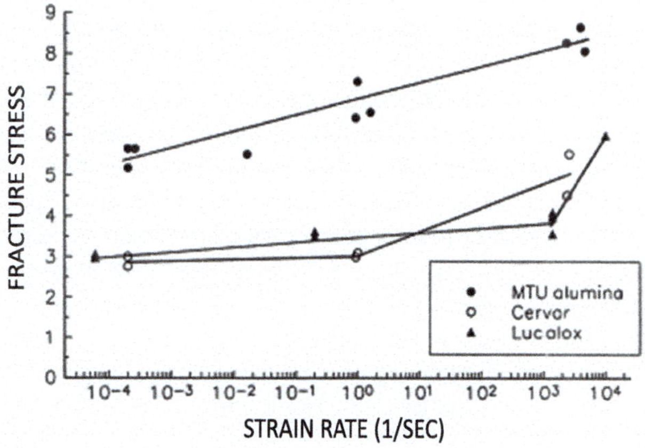

Fig. 4.10 Comparison of the compressive fracture strength of MTU-JS1 alumina with commercial aluminas as a function of strain rate (Staehler et al. [41])

experienced macro-scale fissures in Fig. 4.9d, the post-failure strength should not always be interpreted as an intrinsic property.

Compressive strength measured in an SHPB is a useful way to evaluate material processing parameters. For example, Staeher et al. [41] determined the failure stress (compressive strength) for three different aluminas which were processed differently. The fracture stress of the high pure hot-pressed alumina (MTU-JS) significantly increased at least 1.5 times greater than the commercial aluminas as shown in Fig. 4.10.

The SHPB test is limited, however, in that confining pressures are never more than a small fraction of compressive strength for most brittle materials. Nevertheless, one of the most widely used constitutive models for ceramics, the Johnson–Holmquist model, which is discussed in Chap. 5, requires data at high rates and at high confining pressure. The SHPB is adequate for some cementitious materials and rocks. In 1992, Rajendran [42] reported extensive high strain rate data for metals, ceramics, and concrete in an Air Force technical report. However, for many high-strength ceramics, a test is necessary in which significantly higher confinement is possible. The simplest such test is the plate impact test, the discussion of which begins in the following section.

4.2 Plate Impact Configurations

A planar plate impact test provides a loading path that is very different from the conventional SHPB compression or tension tests. While the deformation is that of *uniaxial strain* in a plate–plate impact test, it is that of a uniaxial stress state in a compressive SHPB test. In other words, confinement is perfect. Under plate impact, before the arrival of any release waves from the lateral boundaries or edges of the plates, the material is subjected to a state of *one-dimensional or uniaxial strain*. The review Chap. 2 described uniaxial strain states with appropriate derivations. In the plate impact test, the mean stress (or pressure) and strain rates are very high compared to SHPB tests. Plate impact experiments are essential for calibrating and validating high strain rate material models that aspire to general applicability. Specifically, plate impact data may be interpreted to infer compression and tensile yield strengths, pressure dependence of elastic moduli, pressure dependence of shear strength, and post-failure parameters. Plate impact produces shock waves in the sample. The physics of shock compression is discussed in Chap. 3.

In the constitutive modelling of engineering materials under extreme loading conditions, the stress tensor is usually viewed as comprised of independent shear and pressure terms ($\sigma_{ij} = S_{ij} + P\delta_{ij}$). Pressure (e.g. mean stress) is considered solely a function of density (e.g. not shear strain) with the relationship of pressure to density expressed through the equation of state. For low strain rate and low-pressure applications, total stresses (σ_{ij}) are entirely determined from the constitutive (or strength) models using "*plasticity*" theories. In many engineering materials, the effect of pressure on strength can be disregarded. Notably, pressure-dependent strength effects are significant in porous solids like rocks, oil shales, soils, porous ceramics, and concrete. At extreme pressures, all materials tend to exhibit pressure strengthening. For certain materials, particularly rocks, there is a transition from brittle (on the meso-scale) to ductile behaviour.

Key fundamental properties such as the yield stress, σ_{HEL}, at the Hugoniot Elastic Limit (refer Chap. 3), σ_{shock} or σ_x, (the peak amplitude of the stress generated by the impact), Hugoniot curve (refer to Chap. 3 on σ_x versus ε_x or shock stress versus shock strain), longitudinal shock wave speed (U_s), and tensile (spall) strength, σ_{spall}, are determined from plate impact test data. Additionally, lateral stresses (σ_y or σ_z) can be measured using stress gauges, from which yield strength can be computed.

4.2.1 *Plate Impact Test*

In the basic plate impact configuration, a high-velocity flyer plate impacts a stationary target plate, as illustrated in Fig. 4.11. Planar plate impact tests offer the ability to control the pulse amplitude, rise time, and duration by adjusting the impedances and dimensions of the flyer and target discs, as well as the impact velocity. Further control over unloading, reloading, or both can be achieved through the use of buffers on the projectile or on the impact or distal sides of the target. Targets are often backed with reference materials (referred to as windows) for optical access or to assist in controlling release or recompression steps.

In a "*symmetric*" plate impact test, both the flyer and target are made of the same materials. Upon impact, compressive stresses are generated and transmitted immediately from the plane of impact to the adjacent stress-free regions of the target and flyer in the form of a stress pulse. As discussed in Chap. 3, the peak particle velocity in the target will be half the impact velocity. Measurement of particle velocities, propagation speeds, or stress amplitudes of the loading pulses determines the load path. These measurements are made before the arrival of release waves from the lateral stress-free surfaces, keeping the strain state strictly one-dimensional while the stress state is 2D axisymmetric. It is crucial that the diameter and thickness of the target plates are selected to prevent lateral release waves from interfering with the central motion until after the primary experimental measurement is complete.

Diagnostics measurements typically include propagation speed, along with either particle velocity or stress, necessary to determine the Hugoniot state (see Chap. 3). Particle velocity is most easily inferred from rear surfaces, with or without a window. The free surface velocity is twice the particle velocity behind the shock wave. The highest time resolution is obtained with velocity interferometry, i.e. a VISAR (Velocity Interferometer System for Any Reflector [43]). Recently, more measurements are conducted with a PDV (Photon Doppler Velocimeter [44]) because it is less expen-

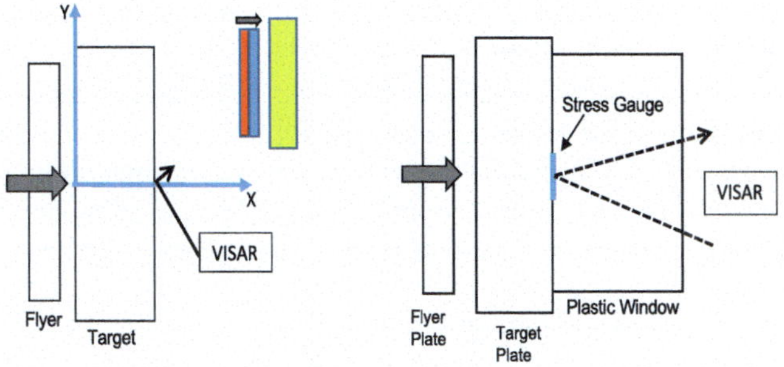

Fig. 4.11 **a** A sketch of plate impact test configuration without any window backing; diagnostic using VISAR **b** with a PMMA or Lithium Fluoride (LiF) window; diagnostics include an embedded longitudinal stress gauge and particle velocity measurement using VISAR. The insert shows a plate impact test configuration in which a backup disc (red) with impedance higher than the flyer disc (blue)

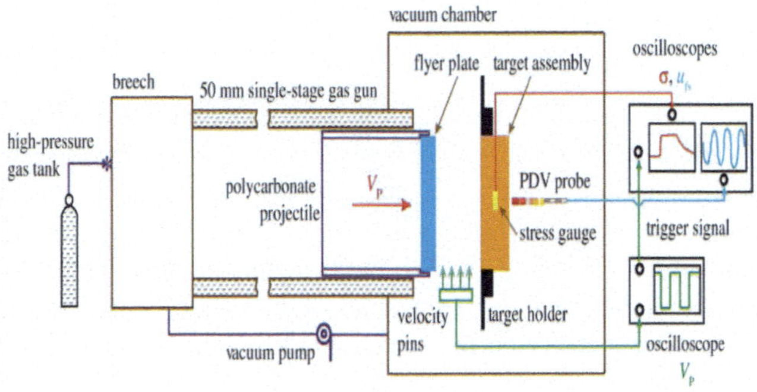

Fig. 4.12 The experimental setup of a plate impact configuration in a laboratory. A polycarbonate projectile sabot carries the thin flyer plate in a 50 mm gas gun. All other devices and diagnostics are indicated in the schematic (Courtesy of Zhang et al. [49])

sive, components are commercially available, alignment is easier, and it has a large dynamic range. Stress is most often determined with piezoresistive gauges [45]. In Fig. 4.11, the VISAR measures the back surface motion of the target plate. These time-resolved measurements are called stress and particle velocity signals or profiles, respectively. The embedded stress gauge measures the shock stress history transmitted into the window, and the stress in the transmitted wave is computed from the impedances of the target and window (see Chap. 3) [45–47]. Any window with a well-known shock response can be used, but lithium fluoride or PMMA has been used most often with piezoresistive gauges in the configuration of Fig. 4.11b. For particle velocity measurements, tests can be performed with and without a transparent window (PMMA—polymethyl methacrylate or lithium fluoride). To obtain the best particle velocity signals from the VISAR, expert plate impact experimentalists often use an optical-quality disc of single crystal lithium fluoride bonded with epoxy to the back of the ceramic target or sample. According to Grady [48], the bonded lithium fluoride surface is first lightly diffused and vapour deposited with about 100 nm of aluminium. It is also important that all critical surfaces are lapped and polished to within a few bands of sodium light.

Plate impact tests are high-precision experiments that require a gun facility along with appropriate diagnostics. In the example in Fig. 4.12, the 50 mm diameter (inner bore) is pressurized by high-pressure gas; helium can produce maximum velocities of about 1000 m/s. Plane shock compression can also be achieved by direct explosive loading, generating triangular pulses, instead of the square pulse from flyer plate impact.

Ahrens [50], Zhernokletov [51], Bourne [52], and Field et al. [53] provide valuable discussions on best practices for plate impact experiments, including various diagnostic techniques. In the example in Fig. 4.12, a plastic (polycarbonate) projectile carries the flyer plate to ensure planar impact with a tilt of less than 1^0. The tilt is also carefully measured through appropriate diagnostics not shown in the figure.

The impact velocity of the flyer is measured by velocity pins, and the back surface motion is determined optically, for example, with a PDV, as shown in Fig. 4.12.

The PDV has been found useful for a broader range of impact scenarios than simple planar plate impact. Its advantages are a large dynamic range and relative insensitivity to noise and amplitude fluctuations, although it has slightly less time resolution than the VISAR. Therefore, it is also useful in multi-axial loading, as demonstrated by Phillip Jannotti et al. [54]. This diagnostic utilizes the Doppler shifts in the reflected laser probe light from the subject surface. They used a PDV in conjunction with two high-speed cameras and measured the back surface motion under multi-axial loading of metallic target plates due to a copper sphere impact at 2200 m/s. This kind of highly instrumented impact configuration provides additional information on the fracturing of the target plate, whether it is metallic or brittle material. Data like these can be employed to validate not only constitutive models but also computational modelling of fracture and fragmentation.

Figure 4.13a shows the target configuration with an embedded lateral stress gauge to measure *lateral stresses*, σ_{yy} during a plate impact test. The specimen is cut into two halves as shown in this figure, and the gauge is glued to one surface. After the gauge installation, the two pieces are carefully rejoined with a very thin epoxy layer. With the direct measurement of both longitudinal and lateral stresses, the shear strength of the target is experimentally determined using the classical relationship in plasticity, $\tau = (\sigma_1 - \sigma_3)/2$, the σ_1 and σ_3 represent the maximum and minimum principal stresses. Stepped target plates with two embedded stress gauges, as shown in Fig. 4.13b, can also be prepared or assembled to measure the shock speed directly from plate impact tests. The gauge record shows the arrival time of the shock wave, allowing the calculation of the wave speed as $d/(t_1 - t_2)$. Since the profile is recorded, in many cases, one can also compute the Lagrangian phase velocity as a function of stress, enabling the integration of stress and velocity to directly compute the stress–strain path (see Chap. 3). Rosenberg et al. [55] used the lateral stress gauge methodology in impact experiments to measure dynamic yield strengths in alumina. We recommend readers refer to the publications of Rosenberg and his colleagues in

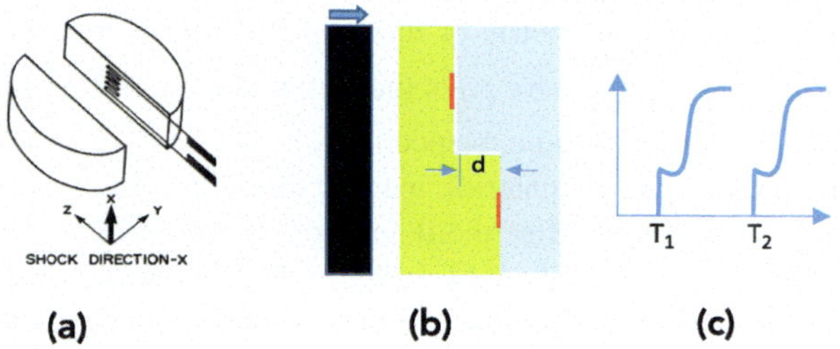

Fig. 4.13 **a** A schematic of a lateral stress gauge glued to a surface normal to the Y-direction. The gauge records the lateral (σ_{yy}) stress history normal to the surface of the stress gauge in the y-direction; **b** two longitudinal gauges to measure shock velocity. **c** The time arrival of the shock wave at two different locations

Israel during the '*80s and '90s* for more reference materials on piezoresistive stress gauges.

4.2.2 *Hugoniot and Hydrostat*

Calculation of stress histories after impact requires data for: (1) a strength/constitutive model and (2) a hydrodynamic equation of state (EOS). In general, the constants for strength and EOS parameters are determined from stress wave-based high strain rate (SHPB) and high-velocity plate impact experimental data. A description of the hydrodynamic EOS can become complex for brittle materials due to various thermodynamics and mechanistic deformation processes that occur in the solids during shock wave loading. For instance, at very high-impact velocities (or pressures), some brittle materials, including some high-strength ceramics, exhibit elastic–plastic transitions, micro-defect evolutions, and phase transformations. The measured velocity profiles from plate impact experiments indirectly reveal these deformation-related complexities. For materials that exhibit such (relatively rare) behaviour, the determination of a *Hugoniot* and *hydrostat* using static (diamond anvil cells with pressure volume instrumentation and dynamic shock wave propagation) experimental techniques will enable one to construct the hydrostat and EOS and give clues to micro- and meso-scale transformations.

Symmetric impacts (employing the same material for flyer and target) are utilized at times because they eliminate the need to measure peak particle velocity. It is simply half the impact velocity. Shock amplitude is given by $\sigma_{shock} = 1/2\rho U_s U_p$, where ρ is the initial material density, U_s is the shock speed, and U_p is the impact velocity. As discussed in Chap. 3, for most materials, the shock speed is a linear function of the particle velocity, $U_s = C_0 + \alpha U_p$ where C_0 or C_L is the longitudinal sound speed at room temperature at low pressures. At very high-impact velocities or with very compressible materials, this relationship often becomes nonlinear with second-order terms in U_p. It is important to note that the shock strain or longitudinal strain, ε_x, is equal to the volumetric strain, e_v, under uniaxial strain state as in the plate impact test configuration, e.g. the volumetric or uniaxial (longitudinal) strain is given by $\varepsilon_x = \left(1 - \frac{\rho_0}{\rho}\right)$ or = $(\frac{V_0}{V} - 1)$ where ρ_0 is the initial density of the material and V_0 and V are initial and current volume, respectively.

For an ideally elastic–plastic material, loading along the Hugoniot path can take place first elastically, and then after the deviatoric stress exceeds the yield stress, plastically; see, for example, Grady [56]. In Fig. 4.14, we consider the compressive strain and compressive pressure as positive for the purpose of plotting the σ_x versus ε_x in the first quadrant. At enough high pressures, both the hydrostat and EOS indexEquation of state behaviours become nonlinear. The shock state from a conventional planar plate impact test provides one data point on the Hugoniot . If several plate impact tests are performed at increasing impact velocities, we can construct the Hugoniot as the locus of all points $(\sigma_x, \varepsilon_x)$. The pressure versus volume (P versus

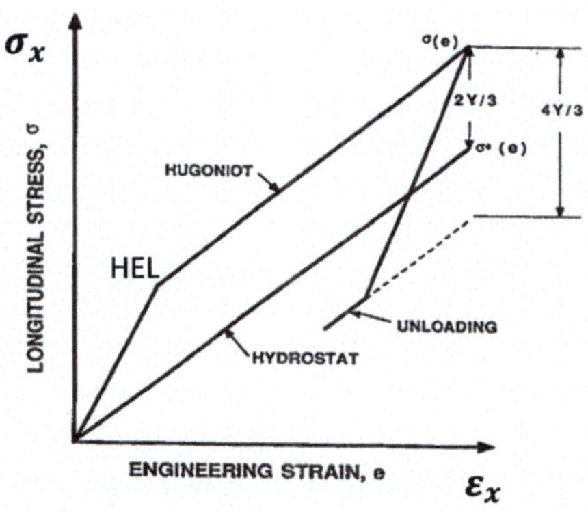

Fig. 4.14 A schematic of Hugoniot, hydrostatic, and compressive strength of the material, Y. Using von Mises yield criterion, a relationship, $P = \sigma_x + \frac{2}{3}Y$, between the pressure, P, shock stress, σ_x, and strength Y is derived in this chapter. Note that $Y = \sigma_1 - \sigma_3$

e_v) relationship, or the "*Hydrostat*" can be determined from the Hugoniot curve. We can correlate the strength of the material, Y to Hugoniot and Hydrostat, as shown in Fig. 4.14. In this figure, the strength is assumed to be constant as in an elastic-perfectly-plastic material. In this simple elastic-perfectly-plastic case the Hugoniot and hydrostat are parallel and offset by $2/3Y$, where Y represents the yield strength.

As explained in Chap. 3, except at sufficiently high stresses, an elastic–plastic Hugoniot such as in Fig. 4.14 produces two shock waves. The first wave, termed the elastic wave, has an amplitude equal to the HEL, while the second represents the maximum shock stress. Determining the HEL establishes the limit of elastic behaviour, signifying the highest amplitude of elastic stress the material can withstand. Measuring the elastic strength in a plate impact is relatively straightforward: conduct an impact where the material fails, and measure the stress behind the fastest wave generated by the impact. Its amplitude indicates the Hugoniot elastic limit.

In earlier shock literature, the distinction between the Hugoniot and hydrostat was often disregarded. During the 1950s and 60 s, shock wave propagation in solids was analysed using the equation of state (hydrostat) exclusively. For instance, when 6061-T6 aluminium with a shear strength (Y) of approximately 0.3 GPa is shocked to 10 GPa, the difference ($2/3Y$) between the hydrostat and Hugoniot is merely 2%. Hence, the EOS determines the stresses due to volume change, as in any hydrodynamic calculation.

Computer codes utilized in shock analysis, whether they're finite difference, finite element, or finite volume, were frequently termed hydrocodes. However, during the 1970s, researchers engaged in modelling and simulation, specifically those involving projectile penetration into thick high-strength metallic or ceramic targets at ordinance velocities ($\approx < 1000$ m/s), recognized the importance of using the Hugoniot for accurately modelling shock amplitudes. For instance, the compressive strength, Y, for Mar-200 steel is approximately 1.3 GPa [42]. The use of the term "*hydrocode*" persists, but in most modern applications, it's a misnomer.

When Y significantly falls below the pressure, P, the shock stress, σ_x, is assumed to be equal to the pressure. This assumption holds for low-strength materials; however, with the advent of high-strength steels like tool steel, MAR-200, MAR-250, and other armour steels, this assumption is often no longer valid. The difference between the Hugoniot and hydrostat for high-strength steel ranges from 8% to 9%. For accurate modelling and simulation, computational and analytical models need to include the strength term, Y, to precisely predict shock propagation in these materials. This is also applicable to high-strength ceramics, such as aluminium oxide, boron carbide, titanium diboride, etc. Nicholas and Rajendran [27] discuss various aspects of the Hugoniot and Hydrostat in the book "*High Velocity Impact Mechanics*" by Zukas [27].

4.2.3 Particle Velocity Profiles

The velocity profiles measured with VISAR in a plate impact test contain numerous significant features related to the elastic–plastic transition, shock stress, release wave, time arrivals of spall signals, reshock characteristics, and absence of spall (indicative of no tensile failure). A comprehensive understanding of these velocity profiles is crucial for interpreting these features, particularly in brittle materials.

For readers unfamiliar with shock wave experiments, gaining insight into the interpretations of VISAR or stress gauge measured profiles in various materials such as rocks, ceramics, concrete, and metals can be invaluable. Therefore, to facilitate this understanding, a wave profile for copper is presented as an example in Fig. 4.15.

A typical free surface velocity plot (profile or signal) for a *metallic* target plate is given in Fig. 4.15. The velocity signal contains various features and information related to HEL, shock stress, unloading wave, spall stress, wave velocity, etc.

If the wave speed is C_L and the target thickness is "h", then the flyer plate impact-generated shock wave arrives at time $t_1 = (h/C_L)$. The impact-generated shock (compressive) wave that travelled into the flyer plate reflects back from the rear surface as an unloading (release) wave and propagates back into the target plate arriving at point "**R**" in Fig. 4.15 and drops the velocity from its peak value. Depending on the impact velocity levels, either the velocity goes to "**B**" or it unloads to "**S**" and then turns up to reach "**D**". When the impact velocity is below V_{HEL}, the impact generates only elastic deformation in both flyer and target plates, and therefore the release wave from the flyer plate's rear surface will bring the peak velocity to zero. In a symmetric impact, the intersection of the unloading waves from the target and flyer plates will reduce the stress in the target to—(minus) the shock stress. If the tensile strength of the target is less than this, the target will spall.

The spallation process creates a distinct signature (often called the spall signal) in the free surface velocity profile. The spall wave arrives at the point "**S**", in Fig. 4.15. The material particle velocity history then follows the solid line (recompression) at "**S**" to "**D**". The signal beyond point "**S**" is denoted as the "*spall signal*". This spall signal is often used in the estimation or calibration of failure model parameters.

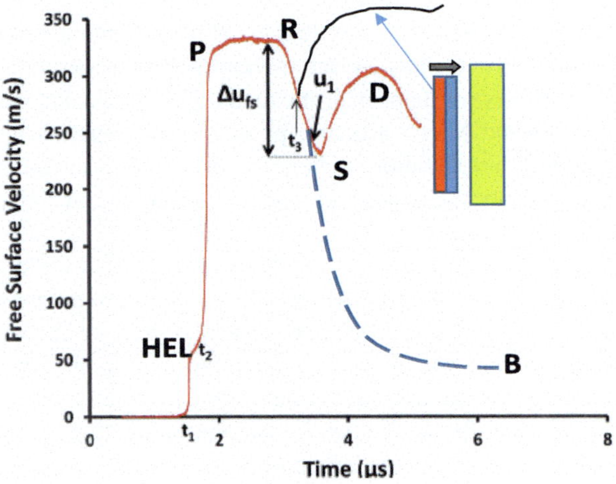

Fig. 4.15 A typical free surface velocity profile from a symmetric impact planar plate impact experiment without a plastic window. The curve between R and B indicates "*no failure*" of the target; between RSD indicates "*Spall Failure*"; the insert shows reshocking (black) due to flyer's high-impedance backup plate (red). The "reshock" aspect is discussed in a later section

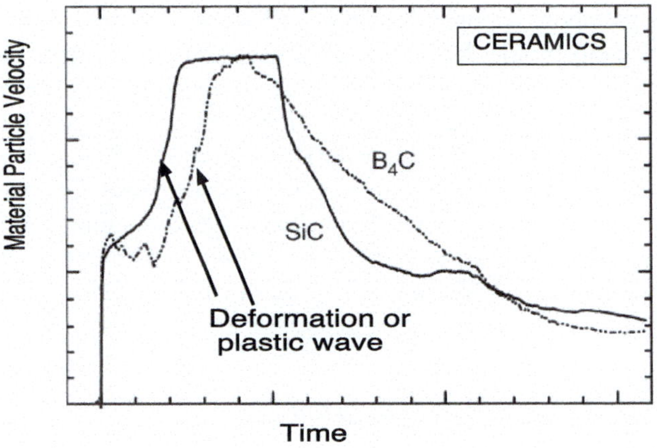

Fig. 4.16 Velocity profiles for silicon carbide and boron carbide measured by VISAR at the interface of the ceramics and a lithium fluoride window

Numerous papers can be found in literature on spall modelling. We discuss further spall signals in brittle materials in a later section of this chapter.

In ceramics, the velocity profile for shocks to the HEL often contains additional features. Unlike metallic solids, the HEL of ceramics is frequently not uniquely identified. The VISAR measured velocity profiles for silicon carbide (SiC) and boron carbide (B4C) are shown in Fig. 4.16. As one can see, the shape and rise time in these two ceramics appear to be very different. While the SiC exhibits characteristics that are similar to metallic material profiles, the B4C exhibits distinctly different behaviour when compared to metals. The interpretation of micro-structural changes at HEL and the deformation wave shape in brittle materials is somewhat speculative due to lack of physical evidence. Micro-structural processes could include heterogeneous yielding, densification, amorphization, and micro-crack growth and coalescence. It is very difficult to recover shocked plates without creating additional damage, although there have been several attempts.

Grady [57] recently conducted an in-depth review of shockwave compression measurements conducted over the past 40 years in B4C and offered hypotheses concerning micro- and meso-scale phenomena. Chen et al. [58] have shown that the loss of shear strength observed in B4C is associated with formation of amorphous zones, shock compression response is also a function of density if the porosity is not very close to zero. In particular, the Hugoniots for porous solids, like cement, concrete, and rock, are sensitive to the volume content of porosity. See, for example, Grady's discussion of results from plate impact tests on B4C with two starting nominal densities of 2.4 g/cm^3 and 1.0 g/cm^3 with implied initial porosities of approximately 3–7% and 22–25%, respectively.

Velocity profiles in some high-strength crystalline ceramics, such as Titanium Diboride (TiB2) and Aluminium Nitride (AlN), exhibit multiple knees or kinks, complicating interpretations within a conventional elastic–plastic framework. Figure 4.17 displays the velocity profiles for boron carbide and aluminium nitride. Both ceramics present anomalies concerning the HEL point. The shape and rise time in boron carbide suggest microstructure and lattice-dependent deformation kinetics. When aluminium nitride is shocked to 35 GPa, well above the HEL value of 8 GPa, the wave profile displays highly nonlinear behaviour with complex characteristics. Grady [48] suggests that the 20% volume strain and large dissipation associated with the shock transition process lead to substantial compression wave splitting and rapid attenuation of the peak wave amplitude, as shown in Fig. 4.16 for three different thick target discs.

4.2.4 $X - t$ (Lagrangian) Diagram

Introduced in Chap. 3, one of the most effective methods for analysing wave propagation in a rod or a plate is through the use of an $X - t$ (Lagrangian) diagram, as illustrated in Fig. 4.18a. This diagram is instrumental in interpreting the time history of material particle velocity. The diagrams can be in either Lagrangian or Eulerian coordinates, with the Lagrangian framework proving more useful and easier to construct in most instances (see, for example, Fig. 4.18), as it eliminates the need to incorporate particle velocity. In plate impact tests, where material particles move in one direction, describing one-dimensional wave propagation becomes feasible through a plot of the distance travelled from the impact plane (X-axis) against time (Y-axis). We presented the $X - t$ diagram for the SPHB test in Sect. 4.1 to interpret the strain gauge signals. Generally, every feature of a gauge record is a result of

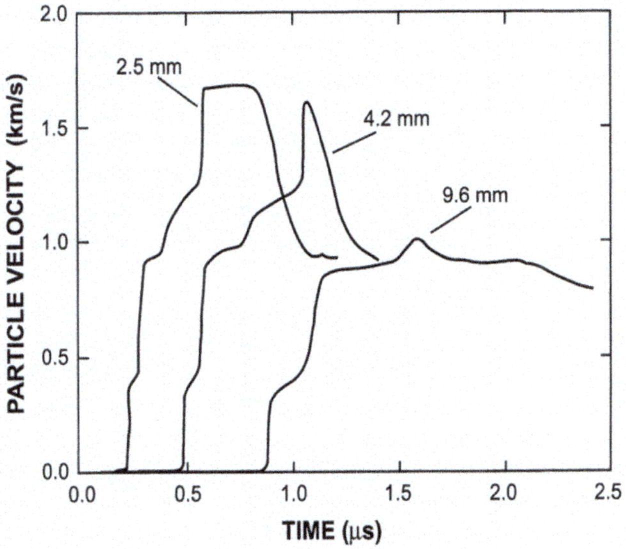

Fig. 4.17 Velocity interferometry wave profiles measured in aluminium nitride for three thicknesses. Courtesy of Reference [48]

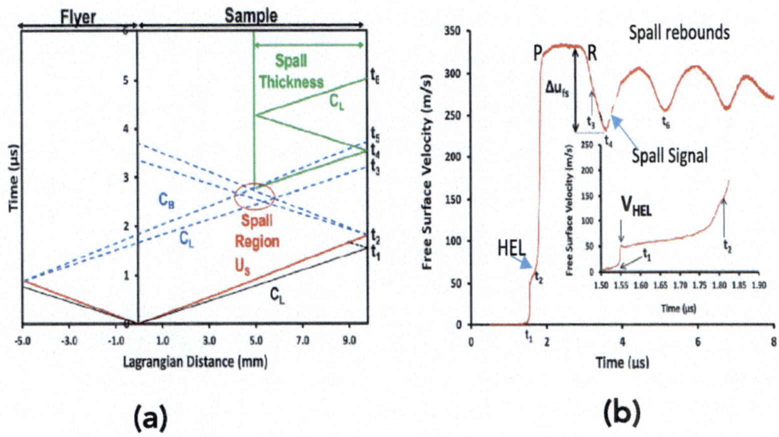

Fig. 4.18 **a** Representative $X - t$ diagram, and **b** VISAR free surface velocity data for an Al 5083-H116 sample impacted near 350 m/s through the thickness of the plate. CL and CB are longitudinal and bulk wave speeds, respectively. The inset of the free surface velocity data better displays the HEL and shock wave arrival [59]

wave arrival. By comprehending the time history of both compression and release waves at the measurement locations, it becomes possible to realistically discern and categorize the types of waves generated at the impact and stress-free rear boundaries of bars or plates. The analysis of $X - t$ diagrams may involve the method of characteristics. However, if the measurement is influenced by waves from the edges and lateral boundaries, the $X - t$ diagram may not prove useful for gaining a better understanding of wave propagation. Additionally, the analysis of propagation based on the $X - t$ diagram under uniaxial stress or strain states could become complex when the material response involves defect evolutions and damage-induced inelastic deformation, such as fracture, fragmentation, and pulverization. It's crucial to note that under multi-axial loading, as seen in projectile penetration into a target plate, a computational modelling approach will likely be more meaningful than relying solely on the $X - t$ diagram.

Figures 4.16 and 4.17 exhibit the absence of spall signals in the velocity profiles for two distinct ceramics. To highlight the difference between a metallic target and a ceramic target, we included a schematic of the plate impact velocity profile for an aluminium alloy 5083-H116 from Reference [59] in Fig. 4.18. When brittle solids (including most high-strength ceramics) are subjected to shock above the Hugoniot Elastic Limit (HEL), their spall strength diminishes to zero, resulting in the absence of spall signals in velocity profiles.

In a study by Whelchel et al. [59], VISAR measurements of stress-free back surface velocities were presented for symmetric planar plate impact experiments. These experiments covered various shock stress levels, ranging from 1.5 to 6.2 GPa. The VISAR profiles and the $X - t$ diagram are illustrated in Fig. 4.18, effectively showcasing shock wave propagation in a plate impact test.

The $X - t$ diagram in Fig. 4.18 displays the time arrivals of compression and release waves in both the flyer and the target. Events in the $X - t$ diagram can be correlated with different points on the velocity profile. The initial elastic shock, travelling at the longitudinal sound speed, C_L, reaches the rear of the target at time t_1. Following this, the slow-moving plastic wave arrives at the rear surface at time t_2. The elastic–plastic transition is visible in the velocity profile at the HEL (see insert in Fig. 4.18).

A characteristic time is involved in the evolution of plastic flows due to dislocation motion and other atomistic mechanisms. When the plastic shock wave reaches the target-free surface, the particle velocity peaks between points **P** and **R**. Subsequently, the elastic release wave from the back surface of the flyer arrives at point **R**, causing the velocity to drop to a lower value at t_3.

The plastic release wave follows the elastic wave from the flyer and reaches the target rear surface just after time $\mathbf{t_3}$, further decreasing the velocity. If the impact velocity is high enough to cause complete or partial tensile failure, a recompression wave emanates from the spall-generated stress-free plane. This wave brings the stress back from its tensile value at spall to zero, the value after the spall has opened. The spall wave arrives at the rear of the target at time t5, and the velocity profile exhibits a spall signal, as seen in Figs. 4.18b and 4.19.

The spall location from the impact plane in the target can be estimated from the wave speeds and thicknesses of the flyer and target using $(H - \eta h)$, where "H" is the target thickness, "h" is the flyer thickness, and η is the ratio of elastic sound speeds

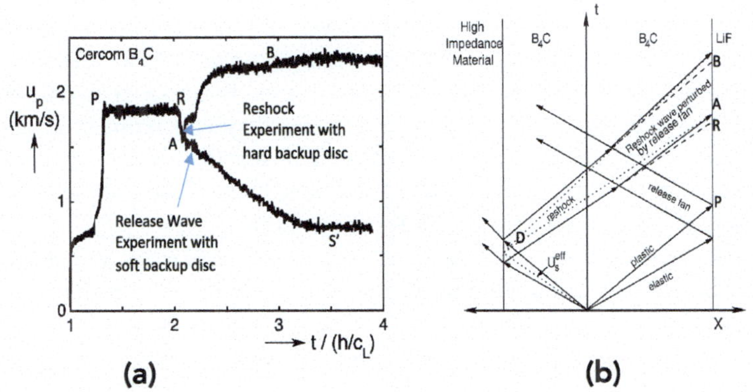

Fig. 4.19 **a** A schematic of the projectile with boron carbide flyer with a soft or hard backup plate and **b** VISAR measured the particle velocity profiles with respect to normalized time (Vogler et al. [61])

of the target and flyer. The "spall thickness" in Fig. 4.18b is given by "ηh". For a symmetric impact and when the target thickness is twice that of the flyer ($H = 2h$), the spall plane will be located at the centre region of the target.

The dotted lines represent elastic and plastic release waves from the rear surfaces of both the flyer and target. Note that when the release wave from the rear of the target reaches the impact plane between 3 and 4 μs, the flyer plate separates from the target plate. After the spall, the waves bounce back and forth in the "spall thickness region" between the spall and target rear surfaces, as shown in Fig. 4.18a. These wave reflections appear in the velocity profile as spall rebounds beyond time t4, as in Fig. 4.18b.

The power of the (X, t) framework is further demonstrated in a paper by the shock physics group at CalTech [60]. They studied glass and were able to identify failure wave nucleation and propagation speeds. While basic plate impact tests can determine stress–strain paths for 1D loading and unloading, relatively straightforward modifications can provide more insight into properties that might be history dependent. Much can be achieved by adding additional driver, buffer, and backing plates. The simple load/unload sequence can be modified to include reshock, partial unloading, shear shocks, divergent shocks, and soft recovery. Interpretation can be unambiguous, but it often requires modelling and simulation of the tests themselves.

Vogler et al. [61] provided examples of experiments involving a backing plate behind the flyer plate. They conducted both release and reshock experiments on boron carbide to investigate the impact of damage (micro-cracking) on the degradation of material strength and Hugoniot. By introducing a backup disc with lower impedance than the flyer disc, they were able to control the release at point **R** in Fig. 4.18b. Through careful adjustment of the longitudinal sound speed in the soft backup and flyer, as well as the thicknesses of these discs, they could control the time of wave arrival at the back of the target material (ceramic). In cases where the flyer backup impedance is higher, a reshock wave is generated and propagated in the target, as illustrated in Fig. 4.19.

In their experiments, the flyer and target were hot-pressed boron carbide with a density of 2.9 g/cm^3. In a conventional symmetric plate impact test, the release wave from the back of the flyer plate unloads the stress to zero as it travels back into the target. However, due to the high-impedance backup plate (copper or tantalum), the flyer's rear surface is no longer stress free. This results in a portion of the wave transmitting into the backup disc, while some reflects back into the flyer disc as a compressive (reshock) wave (see point **D** in Fig. 4.18b)).

This reshock wave travels into the target, meeting the release wave from the back of the target at **E**, and continuing to propagate towards the rear of the target to **A**. Consequently, this increases the particle velocity from its initial value. In Fig. 4.19a, the reshock wave causes the particle velocity to rise to a new level B, surpassing the incidence-wave generated level at P. The reshock wave travels through a damaged ceramic, whose original properties have been modified due to damage evolution during the initial shock propagation and release. Further details on this reshock experiment will be discussed in the modelling section.

The velocity profile, as shown in the reshock configurations, contains rich information on the degradation of shear strength in ceramics. These reshock configurations are often employed to drive materials to high pressures under more isentropic conditions than in a single shock wave, mitigating entropy creation. This approach allows for the determination of an isentropic EOS without the need to correct for shear stress and irreversible heating, which typically causes the offset between a Hugoniot and hydrostat.

4.2.5 *Compressive or Shear Strength*

Dynamic yield strength measurement at very high strain rates ($> 10^4/sec$) often motivates plate impact tests to determine material strength. For elastic compression, the yield strength (Y) is related to the Hugoniot Elastic Limit (HEL), at which deformation becomes plastic or inelastic. Assuming the von Mises yield criterion , the relation is given by

$$Y = \frac{\sigma_{HEL}}{\left(\frac{K}{2G} + \frac{2}{3}\right)} \text{ or } \left(\frac{1 - 2\vartheta}{1 - \vartheta}\right) \sigma_{HEL} \tag{4.2}$$

In the plastic range, for an elastic-perfectly-plastic material, the difference between longitudinal and lateral stresses remains equal to the yield strength. For such materials, the offset between the Hugoniot and the hydrostat is constant and equal to $2Y/3$. Rosenberg and Partom [62] utilized the stress gauge technique to determine the von Mises yield stress in both metals and ceramics. Articles describing the applications of stress/pressure/quartz gauges can be found in the proceedings of the American Physical Society conference on "*Shock Waves in Condensed Matter*" [63]. Using two stress gauges in the same plate impact experiment, one in the lateral (transverse) direction and the other in the longitudinal (shock) direction, Bless et al. [64] mea-

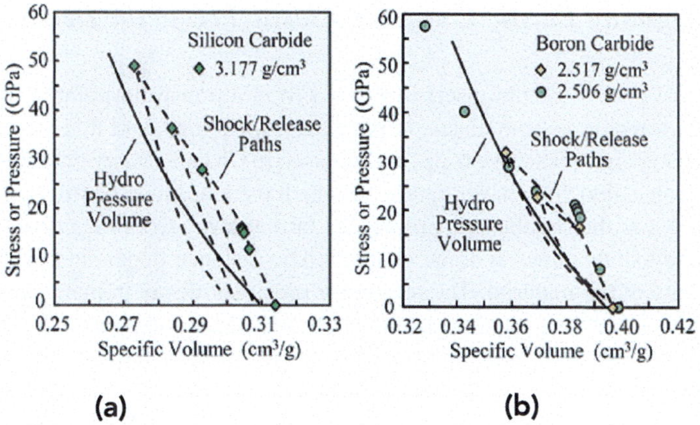

Fig. 4.20 Shear strength of silicon carbide and boron carbide when shocked above the HEL. **a** Retention of shear strength in silicon carbide; **b** Loss of shear strength in boron carbide [65]

sured the yield strength in soda lime glass at various shock stress levels. The yield strength increased linearly with shock stress up to the HEL but decreased at stress levels well above the HEL. Even if Y changes, the offset between the Hugoniot and hydrostat remains $2Y/3$, allowing computation from one to the other with usually negligible uncertainty due to temperature rise. These measurements are extremely useful in the development of constitutive models for brittle materials. Grady [65] discussed the loss of shear strength in boron carbide as shown in Fig. 4.19. While the shear strength of boron carbide is tending to zero (the decreasing differences in the abscissa between the hydrostat (circles and diamond symbols in Fig. 4.20) and hydro pressure volume (EOS)) when shocked above the HEL, the shear strength of silicon carbide however is essentially unchanged when shocked above the HEL as seen in Fig. 4.20a.

The relationships between hydrostat and EOS were discussed in Sect. 4.2.2. In metallic materials, compressive strength Y usually does not degrade after undergoing shock compression. In some metals like copper, it has been shown by Rajendran et al. [66] that the shear strength increases after pre-shocking. However, in most ceramics, the shear strength decreases with increasing shock compression when shocked above the HEL, and in some ceramics, it becomes negligible. This "*shear strength loss*" is an important aspect of brittle material behaviours under shock loading. Experimental measurements and analyses of "*shear strength loss*" in ceramics contribute significantly to the modelling of damaged ceramics under inertial or boundary-induced confinement. For example, when a metal projectile penetrates a thick confined ceramic armour plate, the strength characterization of pulverized ceramics materials directly beneath the projectile has a large influence on penetration rate. Thus, constitutive models used for armour need to take this phenomenon into account.

4.2.6 Hugoniot Elastic Limit and Elastic Precursor Decay

The HEL in Fig. 4.14 or the insert in Fig. 4.17b is a transition point at which the deformation transitions from elastic to plastic or "*inelastic*", and it is manifested as an abrupt change in stress wave form, as discussed in Chap. 3. Many materials exhibit a phenomenon called "*precursor decay*" in which the amplitude of stress or velocity at HEL decays as the shock wave propagates farther away from the impact plane; it reaches steady-state values at some characteristic distance depending on the strain rate sensitivity of the material. The reason for precursor decay in many cases is that the initial shock involves an almost infinite strain rate, and dispersion in this shock causes a slight relaxation in rate until a steady wave is produced.

The presence of precursor decay in metallic materials is well understood and confirmed through extensive data from plate impact tests. Nicholas et al. [67] have presented analyses of the precursor decay phenomenon in metals through constitutive and computational modelling of plate impact experiments to demonstrate the influence of strain rates on the evolution of HEL in metals. The steady-state value of HEL is considered as a material property, σ_{HEL}, also called dynamic compressive strength at very high strain rates.

There are only a very few plate impact experimental data available to confirm the existence of elastic precursor decay in most ceramics [70, 71]. In Fig. 4.21, the HEL (pointed out by the arrows) decays from a value of about 2.7 GPa at 2 mm distance from the impact plane to a steady value of 2.2 GPa at 12 mm. This phenomenon has been reported in several ceramics as well as in amorphous glasses [70]. While a study by Rosenberg et al. [71] reported elastic precursor attenuation in aluminium oxide over sample thicknesses of about 5 mm to 15 mm, the measurements reported by Cagnoux and Longy [72] did not indicate precursor decay. One must pay attention to this if the goal of an experiment is to measure the yield stress, the sample should be thick enough that the HEL has relaxed to a steady value. Later studies in 1990s by Grady [49] indicated the absence of elastic precursor decay in many pure ceramics

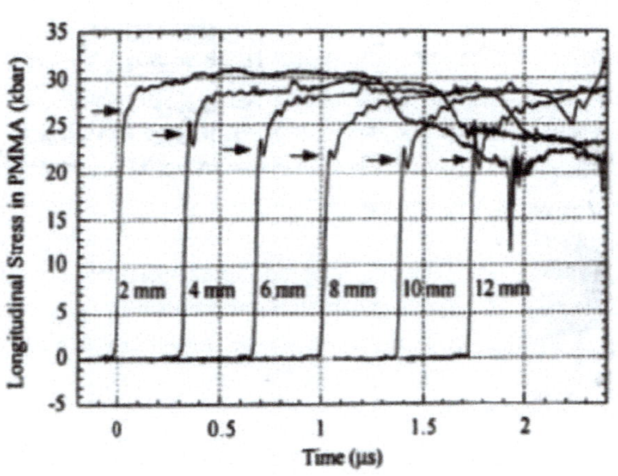

Fig. 4.21 Stress gauge measured shock wave profiles at different distances (or locations) from the impact plane in pure alumina. The stress level at the kink in each profile as shown by the arrow is the HEL value in the plastic window (or PMMA). The HEL saturates to a steady value of about 22 *kbar* or 2.2 GPa [68]

Table 4.1 Precursor decay data for two types of alumina (Rosenberg et al. [69])

$AD-85$				Al_2O_3			
ρ (gm/cm^3)	C_L (mm/μs)	Thickness (mm)	σ_{HEL} (GPa)	ρ (g/cm^3)	C_L (mm/μs)	Thickness (mm)	σ_{HEL} (GPa)
3.7	9.32	5	6.27	3.9	10.52	5.1	10.8
		9.5	4.85			9.95	8.6
		14.1	4.83			14.95	8.5

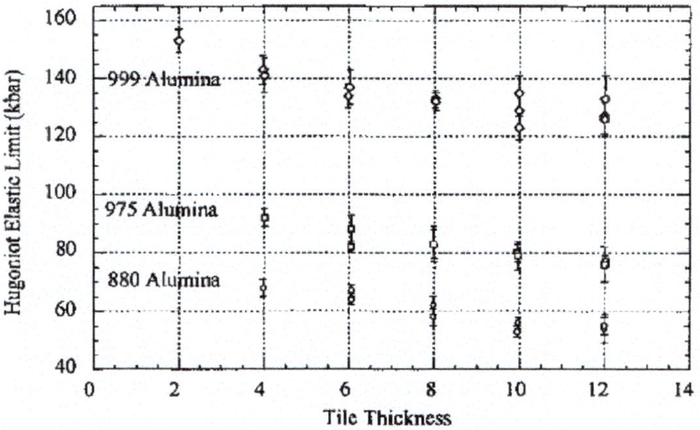

Fig. 4.22 The precursor decay in three different aluminium oxides (Murray et al. [68])

including alumina with the exception of boron carbide. Rosenberg et al. [71] reported HEL values with respect to target thickness for AD-85 (85% alumina) and A12O3 (99% alumina) as given in Table 4.1.

The Manganin stress gauge recorded time history of the shock stress in PMMA measured at the interface of ceramic and PMMA window was shown in Fig. 4.21. The impedance matching solution from Chap. 3 provides a relationship between the stresses in ceramic and PMMA, respectively. Therefore, the stress at HEL in the ceramic can be obtained using the values for the impedances of PMMA and ceramic as

$$\sigma_{HEL} = \frac{Z_{Ceramic} + Z_{PMMA}}{Z_{PMMA}} \sigma_{PMMA} \tag{4.3}$$

Figure 4.22 presents plots of precursor decay for alumina as reported by Murray et al. [68].

To assure that the measured value is a steady-state value for the HEL, data are obtained from three plate impact tests with different thicknesses of the target. The precursor decay phenomena have been observed in both types of alumina with slightly different densities. As we mentioned earlier that the micro-cracking-type defect evo-

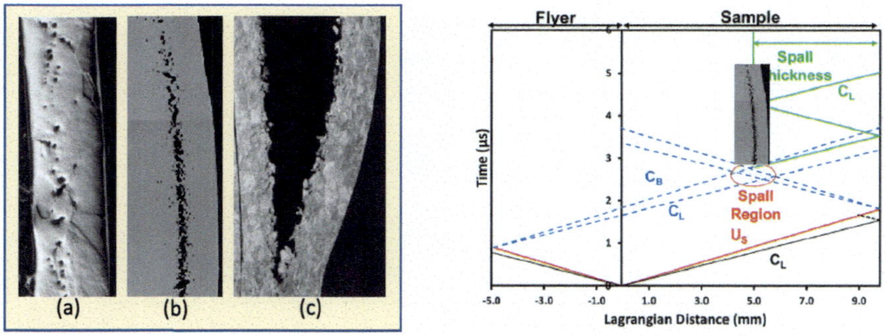

Fig. 4.23 Spallation in plate impact experiments on copper; **a** incipient spall, (2) intermediate spall, and (3) complete spall fracture. A schematic of $X - t$ diagram is also shown in **b** (*Note*: not drawn to scale with respect to spall thickness or location) ([73, 74])

lution in brittle materials often leads to the elastic–inelastic transition at the HEL point in measured data.

Glass is also a brittle material that exhibits precursor decay. Bourne et al. [70] used stress gauges in their plate impact experiments to measure both longitudinal and lateral stresses for ceramics and different glasses. The HEL values for the DEDF glass decreased from about 6.2 GPa for a target thickness of 2 mm to about 4.6 GPa for a 15-mm-thick target.

In general, strain (or loading) rate-dependent micro-mechanical deformation processes causes elastic precursor decay. It has been well established that almost all brittle materials exhibit strain rate-dependent material behaviour under shock and impact loading conditions. The precursor decay phenomenon in cementitious materials, especially concrete, has not been thoroughly investigated. In these relatively low-strength materials, the pore collapse-type mechanisms also could contribute to the HEL point.

4.2.7 *Measurement of Spall Strength*

Since the ceramic targets pulverize when the impact velocity is above HEL, it is not possible to establish any physical evidence to explain the spall phenomenon. To develop a micro-structural understanding of the spall process, as well as for completeness, we present spall results for a metallic material, like copper in Fig. 4.23.

To understand spall-type dynamic tensile failure processes, as discussed in Sect. 4.2.4, $X - t$ diagrams are essential. Referring back to Fig. 4.15, the release (unloading) wave from the rear of the flyer decompresses the materials as it propagates into the target. Under symmetric impact (i.e. same flyer and target materials), the stresses in the target become zero after decompression. The release wave continues to travel towards the rear of target.

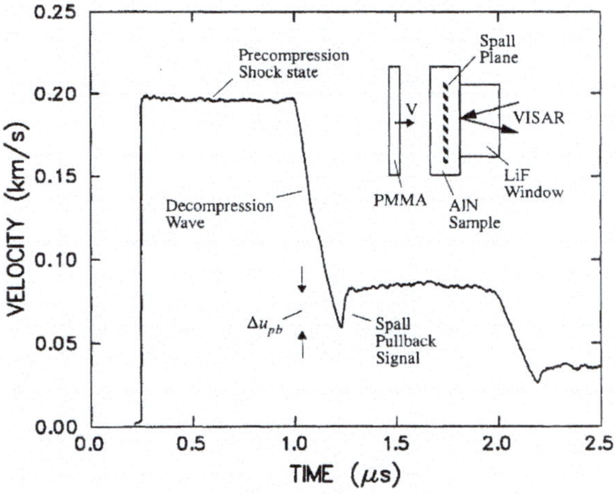

Fig. 4.24 Spall in aluminium nitride when shocked below HEL. This profile was obtained by Kipp and Grady [76] from a plate impact test as shown in the insert

It can be seen from the dotted blue lines in Fig. 4.23 that the release wave from the target's rear surface and the release wave from the flyer cross each other at the centre of the target generating significant triaxial tensile stress state to cause spallation of the target. Spall is caused by nucleation, growth, and coalescence of micro-voids in the material (see, for example, Curran et al. [75]). When the impact velocities are slightly above V_{HEL}, but above some threshold shock stress, micro-voids nucleate and individual voids grow in the spall region as shown in Fig. 4.23a at very high triaxial stresses, $P/\sqrt{3J_2}$, that develop inside the target due to release wave interactions, where P is the pressure and J_2 is the second invariant of the stress deviator. All three principal stresses become tensile.

As impact velocity increases, tensile stress leads to void coalescence (Fig. 4.23b), eventually creating a fracture surface within the target. At sufficiently high-impact velocities or shock stress levels, free surfaces form around the spall region, causing complete failure of the target plate, as seen in Fig. 4.23c. To calculate spall strength, refer back to Fig. 4.15b or 4.15, where it is determined from the pullback velocity Δu_{fs}, using the relationship, $\sigma_{spall} = 1/2\, \rho\, U_s\, \Delta u_{fs}$. Generally, the spall stress (or tensile strength) for metals remains independent of the loading (or pulse) duration or shock pre-compression stress levels, but this is not the case for brittle materials. Spall stress typically decreases with increasing shock stress. When shocked above the HEL stress, the spall strength of most ceramics is usually zero or negligible due to damage induced by micro-cracking under the initial compressive shock wave. However, a spall signal can still be observed in ceramics when the impact velocity well below V_{HEL}, as shown in Figs. 4.24 and 4.25a. In contrast to metals, the spall strength of ceramics is typically affected by the magnitude of the compressive wave, leading to non-ideal spall signals in the velocity profile.

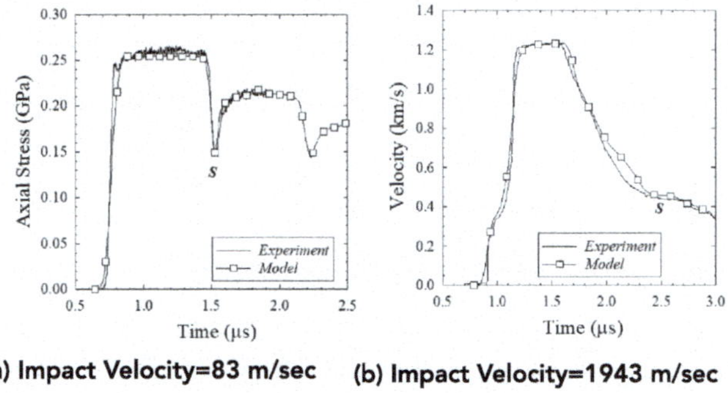

Fig. 4.25 Spall signals in Al_2O_3 **a** shocked below the HEL **b** shocked above the HEL

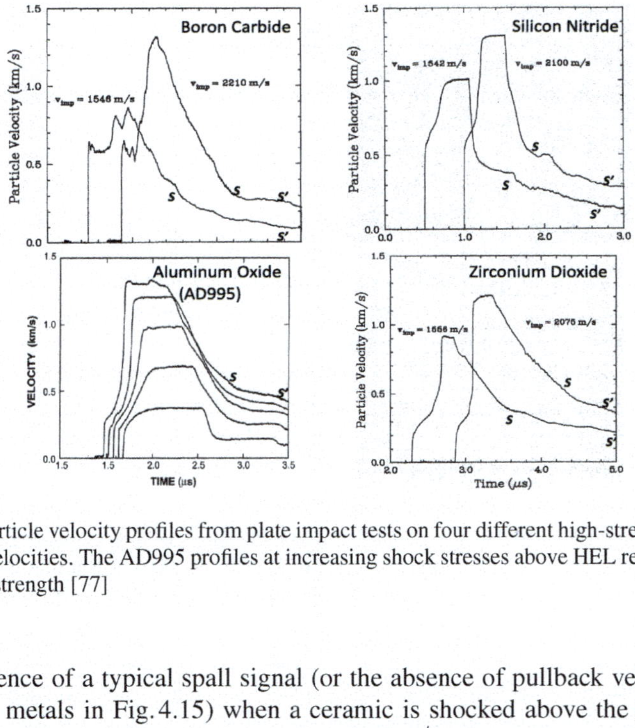

Fig. 4.26 Particle velocity profiles from plate impact tests on four different high-strength ceramics at different velocities. The AD995 profiles at increasing shock stresses above HEL reveal complete loss of spall strength [77]

The absence of a typical spall signal (or the absence of pullback velocity Δu_{fs} as seen for metals in Fig. 4.15) when a ceramic is shocked above the HEL, along with the slow stress release between points **S** and **S**$'$ indicates zero spall strength in these materials, as shown in Fig. 4.26. This loss of spall strength for four different ceramics when shocked above the HEL is shown in Fig. 4.26. Several experimental investigations conducted by various researchers have confirmed the loss of tensile strength above HEL, such as Grady [48] in high-strength ceramics.

Grady and Wise [78] reported spall studies on high-purity and fine-grained alumina subjected to two different shock stress levels: one slightly above the HEL and the other well above the HEL. The spall stress calculated from the pullback velocity in the VISAR measured velocity profiles were about 0.5 GPa and 1.2 GPa, respectively. With the lack of sufficient number of tests on the fine-grained alumina, they suggest that their study could also be inconclusive. However, we can also speculate that the presence of spall strength in alumina when shocked above the HEL is due to the suppression of micro-cracking due to dislocations (or micro-plasticity) activities during shock compression. According to Grady and Wise [78], the lack of significant micro-cracking could lead to the substantial dynamic tensile strength observed in their plate impact experiments. Based on these shock studies on different ceramics, it is important to recognize that the tensile strength could be significant in certain ceramics when shocked above HEL due to microstructure homogeneity. The conclusion is that the loss of spall strength for shocks above the HEL is not a general feature of ceramics, but needs to be investigated for each ceramic of interest.

Since plate impact experiments require specialized facilities and are expensive to conduct, new design concepts can be explored through two-dimensional simulations using finite element methods (FEM) capable of simulating shock wave propagation in solids. With realistic equations of state (EOS) and strength models for the materials involved, a robust FEM or other computational codes can be used to calculate stresses for shock wave propagation analysis. Further details about numerical or computational codes for shock wave analysis are discussed in Chap. 6, and examples of using computations to design experiments are provided.

Characterizing the strain rate induced by plate impact is challenging due to its high and difficult-to-control nature. Researchers have sought techniques to control the rise time of stress pulses, as illustrated in Fig. 4.27. For example, a grooved flyer plate, as discussed in a recent paper by Forquin and Zincsner [79], can influence stress spread and rise time. They performed FEM simulations of wavy-machined flyer plates and buffers to achieve the desired increase in the rise time of the pulses. This exercise in designing planar plate impact experiments can assist shock wave experimentalists in reducing developmental time and cost.

The study involved modelling and simulating three different configurations: (1) a traditional planar plate impact test, (2) a grooved flyer plate (Fig. 4.27d), and a grooved buffer plate ((Fig. 4.27g)). The loading pulse duration for the ceramic is restricted by the thickness of the buffer plate. When the shock wave generated by the impact of the flyer plate travels through the buffer plate, the release waves from the lateral boundaries also move towards the central regions of the ceramic plate.

In a real laboratory experiment employing a gun barrel, the time available for measurements under a genuinely uniaxial strain state is limited by the diameter of the flyer plate and the thickness of the buffer plate. To illustrate, if the wave speed is 5 mm/μs, and the radius and thickness are 20 mm and 5 mm, respectively, for both the flyer and buffer plates, the lateral wave would roughly catch up with the initial shock wave travelling through the buffer plate in less than 4 μs. Naturally, in simulations, we can always apply appropriate lateral boundary conditions to ensure that the analysis is conducted under a uniaxial strain loading condition.

Figure 4.28a illustrates the computed axial stress history at the interface between the buffer and silicon carbide, while Fig. 4.28b shows the particle velocities for the three configurations. These plots aim to analyse the shape and characteristics of the wave pulse generated by the flyer plate impact at a velocity of 1000 m/s. Both the

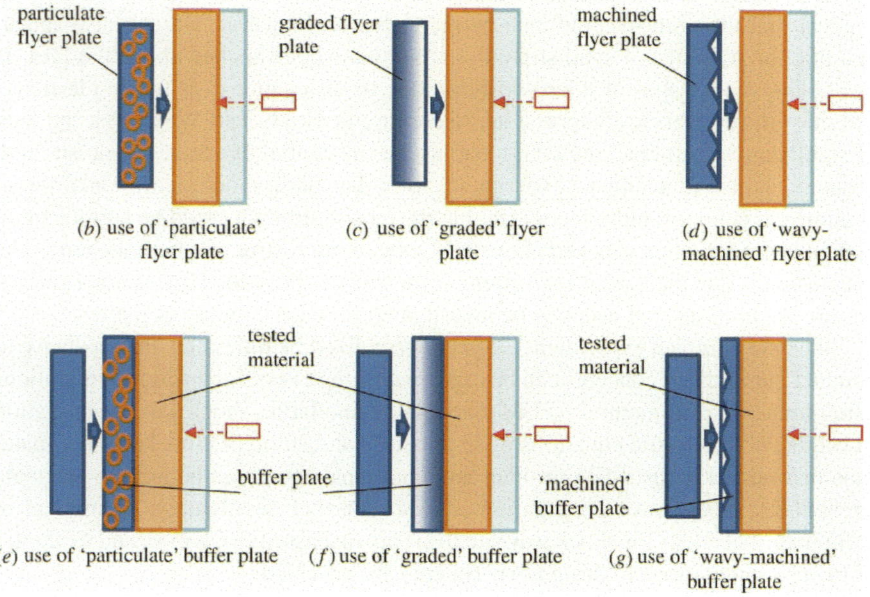

Fig. 4.27 **a** The conventional plate impact configuration was shown in Fig. 4.9 and **b–g** are the various conceptual plate impact configurations for wave shaping techniques to smooth the wave pulse [79]

wavy surfaced flyer and the wavy surfaced buffer demonstrated very similar histories. The use of "wave shaping" techniques appears to achieve the intended increase in the rise time of the pulses.

The computed axial stress history at the interface between the buffer and silicon carbide is plotted in Fig. 4.28a and the particle velocities in Fig. 4.28b for these three configurations to study the shape and other features of the wave pulse generated by the flyer plate impact at 1000 m/s. The wavy surfaced flyer and the wavy surfaced buffer both produced very similar histories. The effects of these "*wave shaping*" technique seem to produce the desired increase in the rise time of the pulses. This computational exercise on designing a planar plate impact experiment to study the effects of varying impedance between flyer and a layered target configuration on the pulse shape and duration will enable shock wave experimentalists to reduce developmental time and cost. As discussed in Chap. 3 on shock wave propagation, impedance analysis, which is based on the EOS curves, can be used to compute the amplitudes of transmitted and reflected waves generated by material interfaces.

The plate impact data presented so far under this section have been used towards the development and validation of constitutive and failure models. However, it is often the case that the application of model parameters to realistically predict shock and impact response of brittle materials under multi-axial loading conditions that occur in most scenarios of engineering importance will require further validation and recal-

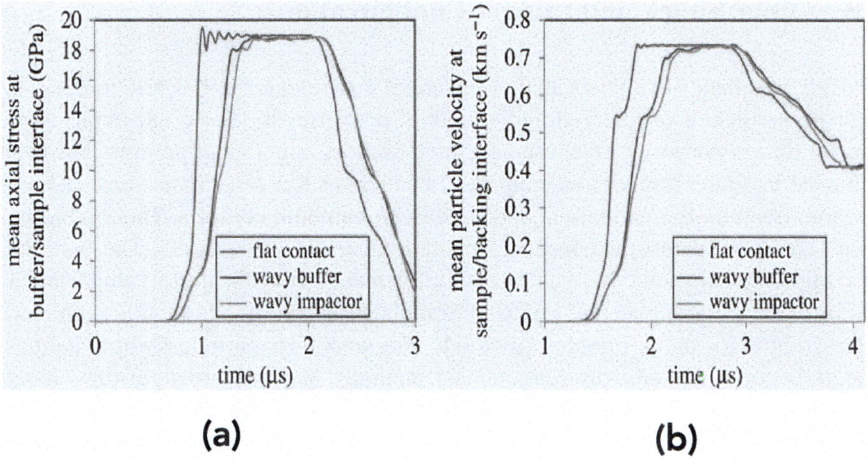

Fig. 4.28 **a** The average axial stress at the interface of buffer and ceramic and **b** the average particle velocity at the interface of ceramic and lithium fluoride window. The stress and velocity histories are shown for three configurations [79]

ibration of the model parameters. For this purpose, well-controlled additional or new impact and shock experiments other than the uniaxial stress and uniaxial strain states may be required. The next section discusses various alternative impact experimental setups and measurement techniques to acquire data for improving models and their predictive capabilities.

Shock wave experimentalists can potentially reduce developmental time and costs through this computational exercise to design a planar plate impact experiment that studies the effects of varying impedance between the flyer and a layered target configuration on the pulse shape and duration.

As discussed in Chap. 3 on shock wave propagation, impedance analysis, which relies on the EOS curves, allows for the computation of the amplitudes of transmitted and reflected waves generated by material interfaces. The plate impact data presented in this section have been instrumental in the development and validation of constitutive and failure models. However, realistically predicting the shock and impact response of brittle materials under multi-axial loading conditions, prevalent in various engineering scenarios, often requires further validation and recalibration of the model parameters. To achieve this, well-controlled additional or new impact and shock experiments, beyond the uniaxial stress and uniaxial strain states, may be necessary. The subsequent section discusses various alternative impact experimental setups and measurement techniques aimed at acquiring data to enhance models and their predictive capabilities.

4.3 Other Shock and Impact Configurations

The advancement of sophisticated commercial and research codes has enabled the realistic prediction of material and structural responses to shock and impact, crucial for design purposes. Over the past three decades, significant progress has been achieved in addressing various numerical challenges like large distortion, material erosion, fracture, fragmentation, and smooth/non-smooth contacts. These advancements have considerably enhanced the predictive capabilities of codes. The evolution of computational methods—such as finite element methods, finite volume methods, arbitrary Lagrangian and Eulerian formulations, cohesive laws, fluid–structure interaction algorithms, meshless methods like smoothed particle hydrodynamics, materials point methods, discrete element methods, and fracture algorithms based on peridynamics—has substantially improved design analyses. This progress allows for more realistic predictions in crashworthiness studies and events involving shock and impact, such as bird impacts on airplane structures or debris impacts on space structures.

An essential aspect in computer simulations is the accurate description of the nonlinear responses of brittle solids through material constitutive (strength) equations. The dynamic strength of materials is influenced by multiple factors, including pressure, strain rate, temperature, and loading histories. Material strength model parameters are often derived or estimated using data from both quasi-static and dynamic tests under uniaxial stress (utilizing SHPB tests) and uniaxial strain (using plate impact tests). However, when these parameters are applied in simulating impact problems, discrepancies between simulation and experimental results can arise. This is evident in scenarios such as projectile penetration into thick plates, structural responses to blasts, high-speed car crashes, oil exploration drilling, and modelling deeply buried concrete targets.

To improve the predictive capabilities of design codes, it's essential to conduct verification and validation of material models under loading conditions that extend beyond uniaxial stress or strain. In this section, we briefly introduce several experimental techniques that aid in refining and improving model parameters. While these techniques have primarily been utilized in characterizing high-strength armour ceramics, they can also be applied to the characterization of brittle materials like cementitious materials and rocks. The subsequent sections will explore a selection of experimental techniques to evaluate parameters not directly obtainable from simple laboratory experiments.

4.3.1 Oblique Plate Impact Experiment

Oblique impact tests are intended to generate intense shear waves. There are two types: shear wave propagation and anvil compression. Their motivations are: (1) to obtain time history of compression and shear wave motions of the stress-free back surface of the target disc or plate and (2) to determine stress–strain response at very high strain rates under a combined pressure and shear loading condition. Both plate configurations were developed at Brown University by Rodney Clifton and his co-workers [80, 81]. Both involve impacts where the velocity vector is not normal to the plate surface, e.g. impact is oblique. Thus both impose normal stress and shear on the impact plate. The first type is a simple wave propagation measurement. The second gives stress–strain history, similar to a SPHPB. Nathenson et al. [82] extended the oblique impact technique to study brittle materials, like silicon nitride and evaluated the dynamic response under combined pressure and shear impact.

Abou-Sayed et al. [80] first developed the shear wave propagation configuration in which an oblique flyer disc impacts a stationary oblique target disc at high velocities generating a combined pressure and shear waves as shown in Fig. 4.29. Taking advantage of the wave speed difference between the compression and shear waves, the angle of tilt can be varied to obtain different stress states and non-proportional loading paths. For instance, the longitudinal wave speed and shear wave speed in pure alumina (AD995) are about 10.56 mm/μs and 6.24 mm/μs, respectively. In silicon carbide, they are 12 mm/μs and 7.67 mm/μs, respectively, as reported by Grady [48].

In the anvil compression version, an inclined flyer plate carrying a thin layer of specimen material impacts a similarly inclined anvil plate such that the specimen is loaded under a combined pressure and shear loading condition. Validity of data requires that the flyer and anvil remain elastic and that measurements are made at the centre of the anvil prior to arrival of release waves from edges and lateral boundaries. Just as in a SHPB, but at much higher rates, the specimen will undergo several reverberations of the compression and shear waves eventually attaining stress equilibrium. The measured compression (normal) wave and shear wave profiles recorded at the rear surface of the target anvil are used to determine the strength. The pressure and shear waves in the elastic anvil are uncoupled under conditions of superimposed hydrostatic pressure. Therefore, data from this oblique plate impact test can be used to extend SHPB data to higher rates and higher pressures.

Despite their potential, oblique impact tests have not been as extensively utilized as normal impact tests due to their complexities in alignment. For instance, consider the setup used by Yuan et al. [83] (Fig. 4.29), quite similar to the planar test (as shown in the schematic of Fig. 4.12) but featuring a gun barrel with a keyway enabling the projectile to travel inside the gun without tilt or rotation.

Yuan et al. [83] launched the tilted or inclined flyer plate with a fibreglass projectile with a sealing O-ring and a Teflon key that slides in a keyway inside the gun barrel to prevent any rotation of the projectile (see Fig. 4.29). Both the inclined projectile and target plates are parallel to each other, and the particle velocity in the target has components both normal and parallel to the plane of impact as shown in Fig. 4.28. The impact velocity V has two components, v_0 (normal) and u_0 (shear).

As utilized by Abou-Sayed et al. [80], the normal velocity component was recorded using a laser velocity interferometer, and light scattered at low angles from the surface was monitored with transverse displacement interferometer (TDI). Yuan et al. [83] employed a laser interferometer, based on the multi-beam VALYNTM™

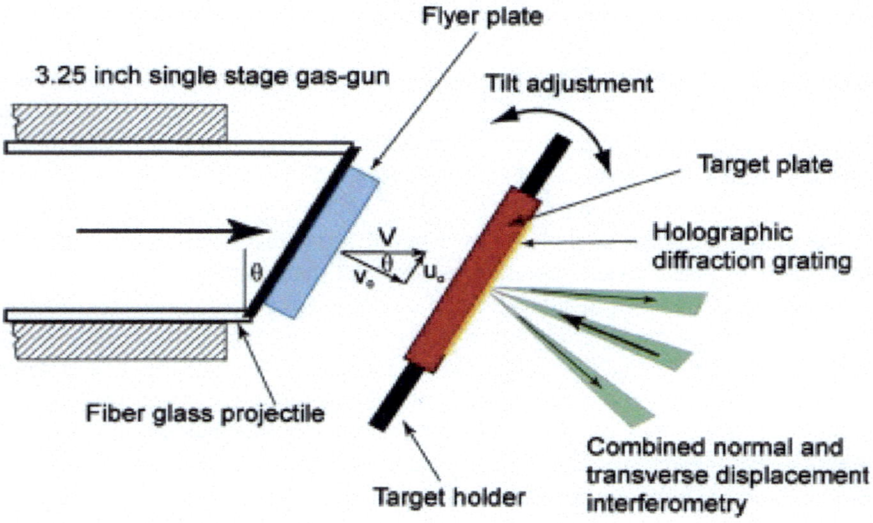

Fig. 4.29 A schematic of oblique plate impact experiment using 82.55 mm gun and multi-beam VALYNTM™ VISAR to measure the v_o (normal) and u_o (shear) particle velocity time histories [83]

VISAR [82], to measure the combined normal and transverse particle velocities at target's rear surface. The TDI requires a diffraction grating on the rear surface as discussed by Abou-Sayed et al. [80] and Yuan et al. [83] in their respective references. Due to the fragility of polymer diffraction gratings and the difficulty of etching gratings directly into the ceramic, Klopp and Shockey [84] adapted a Laser Doppler velocimeter (LDV) system that is often used in fluid-mechanics experiments, to measure the transverse motion of the rear surface.

The transverse and normal velocity profiles from the oblique plate impact experiment provide otherwise hard-to-obtain validation for large shear constitutive models. The interpretation of the profiles will require constitutive description of the brittle material. Though the method was developed for metals, very few investigators have reported data from the combined pressure-shear experiment on brittle materials. As we mentioned earlier, this is a complex test to perform, and requires an expensive modification of the launch tube. Data for ceramics from this highly precision experimental technique are scarce.

4.3.2 Double Flyer Plate Impact

The double flyer plate impact method, introduced by Yaziv [85], was initially intended for examining the behaviours of damaged metals resulting from tensile failure. The experiment's principle involves inducing spall in a metal through the impact of a first flyer. Subsequently, a second flyer generates a timed strong shock in the same experiment to collapse the spall-nucleated voids in the target caused by the first flyer's impact. This technique has been extended to ceramic materials to gain a better understanding of how shock waves travel in an inertially confined cracked ceramic and to calibrate parameters for damage models. This plate impact configuration is particularly useful for determining wave speeds in micro-cracked or fragmented ceramics under lateral confinement. However, its optimal use involves a combination of computational and constitutive modelling because its goal is to examine the evolution and effects of micro- and meso-scale damage. The successful matching of the measured time history of the particle velocity helps verify assumptions on constitutive equations. Although data from double flyer impact tests on most high-strength ceramics or cementitious materials are generally not available, this plate impact test is both highly precise and time-consuming, especially in the case of double flyer and oblique impact configurations. Bless and colleagues [86, 87] obtained double flyer impact data at the University of Dayton Research Institute, initially for an OFHC copper target and later for alumina. The schematic representation of the two flyer plates separated by a small gap is illustrated in Fig. 4.30a.

The first flyer (aluminium) has a lower shock impedance than the second flyer (copper). Yaziv [85] and Yaziv et al. [86] used target thicknesses between 6.35 and 6.5 mm and the flyer thickness of 2 mm to 4 mm. The impact velocities were between 100 and 300 m/s. Since the measured spall strength for AD85 (85% alumina) was around 0.4 GPa (4 $kbars$), the tests were performed well below the HEL to generate spall. Stress histories at the interface of the target and PMMA were measured with Manganin stress gauges. Yaziv also used a VISAR to measure the particle velocity at this interface. Figure 4.30b shows the velocity history for the crystalline ceramic

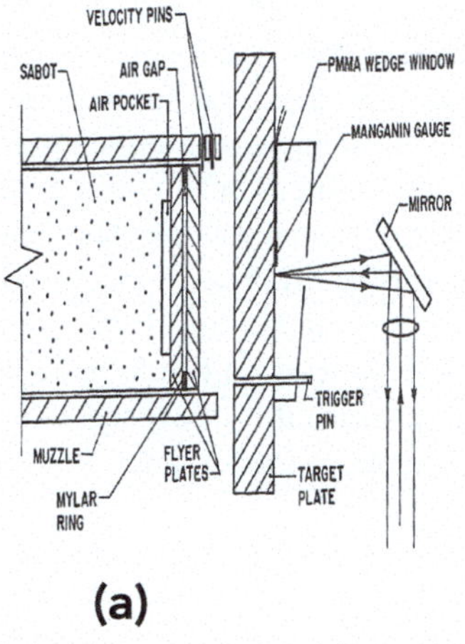

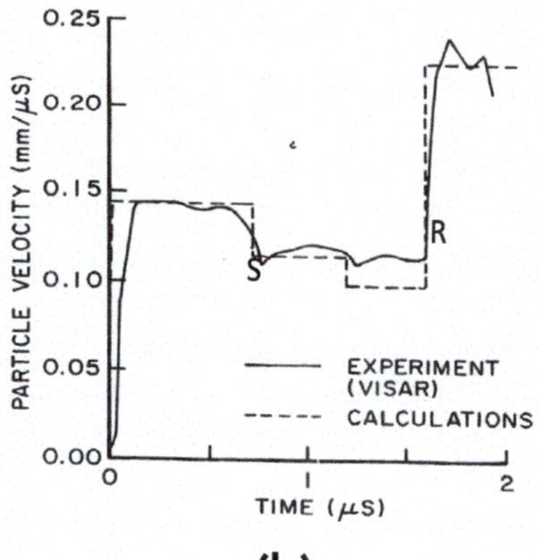

Fig. 4.30 **a** Double flyer impact configuration and **b** the particle velocity history at the interface of PMMA window and the target plate [86]

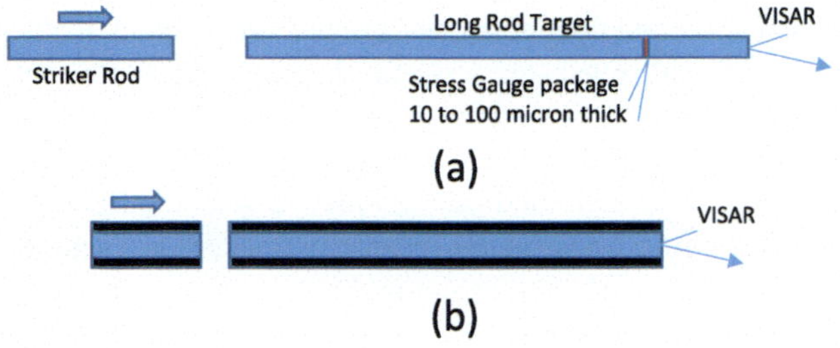

Fig. 4.31 A schematic of long rod experiment: Rod-on-rod impact configuration with stress gauge and VISAR diagnostics; **a** without confinement and **b** with confinement

AD85. The first shock generated a spall signal at **S** due to spall occurring inside the target. Based on the wave speeds, thicknesses of the flyers, the air gap between them, and the target, the reshock caused by the high-impedance second flyer impact reaches the interface between the target and PMMA window at approximately 1.7 μs. Calculations by Yaziv et al. [86], using $(X - t)$ analysis, revealed that the increased density and low sound speed delayed the reshock arrival at the ceramic/window interface. In summary, data from double flyer experiments can be valuable for applications where the tensile-damage properties of ceramics influence engineering performance.

4.3.3 Rod-on-Rod Impact

The framework of wave propagation in 1D strain that is used for plate impact tests can also be used to analyse impacts with 1D stress boundary conditions, e.g. impacts onto long bars. In the simplest interpretation, the impact fails the target bar and launches a wave whose amplitude is the largest elastic stress the bar can support, which is, by definition, Y. However, imperfect impact conditions may result in a stress lower than this, especially in brittle materials. In addition, a brittle ceramic may fail in hoop tension instead of principal compressive stress. In many constitutive models, shear failure is a function of mean stress, and thus the failure in a bar, which takes place at a mean stress of one-third of the principal stress, can be an important datum in model calibration. Since bar strength measurement is based on wave amplitudes, it avoids some of the problems of stress equilibrium which can inject uncertainty into interpretation of SHPB data. However, as with most tests on brittle materials, the statistical spread in the measured data can be influenced by how the rods or bars were manufactured or processed from the stock materials as well as test-to-test variation.

This rod-on-rod experimental technique shown in Fig. 4.31a was pursued for ceramics by Bless and his co-workers [86, 87]at the University of Dayton Research Institute, Dayton, Ohio during 1980s and later at the University of Texas, Austin, Texas. During the 90 s, researchers in Sandia National Laboratory, USA and Cavendish Laboratory at Cambridge, UK used this technique to develop a better

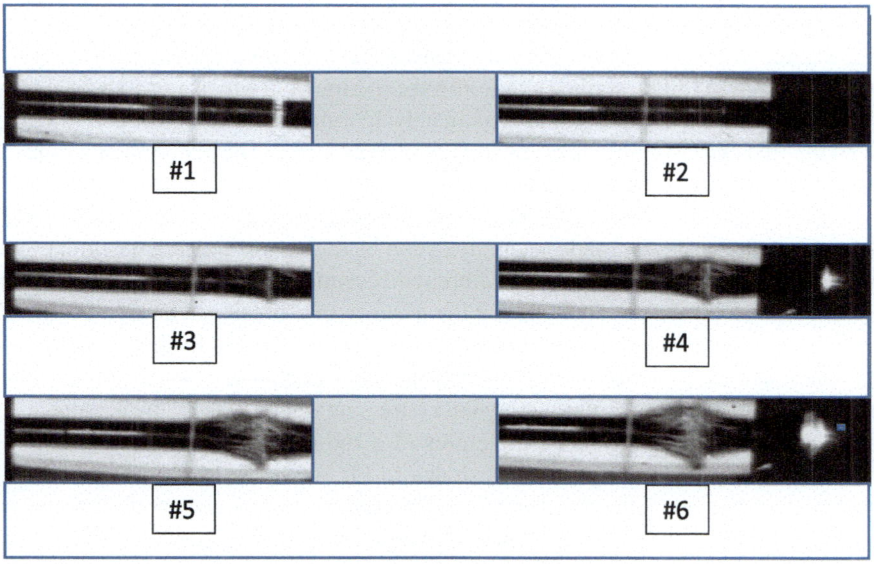

Fig. 4.32 High-speed photographs at a framing rate of $105/s$ from a 12.7 mm diameter AD-85 bar-on-bar impact at 179 m/s. The rods were painted black for better resolutions. Each frame is at an interval of 10 μs [88]

understanding of high strain rate response of high-strength ceramics. Several variations have been introduced in the rod impact configuration, including the use of metallic plate as the impactor and wave shapers, while the target had been always the slender long rod.

In the rod-on-rod impact configuration as shown schematically in Fig. 4.31, a short slender ceramic rod impacts a stationary long ceramic rod having a length-to diameter ratio of about 8. Both rods are made of the same material. The target rod experiences strain rates in the range of $10^2 - 10^3/s$. A stress gauge is typically embedded in the target rod (several diameters away from the impact end) to record the stress history. The rods also can be confined in sleeves as shown in Fig. 4.31b to avoid cracking up the impact end and also to study the effects of confinement on the peak value of the measured stress or particle velocity. Brar et al. [87] reported stress history data from the sleeved confined long rod tests. They investigated the confinement effect on the peak value of the stress.

A VISAR or PDV can also be used to record both longitudinal velocity histories of the target rod's free surface. One can also measure the lateral motion of the rod using VISAR at a location close to the free end. Failure of ceramics can be studied through these instrumented rod-on-rod impacts and high-speed photography. At very early times the loading state is one-dimensional strain, so the initial amplitude of the transmitted stress pulse is comparable to the HEL of the ceramic.

It is highly challenging to measure the stress histories very close to the impact plane. In past rod-on-rod tests, the stress gauge was embedded about 25 mm from the free end of the long rod as shown in Fig. 4.32. The striker rod was 50 mm long and

12.5 mm in diameter ($L/D = 4$), while the target rod was 100 mm long and 12.5 mm in diameter ($L/D = 8$); both rods were made of AD995 ceramic. For diagnostic purposes, Rajendran [88] described a rod-on-rod experiment in which a stress gauge was embedded inside the target rod approximately 75 mm from the impact end (25 mm from the free end) to record the time history of axial stress. The experimental impact velocities were 175 m/s and 278 m/s.

When examining points very close to the impact end (distances less than two rod diameters), the peak amplitude is anticipated to be under uniaxial strain, similar to a plate impact test. However, this peak amplitude eventually relaxes to the yield stress for uniaxial stress deformation. Given the absence of a precise definition of "*yield stress*" for ceramic materials, particularly under uniaxial stress loading conditions, it is often defined as the peak amplitude of the measured pulse. As a result, the sole useful information in the time-history data (stress or velocity) is the peak value.

A series of high-speed camera pictures of a rod-on-rod impact experiment on AD-85 ceramic taken every 10 μs is shown in Fig. 4.32. These pictures provide a qualitative understanding of the macro-scale fracture process occurring in the rods and the extent of the fracture as a function of time. In the high-speed camera pictures, one can observe several splitting-type macro-cracks forming and propagating towards the stress free end of the long rod. Simha [88] measured particle velocities and also obtained high-speed camera pictures for rod-on-rod tests on AD-995 alumina. Though high-speed camera pictures are obtained for the rod-on-rod tests, we will see later in the modelling chapter that it has not been possible to uniquely quantify any scalar damage model parameters that may be used in computational modelling. Therefore, attempting to model this macro-crack propagation using a continuum mechanics finite element code will not be accurate and is futile. It has been possible to match the maximum (peak) value of the stress between the model and data [89]. Also, the data from rod-on-rod experiments can be used in a qualitative sense to validate the model parameters under a uniaxial stress state.

The fracture initiates at the impact end, leading to several axial splitting-type macro-cracks to form and propagate towards the gauge location. Simha et al. [90] have discussed the validity and interpretation of stress measurements in the bar experiments. They concluded that the peak stress measured by the stress gauge cannot be correlated to the intrinsic strength of the ceramic material since its value is influenced by the impact velocity, striker rod material, location of the gauge (relative to the impacted end of the target rod), and the target rod diameter. Since the striker and target bars are brittle materials, severe cracking occurs instantaneously at the impact end. These apparently can lead to failure and unloading before the theoretical peak 1D yield is achieved. At higher impact velocities, the cracking becomes more intense and the damage zones extend further into the long target rod to reduce the peak value even lower [91]. The velocity and stress histories for AD995 alumina are shown in Fig. 4.33.

Figure 4.33a presents VISAR-measured longitudinal particle velocity histories from two similar tests conducted at about 112 m/s, depicting nearly identical velocity trends with a marginal difference of approximately 10%. In Fig. 4.33b, the peak stress (2.8 GPa) measured by the Manganin gauge at a higher velocity of 278 m/s was found

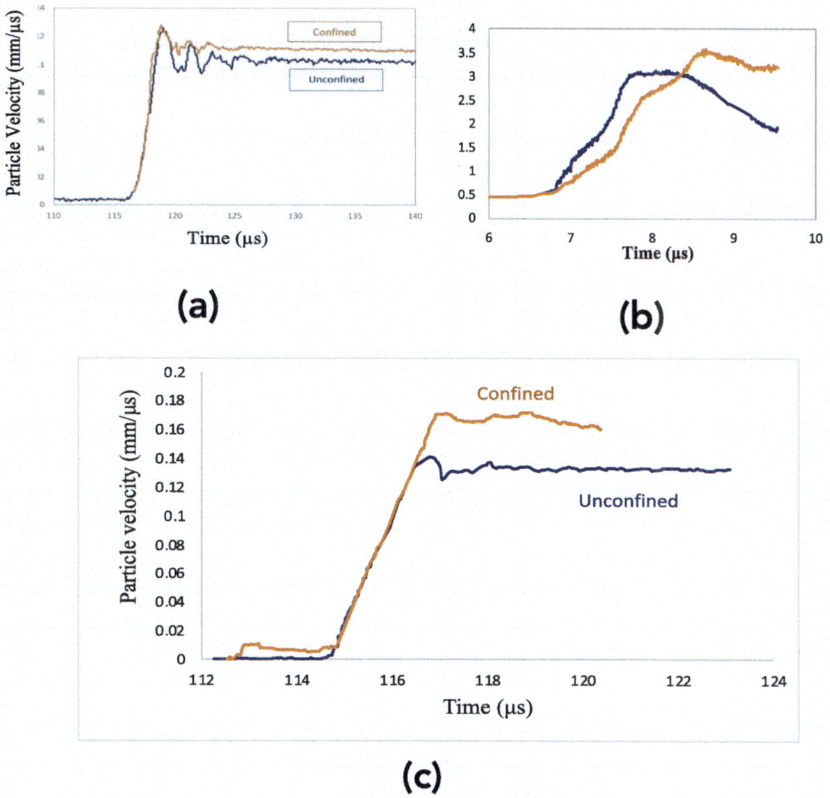

Fig. 4.33 Free surface velocities for two tests on AD-995 alumina bar at impact velocity about **a** 112 m/s; **b** at about 297 m/s and **c** comparison between confined and unconfined bars

to be lower than the peak stress (3.3 GPa) observed for the test at 175 m/s. Brar [91] reported that doubling the impact velocity did not double the maximum stress but indicated a slight increase. The Manganin gauge stress data for AD995 alumina, reported by Simha [88] and Brar et al. [87], suggests that the peak stress (or velocity) is not solely a material property. Their data reveals that the peak value is influenced by various factors, such as impact velocity, striker rod material, gauge location, and target rod diameter.

Bless [92] summarized measurements of bar impact strength for several ceramics, establishing failure stress (Y_B) under uniaxial stress loading. For instance, Y_B for alumina, titanium diboride, silicon carbide, and other brittle materials like borosilicate glass were identified. Sintered alumina showed Y_B values between 3.6 and 4.2 GPa, while hot-pressed alumina was at 4.1 GPa. Bless also included high-speed images of fracturing rods in the article "*Using Bar Impact to Determine Dynamic Properties of Ceramics*", often revealing different failure modes in rods that could be valuable in formulating failure models. A preliminary conclusion suggests that the strength measured in a bar impact test might only represent a lower bound on Y.

Since the unconfined edges of the ceramic bars (or rods) develop tensile cracking immediately after the impact and create release waves at the impact plane, one method to suppress such cracking is to confine both the striker and target rods by confining them in a steel or tantalum sleeves. The particle velocity data from both unconfined and confined tests at similar velocities, as shown in Fig. 4.33c, further validates the rod-on-rod test. Due to increased cracking at the impact end in the unconfined rod, the higher velocity test (278 m/s) produced lower peak velocity compared to the lower velocity (175 m/s) test. However, when the rod is confined, due to the suppression of cracks at the impact end, the peak velocity at a higher velocity was higher than the peak velocity value in the lower velocity test. In fact, the high-speed camera images obtained from tests on unconfined rods support these findings.

The time histories of both the VISAR and Manganin gauge measurements from the rod-on-rod tests enable strength or constitutive model developers to not only fine-tune certain damage model parameters but also to validate and introduce modification to the models themselves. Our ability to model both the rod-on-rod tests under uniaxial stress state and the plate impact test under uniaxial strain state using one set of model parameters is important in the realistic prediction of impact problems under multi-axial loading, such as the projectile penetration into thick armour plates or blast loading on protective structures.

4.3.4 Single-Density Metal Plate Impact on a Long Ceramic Rod

While the repeatability of the rod-on-rod experiments has been confirmed by various researchers, issues persist regarding uncertainties and data scattering. To mitigate these uncertainties and minimize cracking at impact points, researchers introduced a new experimental setup involving a metal plate impacting a ceramic rod. The concept behind this test is to enable the metal flyer to flow around the high-strength ceramic rod, allowing for a characteristic dwell time. This, in turn, results in a stress pulse that is somewhat ramped, similar to the effect of pulse shapers in SHPB tests. Careful consideration is necessary in choosing the metallic material for the flyer. If the flyer material is too soft, the issue could swiftly transform from a stress wave problem into a projectile penetration problem.

Rosenberg et al. [93] reported an experiment in which a metal flyer plate was made to impact a long slender aluminium rod. Constantan and Manganin gauges were used to measure the stress-time history in the rod. Simha [88] reported several plate-on-rod experiments on AD995; he investigated the influence of the flyer material, rod diameter, and several other parameters on the measured stress profiles. The maximum stress measured in the rod can be interpreted as the compressive strength under high strain rate loading. Wise and Grady [94] extended this impact configuration in a test

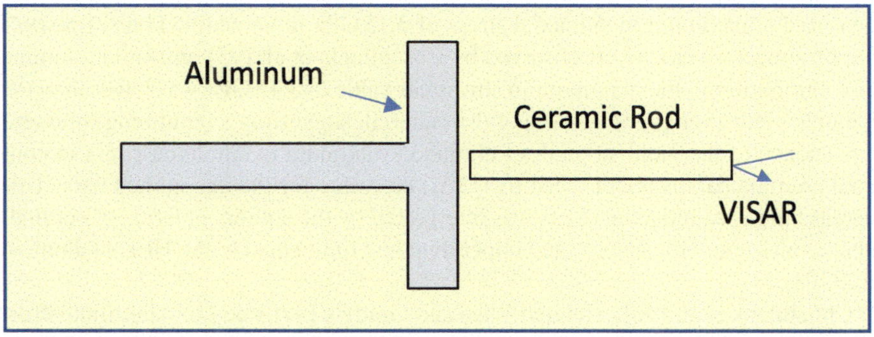

Fig. 4.34 A 6061-T6 aluminium metal plate with a long tail rod impacting a stationary AD995 (Alumina) long rod target at 1035 m/s. VISAR measures the free surface velocity at the rear end of the AD995 ceramic rod

in which a flat 6061-T6 aluminium plate (disc) with a long trailing rod was made to impact a stationary long ceramic rod. The rod section of the flyer plate maximizes the duration of high compressive stresses at the impact end of the ceramic rod. As shown in Fig. 4.34, this configuration permits a time-resolved investigation of the dynamic response of materials under a multi-axial strain state. The stress state in the ceramic rod well away from the impact face is uniaxial. During the experiment, time-resolved particle velocity data were obtained at the free end of the target rod.

The bar on rod configuration has remained non-conventional. In bar impact tests on ceramics, most researchers have used embedded stress gauges. These stress gauge signals can be complicated by unloading waves due to fracture (macro-crack propagation) at the gauge location. Rear free surface measurements may be easier to interpret. Even so, the peak free surface or stress value may not be a material property, since its value is influenced by the impact velocity, striker rod material, location of the gauge (relative to the impacted end of the target rod), and the target rod diameter. On the other hand, in the study by Brar et al. [87], in which 12.7 mm diameter and 300 mm AD 998 and AD94 long rods were impacted by thick 15 mm steel impactor (or flyer) at velocities ranging from 96 m/s to 560 m/s, it was found that after a short peak the stress levels relaxed to the static values. This is an expected behaviour.

Simha [88] presented data from Manganin gauge stress measurements for AD995 alumina for two configurations: (1) a 4340 steel plate impacting a long AD995 rod and (2) a short AD995 rod impacting a long AD995 rod. In these experiments, fracture initiated at the impact end, with several splitting-type macro-cracks forming and propagating towards the gauge location. Attempting to model this macro-crack propagation using a continuum mechanics finite element code will not be accurate. However, it is still true that the peak measured stress represents an elastic stress which the material can support; thus the strength must be at least that high.

Wise and Grady [94] performed impact tests on unconfined and confined (in close-fitting tantalum sleeve) Coors 99.5% alumina bars with aluminium flyer plates launched at velocities ranging from 1035 to 2182 m/s. The magnitude of confinement stress was not quantified since the bars were not shrunk fit into metal sleeves. They measured a maximum in situ axial stress of 3.15 GPa in unconfined bars, irrespective of impactor velocity, as observed by Cosculluela et al. [95] on similar alumina bars. Furthermore, the maximum in situ stress value of 3.15 GPa was lower than the dynamic yield strength of 4.3 GPa of the material, suggesting a (confining) pressure dependence of the yield strength of alumina. Maximum in situ axial stress in confined alumina bars is about twice (6.1 GPa) the value for the unconfined bars. This increase in measured axial stress was interpreted by the authors in terms of confinement, consistent with an expected upper dynamic limit equal to the HEL of alumina (6.2 GPa).

Chhabildas et al. [96] extended Wise and Grady's [94] study to determine effect of confinement (in shrink fit steel cylinder) on in situ axial stress for 99.5% alumina bars impacted by steel plates at 300–366 m/s. Measured in situ axial stress in confined alumina bars was in the range of 4.6 to 5.1 GPa, significantly higher than in the case of unconfined bars, and in agreement with earlier studies. Tensile waves generated as a result of lateral release in a sleeved bar target are totally eliminated due to confinement during the initial loading stress wave. Simha [88] and Simha et al. [90] reported an increase in measured in situ axial stress (4.2 GPa) in 12.7 mm diameter confined (with a steel sleeve) Coors©alumina.

The straightforward interpretation of these bar impact results may be hindered by uncertainties associated with fracturing. At low confining pressure, failure could potentially be attributed to localized tensile forces, possibly influenced by the wing crack mechanism. The challenges in data interpretation do not stem from experimental issues; instead, they are more likely linked to an insufficient conceptual framework. Addressing these challenges may necessitate simulations incorporating enhanced constitutive models.

4.3.5 Graded-Density Plate on Rod Impact

It is possible to achieve high strain rates above 10^3 *s* in the rod-on-rod tests for ceramics that are slightly higher than in SPHB, but lower than in conventional plate impact experiments. But premature fractures at the impact end makes the interpretation of data difficult. The use of single-density metal flyer plate to impact a long slender ceramic rod improved the data and increased the strain rates in tests. It is also possible to spread the stress pulse in the rod by impacting through a flyer whose density is graded from low to high. During the early time of the impact, the soft layer of the flyer plate will flow around the rod and this may reduce excessive damage initiation and propagation.

Chhabildas et al. [96] designed an experiment as shown in Fig. 4.34 to introduce a controlled stress pulse into confined as well as unconfined ceramic rods. The main focus in the design was to be able to obtain data at intermediate strain rates between the SPHB and planar plate impact experiments. They reported an experimental configuration in which an AD995 rod (19 mm in diameter and 75 mm long) was impacted by a graded-density flyer plate. The thicknesses of the four layers (steel, titanium, aluminium, and TPX) in the graded-density flyer plate were 19.04, 1.097, 1.199, and

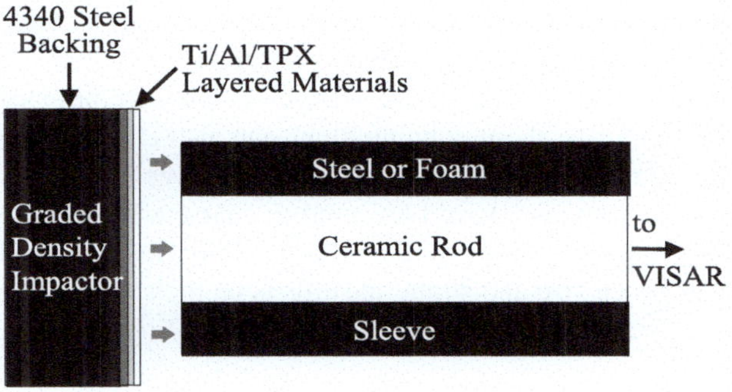

Fig. 4.35 A schematic of the graded-density flyer plate impacting a ceramic rod assembly

1.034 mm, respectively. AD995 ceramic rods (sleeved or unsleeved) were impacted at velocities between 300 and 366 m/s.

A VISAR was used to record the axial particle velocity of the free end of the target rod. A schematic of the experimental configuration is shown in Fig. 4.35. To shape the stress pulse and to make the impact less hard on the long AD995 alumina and silicon carbide ceramic rods, a graded-density flyer plate was used by Chhabildas et al. [97] in their gas gun experiments. The strain rate-dependent failure stress was determined from the graded-density plate–rod experiments. Later, Chhabildas et al. [96] reported high strain rate data for about a 9 mm diametre boron carbide with a length of ≈ 9.42 mm. The impact velocities ranged between 300 m/s and 945 m/s. Graded-density materials were used to impact both bare sleeved rods.

According to Chhabildas et al. [96], the measurements from these experiments demonstrate the following unique features of this test methodology: (1) the time-dependent stress pulse generated during impact allows a smooth and efficient transition from the initial uniaxial strain loading to a uniaxial stress state as the stress pulse propagates through the rod and (2) intermediate loading rates obtained in this configuration lie in a region which is not achieved easily by either SHPB bar or shock-loading techniques. Availability of such measurements enhances the validation of constitutive/damage models.

Cazamias et al. [98] conducted impact of graded-density impactors on unconfined and confined AD995 alumina rods of 12 and 19 mm diametre to investigate the behaviour of ceramics under one-dimensional stress loading conditions. Manganin gauge and VISAR diagnostics were employed in their investigation. They employed target rod of different lengths and found that the larger rods showed higher strength of about 4.5 GPa. For both confined and unconfined rods, the larger diameter rods exhibited an increased strength over the smaller diametre rods. They attributed this strength variation to differences in the stock material. Thus, the difference in strength might be a true scale effect rather than a rate effect. Material variability might also be attributed to the fact that the small-scale rods were cut perpendicular to the face of the plate, while the large-scale rods were cut parallel. Note that this is counter to a common hazard in static tests on brittle materials whose strength may be flaw dependent: larger samples tend to be weaker.

Although these shock wave propagation experiments were primarily developed for high-strength ceramics, these configurations can also characterize other brittle materials such as glass, rocks, cement, and concrete. Over the past three decades, most experiments for ceramics have been used to characterize transparent glasses. Especially noteworthy is the work by B.E. Martin, W.F. Heard, C.M. Loeffler, and X. Nie, on the specimen size and strain rate effects on the compressive behavior of Concrete [99].

Summary

This chapter offers an extensive overview of experimental techniques available for testing and characterizing the behaviour of brittle materials under dynamic loading conditions. Constitutive and failure models, whether derived from physics-based theories or empirical studies, comprise numerous parameters. In engineering applications, constitutive models for brittle materials generally necessitate an equation of state, a model for shear strength, and a model for tensile strength. These crucial parameters can be substantially determined through well-designed high strain rate experiments.

Furthermore, it's often imperative for brittle materials to have measurements of damage and an understanding of its impact on elastic and strength properties. In all applications, it's vital to conduct impact tests in various geometries. Ideally, laboratory tests should provoke similar failure modes as those likely to occur in the intended engineering application. However, it's equally important to assess the specific requirements of each engineering problem, as not all practical scenarios demand precise specification of all constitutive parameters.

Determining equations of state typically involves measuring Hugoniot curves and shear strength along these curves. Hugoniot curves can be computed when any two of the four Hugoniot parameters are measured, such as principal stress and particle velocity. Shear strength during shock compression can be measured using transverse stress gauges, and conventional plate impact techniques are well suited for these measurements.

Additionally, shear strength can be independently measured by comparing uniaxial stress measurements, which can be achieved using a bar impact geometry or Hopkinson bar, providing yield strength (twice the shear strength). Another technique involves shocks produced by inclined flyer plate impacts.

Exploring the effects of damage can be achieved through multiple impact experiments, where a material is first shocked and then either unloaded or reloaded. Using multiple flyer plates, either in contact or with gaps, allows for a wide range of conditions to be examined.

Tensile strength can be most easily measured in uniaxial strain by conducting spall experiments. These experiments are relatively straightforward with plate impact facilities. Direct loading by explosives can also measure spall strength, while failure in 1D stress can be investigated using suitably designed bar impact experiments. Properties after experiencing tensile damage can be determined by using more than one flyer plate.

Damage effects can be probed by using multiple impact experiments, in which a material is first shocked and then unloaded or reloaded. By using multiple flyer plates, either in contact or with gaps, a wide range of conditions can be examined.

Certain parameters, such as the effects of evolving triaxial stress states, cannot be directly measured by simple impact experiments. In such cases, it's necessary to conduct experiments that, to some extent, simulate scenarios of significant engineering interest. Additional instrumentation can be deployed to verify that numerical codes accurately produce stress and particle velocity during intermediate states, providing the correct end result (such as ballistic limit or depth of penetration). Only then can a numerical code be utilized for extrapolation or prediction to new impact parameters.

References

1. Chen WW, Song B (2011) Split hopkinson (kolsky) bar. Springer Publishers
2. Othman R (2018) The kolsky—hopkinson bar machine. Springer International Publishing
3. Walley SM (2018) The origins of the Hopkinson bar techniques. In: Othman R (ed). Springer International Publishing, pp 1–26
4. Kolsky H (1949) An investigation of the mechanical properties of materials at very high rates of loading. Proc Phys Soc Sect B 62(11):676–670
5. Ravichandran G, Subhash G (1994) Critical appraisal of limiting strain rates for compression testing of ceramics in a split Hopkinson pressure bar. J Am Ceram Soc 77:263–267
6. Chen W, Subhash G, Ravichandran G (1994) Evaluation of ceramic specimen geometries used in split Hopkinson pressure bar. DYMAT J 1:193–210
7. Li D, Ngai L, Wong Y (2013) The Brazilian disc test for rock mechanics applications: review and new insights. Rock Mech Rock Eng 46:269–287. https://doi.org/10.1007/s00603-012-0257-7
8. Chen WW, Rajendran AM, Song B, Tie N (2007) Dynamic fracture of ceramics in armour applications. J Am Ceram Soc 90(4):1005–1018
9. Lee OS, Cho KS, Kim SH, Han YH (2006) Deformation behavior of soft material using SHPB technique and pulse shaper. Int J Mod Phys B 20(25, 26 & 27):3751–3756
10. Gray III GT (2000) Classic split Hopkinson pressure bar testing. In: ASM handbook
11. Chen W, Lu F, Frew J, Forrestal MJ (2002) Dynamic compression testing of soft materials. ASME Trans J Appl Mech 69:14–223
12. Song B, Chen W (2004) Dynamic stress equilibration in split Hopkinson pressure bar tests on soft materials. Exp Mech 44:300–312
13. Chen W, Song B (2009) Dynamic characterization of soft materials. In: Dynamic failure of materials & structures, pp 1–28
14. Chen WW (2016) Experimental methods for characterizing dynamic response of soft materials. J Dyn Behav Mater 2:2–14
15. Follansbee PS, Franz CE (1983) Wave propagation in the split Hopkinson pressure bar. Trans ASME J Eng Mater Technol 105:61–66
16. Franz CE, Follansbee PS, Wright WJ (1984) New experimental techniques with the split Hopkinson pressure bar. In: 8th international conference on high energy rate fabrication. Pressure Vessel & Piping Division, ASME, Eds J. W. Schroeder, San Antonio, Texas, June 17–21

17. Nemat-Nasser S, Isaacs JB, Starrett JE (1991) Hopkinson techniques for dynamic recovery experiments. Proc R Soc A 435(1894):371–391
18. Xia K, Yao W (2014) Dynamic rock tests using split Hopkinson (Kolsky) bar system: a review. J Rock Mech Geotech Eng 7:27–59
19. Frew DJ, Forrestal MJ, Chen W (2002) Pulse shaping techniques for testing brittle materials with a split Hopkinson pressure bar. Exp Mech 42(1):93–106
20. Forrestal MJ, Warren TL (2021) A conical striker bar to obtain constant true strain rate for Kolsky bar experiments. J Dyn Behav Mater 7:161–164
21. Grote DL, Park1 SW, Zhou M (2001) Dynamic behavior of concrete at high strain rates & pressures: I. Experimental characterization. Int J Impact Eng 25:869–886
22. Zhou J, Zhao G, Meng X, Liu C, Ma L, Xu W, Cheng X (2023) Experimental investigation of the size effect of rock under impact load. Minerals 13(1):43
23. Forrestal MJ, Wright TW, Chen W (2007) The effect of radial inertia on brittle samples during the split Hopkinson pressure bar test. Int J Impact Eng 34:405–411
24. Gorham DA (1983) A numerical method for the correction of dispersion in pressure bar signals. J Phys E Sci Instrum 16:477
25. Ramesh KT (2008) High rates and impact experiments. In: Sharpe WN (ed) Springer handbook of experimental solid mechanics. Springer, New York, USA, pp 929–960
26. Davies EDH, Hunter SC (1963) The dynamic compression testing of solids by the method of the split Hopkinson pressure bar (SHPB). J Mech Phys Solids 11:155–179
27. Nicholas T, Rajendran AM (1990) High strain rate behavior of materials. Chapter 3 in high velocity impact dynamics. In: Zukas J (ed) John-Wiley & Sons, Inc., pp 127–296
28. Adam KM (1998) Advancements in the split Hopkinson bar test. Thesis, Masters of Science in Mechanical Engineering, Virginia Polytechnic Institute and State University, Blacksburg, Virginia
29. Frew DJ, Forrestal MJ, Chen W (2001) A split Hopkinson pressure bar technique to determine compressive stress-strain data for rock materials. Exp Mech 41(1):40–46
30. Rosenberg Z, Dawicke D, Strader E, Bless SJ (1986) A new technique for heating specimens in Split Hopkinson bar experiments using induction-coil heaters. Exp Mech 26:275–278
31. Lennon AM, Ramesh KT (1998) A technique for a technique for measuring the dynamic behavior of materials at high temperatures. Int J Plast 14(12):1279–1292
32. Lankford J (1981) Temperature-strain rate dependence of compressive strength and damage mechanisms in aluminium oxide. J Mater Sci 16:1567–1578
33. Chen P, Guo B, Chen J (2018) Dynamic Brazilian test using the Kolsky-Hopkinson bar machine. In: Othman R (ed) Chapter 4, the Kolsky-Hopkinson bar machine. Springer, Cham, pp 121–141
34. International Society for Rock Mechanics (ISRM) (1978) Suggested methods for determining tensile strength of rock materials. Int J Rock Mech Min Sci & Geomech 15:99–103
35. ASTM D 3967-08 (2008) Standard test method for splitting tensile strength of intact rock core specimens. ASTM International, West Conshohocken, USA
36. Diyuan L, Yuen Wong LN (2012) The Brazilian disc test for rock mechanics applications: review and new insights. Rock Mech Rock Eng 46:269–287
37. John R, Antoun T, Rajendran AM (1991) Effect of strain rate and size on tensile strength of concrete. In: Proceedings of the American physical society conference on shock compression of condensed matter- 1991. Elsevier Science Publishers, pp 501–504
38. Wang QZ, Li W, Xie HP (2009) Dynamic split tensile test of flattened Brazilian disc of rock with SHPB setup. Mech Mater 41:252–260
39. Dai F, Xia K (2010) Loading rate dependence of tensile strength anisotropy of barre granite. Pure Appl Geophys 167(11):1419–1432
40. Nemat-Nasser S, Horii H (1984) Rock failure under compression. Int J Eng Sci 22(8–10):999–1011
41. Staehler JM, Predebon WW, Pletka BJ, Subhash G (2000) Micromechanisms of deformation in high-purity hot-pressed alumina. Mater Sci Eng A 291:37–45
42. Rajendran AM (1992) High strain rate behaviours of metals, ceramics, and concrete. Wright Laboratory Technical Report, WL-TR-92-4006

43. Barker LM, Hollenbach RE (1972) Velocity interferometer for measuring the velocity of any reflecting surface. J Appl Phys 43:4669–4680
44. Strand OT, Goosman DR, Martinez C, Whitworth TL, Kuhlow WW (2006) Compact system for high-speed velocimetry using heterodyne techniques. Rev Sci Instrum 77:083108–8
45. Rosenberg Z (1981) The use of Manganin gauges in shock reverberation experiments. J Appl Phys 52(6):4000–4002
46. Gupta S, Gupta Y (1984) Piezoresistance response of longitudinal and lateral oriented ytterbium foils subjected to impact and quasi-static loading. J Appl Phys 57(7):2464–2473
47. Gupta S, Gupta Y (1987) Experimental measurements and analysis of the loading and unloading response of longitudinal and lateral manganin gauges shocked to 90 kbar. J Appl Phys 62(7):2603–2609
48. Grady DE (1994) Dynamic properties of ceramic material. Sandia National Laboratories Report. SAND94-3266
49. Zhang QB, Braithwaite CH, Zhao J (2016) Hugoniot equation of state of rock materials under shock compression. Philos Trans R Soc Publ A375
50. Ahrens TJ (1987) Shock wave techniques for geophysics and planetary physics. Methods Exp Phys 24, part A (Eds. G.S. Charles and L.H. Thomas). Academic Press, New York, NY, pp 85–234
51. Zhernokletov MV (2006) Methods and devices for producing intense shock loads. In: Zhernokletov MV, Glushak BL (eds) Material properties under intensive dynamic loading. Shock wave and high pressure phenomena. Springer, pp 33–71
52. Bourne N (2013) Materials in mechanical extremes: fundamentals and applications. Cambridge University Press, UK
53. Field J, Walley S, Proud W, Goldrein H, Siviour CR (2004) Review of experimental techniques for high rate deformation and shock studies. Int J Impact Eng 30:725–775
54. Jannotti P, Doney R, Schuster B (2017) Time-resolved measurement of deformation of metal plates due to high-velocity impacts. Procedia Eng 204:276–283. 14th Hypervelocity Impact Symposium, (2017) HVIS2017, Canterbury, Kent, UK
55. Rosenberg Z, Yaziv D, Yeshurun Y, Bless SJ (1987) Shear strength of shock-loaded alumina as determined with longitudinal and transverse manganin gauges. J Appl Phys 62(3):1120–1122
56. Grady DE (1998) Shock-wave compression of brittle solids. Mech Mater 29:181–203
57. Grady DE (2014) Hugoniot equation of state and dynamic strength of boron carbide. J Appl Phys 117:165904
58. Chen M, McCauley JW, Hemker KJ (2003) Shock-induced localized amorphization in boron carbide. Science 299(5612):1563–6
59. Whelchel RL, Kennedy GB, Dwivedi SK, Sanders TH, Thadhani NN (2013) Spall behavior of rolled aluminum 5083–H116 plate. J Appl Phys 113:233506
60. Joshi A, Ravindran S, Gandhi V, Ravischandran G (2022) Probing the properties and mechanisms of failure waves in soda-lime glass. J Appl Phys 131:205902
61. Vogler TJ, Reinhart WD, Chhabildas LC (2004) Dynamic behavior of boron carbide. J Appl Phys 95:4173
62. Rosenberg Z, Partom Y (1984) Longitudinal dynamic stress measurements with in-material piezoresistive gauges. J Appl Phys 58(5):1814–1818
63. Kipp ME, Grady DE (1990) Shock compression and release in high-strength ceramics. In: Shock waves in condensed matter 1989. AIP, New York, pp 377–80
64. Bless SJ, Brar NS, Kanel G, Rosenberg Z (1992) Failure waves in glass. J Ceram Soc 75(1):1002–1004
65. Grady DE (1994) Shock-wave strength properties of boron carbide and silicon carbide. In: Sandia national laboratory, SAND-94-0300-C
66. Rajendran AM, Brar NS, Khobaib N (1989) Effects of dynamic pre-strain on the subsequent tensile flow strength. In: Schmidt SC, Johnson JN, Davison LW (eds) Proceedings of the American physical society conference on shock compression of condensed matter. Elsevier Science Publishers, pp 401–404

67. Nicholas T, Rajendran AM, Grove DJ (1987) Analytical modelling of precursor decay in strain-rate dependent materials. Int J Solids Struct 23:1601–1614
68. Murray NH, Bourne NK, Rosenberg Z (1996) Precursor decay in several aluminas. AIP Conf Proc 370:491
69. Rosenberg Z, Brar N, Bless SJ (1988) Elastic precursor decay in ceramics as determined with manganin stress gauges. J Phys Colloq 49(C3):C3-707–C3-711
70. Bourne N, Millett J, Rosenberg Z, Murray N (1998) On the shock induced failure of brittle solids. J Mech Phys Solids 46(10):887–1908
71. Rosenberg Z, Yaziv D, Yeshurun Y, Bless SJ (1987) Shear strength of shock loaded alumina as determined with longitudinal and transverse manganin gauges. J Appl Phys 62:1120–1122
72. Cagnoux J, Longy F (1988) Spallation and shock-wave behaviour of some ceramics. In: DYMAT 88-international conference on mechanical and physical behaviour of material under dynamic loading, Le editions physique, C, pp 3–3
73. P. Peralta, S. DiGiacomo, S. Hashemian, S.N. Luo, D. Paisley, R. Dickerson and H. D'Armas, (2009) Characterization of incipient spall damage in shocked copper multi-crystals. Int J Damage Mech 18(4):393–413
74. Hashemian S, Digiacomo S, Peralta P, D'Armas H, Luo SN, Loomis E, McCllelan KJ (2008) Effects of microstructure on spall damage localization in shock loaded copper multi-crystals. TMS Annu Meet 1:319–325
75. Curran DR, Seaman L, Shockey DA (1987) Dynamic failure of solids. Phys Rep 147(5–6):253–388
76. Kipp ME, Grady DE (1989) Shock compression and release in high-strength ceramics. Sandia National Laboratory Report: SAND89- 1461
77. Grady DE, Moody R (1996) Shock compression profiles in ceramics. Sandia National Laboratories Technical Report. SAND96-0551
78. Grady DE, Wise J (1993) Dynamic properties of ceramic materials. Sandia National Laboratory Report: SAND93-0610
79. Forquin P, Zinszner J (2016) A pulse-shaping technique to investigate the behaviour of brittle materials subjected to plate-impact tests. In: Forquin P (ed) Theme issue experimental testing and modelling of brittle materials at high strain rates. Philos Trans R Soc A 375:20160333
80. Abou-Sayed AS, Clifton RJ, Hermann L (1976) The oblique-plate impact experiment. Exp Mech 16:127–132
81. Espinosa H, Clifton RJ (1991) Plate impact experiments for investigating inelastic deformation and damage of advanced materials. In: Kim KS (ed) Experiments in micro mechanics of failure resistant materials. Book number H00723, ASME, pp 37–56
82. Nathenson DI, Prakash V, Dandekar DP (2005) Dynamic response of silicon nitride under combined pressure and shear impact. In: 2005 SEM annual conference and exposition on experimental and applied mechanics. Society of Experimental Mechanics, Portland, Oregon, USA
83. Yuan F, Liou NS, Prakash V (2009) High-speed frictional slip at metal-on-metal interfaces. Int J Plast 25(4):612–634
84. Klopp RW, Shockey DA (1991) The strength behavior of granulated silicon carbide at high strain rates and confining pressure. J Appl Phys 70:7318
85. Yaziv D (1984) Shock fracture and recompaction of ceramics. PhD thesis, University of Dayton, Dayton, Ohio, USA
86. D. Yaziv, S.J Bless SJ and Z. Rosenberg (1985) Study of spall and recompaction of ceramics using a double-impact technique. J Appl Phys 58:3415–3418
87. Brar NS, Bless SJ, Rosenberg Z (1988) Brittle failure of ceramic rods under dynamic compression. J Phys 49(C3):607–612
88. Rajendran AM (2003) Critical measurements for validation of constitutive equations under shock and impact loading conditions. J Opt Lasers Eng 40:249–262
89. Grove DJ, Rajendran AM (2000) Modelling of micro-crack density based damage evolution in ceramics rods. In: Chhabildas F, Hixson (eds) Proceedings of the American physical society conference on shock compression of condensed matter, pp 619–622. APS Press

90. Simha CM, Bless SJ, Bedford A (2000) What is the peak stress in ceramic bar impact experiments?. In: Furnish MD et al (eds) Shock compression of condensed matter. AIP Conf Proc 505:615–618
91. Brar NS (2004) Confined alumina bar-on-bar impact experiments. Technical Report, University of Dayton Research Institute, Dayton, Ohio, USA, UDR-TR-2004-00157, October
92. Bless SJ (2002) Using bar impact to determine dynamic properties of ceramics. In: McCauley J, Crowson A, Gooch W, Rajendran AM, Bless SJ, Logan KV, Normandia M, Wax S (eds) Ceramic armour materials by design. Ceram Trans 134:225–232
93. Rosenberg Z, Mayseless M, Partom Y (1984) Axial stress-time history measurements in impulsively loaded aluminium rods using the double gauge technique. ASME Trans J Appl Mech 52(4):977–979
94. Wise JL, Grady DE (1992) Dynamic multi-axial impact response of ceramic rods. In: Schmidt SC et al (eds) High pressure science and technology. AIP Conference Proceedings, vol 309. New York, pp 777–780
95. Cosculluela A, Cagnoux J, Collombet F (1992) Two types of experiments for studying uniaxial dynamic compression of alumina. In: Schmidt SC et al (eds) Shock compression of condensed matter. Elsevier Science Publishers, pp 951–954
96. Chhabildas LC, Furnish MD, Reinhart WD, Grady DE (1998) Impact of AD995 alumina rods. In: Schmidt SC et al (eds) Shock compression of condensed matter. AIP Conference Proceedings, vol 429. New York, pp 505–508
97. Chahbildas LC, Reinhart WD, Dandekar DP (2002) Taylor-impact experiments for brittle ceramic. In: McCauley J, Crowson A, Gooch W, Rajendran AM, Bless SJ, Logan KV, Normandia M, Wax S (eds) Ceramic armour materials by design. Ceramic transactions, vol 134, pp 269–278
98. Cazamias JU, William D, Reinhart WD, Konrad CH, Chhabildas LC, Bless SJ (2002) Bar impact tests on alumina (AD995). AIP Conf Proc 620:787
99. Martin BE, Heard WF, Loeffler CM, Nie X (2018) Specimen size and strain rate effects on the compressive behavior of concrete. Exp Mech 58:357–368

Chapter 5
Shock and High Strain Rate Modelling

Prologue

With the rapid advancement of high-performance computing capabilities over the past three decades, the study of impact-resistant protective structures has increasingly leaned on advanced design codes, commonly known as "hydrocodes". A key attribute of hydrocodes is their integration and utilization of accurate and realistic material models. These hydrocodes encompass a wide array of empirical models, often organized into a model library, covering a diverse range of materials, including composite laminates.

Despite the presence of user-friendly empirical strength and failure models in these codes, a persistent challenge emerges in determining realistic values for material model parameters, including the equation of state (EOS). This challenge poses a barrier to achieving realistic and accurate solutions for shock and impact problems.

This chapter provides a historical overview of various approaches taken by researchers to describe the brittle fracture of polycrystalline ceramics, heterogeneous concrete, and rocks. Dynamic fracture phenomena involve the intricate interplay of millions of microdefects and their evolution, compromising material strength and stiffness. While many existing models are generally applicable to these scenarios, occasional modifications may be necessary to account for the effects of microstructural features on strength and damage evolution. Nevertheless, when dealing with the isotropic and homogeneous response of any material containing pores, micro-cracks, and inclusions (such as aggregates in concrete), the theoretical treatment tends to be quite similar.

The high strain rate and high-pressure responses of brittle materials often depend on analogous empirical models. Although material properties like elastic moduli, quasi-static compressive strength, fracture toughness, Hugoniot Elastic Limit (HEL), spall stress, etc., can be directly measured through high strain rate and shock experiments discussed in Chap. 4, determining or calibrating model parameters necessitates the modelling and simulation of these experiments themselves.

A. M. Rajendran et al., *Failure of Brittle Materials Under Shock and Impact*, Materials Horizons: From Nature to Nanomaterials, https://doi.org/10.1007/978-981-97-7026-7_5

5.1 Deformation and Damage Mechanisms

In 1971, Hockey [1] reported direct evidence of plastic deformations in both single and polycrystalline alumina ceramic due to microindentation at room temperature. Transmission electron microscopic (TEM) studies revealed high densities of dislocations and mechanical twinning at the surface of the indented specimens. Shockey et al. [2] observed evidence of networks (or systems) of micro-cracking in laterally confined ceramic targets recovered from ballistic impact. Merala et al. [3] conducted micro-structural investigations to evaluate microdefects in three different ceramics under extremely high pressures, reporting the nucleation of a significant amount of dislocations, micropores, and micro-cracking in the shocked samples.

Espinosa et al. [4] employed a combined experimental, micro-structural, and numerical analyses to investigate the source of damage initiation in ceramics. SEM and TEM analyses on the soft-recovered plate impact samples revealed the presence of micro-cracking and micro-plasticity at the triple point grain boundaries. They suggested that the inelastic strains under shock compression are due to these microdefects.

TEM examinations on confined 99.5% pure alumina (AD995) specimens from SHPB tests indicated extensive evidence of plastic flows, according to Lankford et al. [5]. They reported dislocation pileups at grain boundaries further propagated to other adjacent grains. Similar observations have been reported by many researchers regarding the presence of micro-structural features associated with plastic strains in high-strength ceramics. All these studies confirm that, in addition to the strains due to micropores and micro-cracks, it is important to include microplasticity-induced plastic strain descriptions in ceramic constitutive models.

Modelling the dynamic response in terms of stress–strain behaviours of concrete and rocks under unconfined uniaxial stress loading conditions is complex due to damage-induced post-yield behaviours. Ductility is defined based on the amount of strain measurable beyond the peak value of stress. In metals, ductility is directly measured and determined from the stress–strain curve. This is not the case with brittle materials since the amount of strain beyond the peak stress value is highly statistical due to sample-to-sample variation and test parameters, such as specimen size, aspect ratio, geometry, etc.

Microscopic studies and micro-mechanical modelling of rock-type materials [6–12] often help us understand and evaluate mechanisms of macroscopic failure due to nucleation, growth, and interactions of micro-cracks under multi-axial dynamic loading conditions. The sites for nucleation include pre-existing flaws, grain boundaries, interfaces of two different materials (matrix and the second phase particle), and pre-existing micro/macro voids (or pores). These mechanisms contribute to inelastic strains in brittle solids. For instance, pore collapse due to high compressive pressure contributes to inelastic strains and influences the "*plastic*" wave characteristics of shock waves, as observed in particle velocity and stress profiles measured in plate impact experiments on concrete and porous ceramic materials. The anisotropic and heterogeneous rock failure mechanics may be applicable to concrete; however, extending it to crystalline ceramics will be speculative.

5.1.1 Micro-cracking in Brittle Materials

Brittle materials, such as rocks, concrete, glass, and ceramics, undergo fracture and fragmentation under high-velocity impact loading conditions, resulting in pulverization. These macroscopic failures, including spallation due to tensile loading, are caused by various types of micro/mesoscopic damage phenomena. Given that ceramics completely pulverize upon high-velocity impact, recovering shocked ceramics for microscopic studies is challenging. Therefore, establishing clear physical evidence of the microscopic failure processes in ceramics necessitates carefully designed recovery experiments. Intuitively, several of the quasi-statically observed mechanisms can be incorporated into failure model formulations. For instance, during the 1980s and 90s, Nemat-Nasser and his coworkers [10–12], through a combined modelling and experimental study, demonstrated the initiation of micro-cracks at the tip of pre-existing flaws (see Fig. 5.1) in brittle materials to generate splitting and faulting-type macroscopic failures.

The crack growth under compression has been analysed by Horri and Nemat-Nasser [13, 14]. They found that tensile wing cracks initially grow at a large angle with respect to the direction of axial compressive stress and then rapidly grow parallel to the applied stress. The approximate expression to estimate dynamic crack growth derived by Hori and Nemat-Nasser was adopted by Ning et al. [15]. Nemat-Nasser and Deng [11] developed and illustrated a model in which an array of interacting, dynamically growing wing cracks was used to simulate the rate-dependent dynamic damage evolution and subsequent brittle failure of solids under compression. Based on their model simulations under both uniaxial stress (SHPB) and uniaxial strain (planar plate impact), they illustrated that the strain rate-dependent behaviour of compressive failure stress is due to compression-induced, tensile micro-cracks.

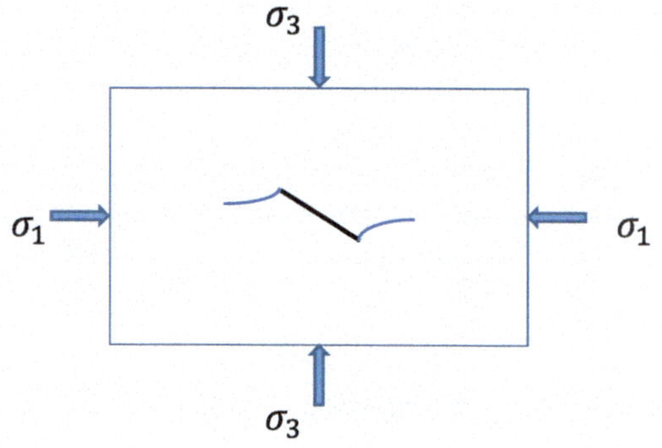

Fig. 5.1 A schematic of wing cracks emanating from the tips of pre-existing flaws (sliding cracks). The maximum principal stress is σ_1 and minimum principal stress is σ_3

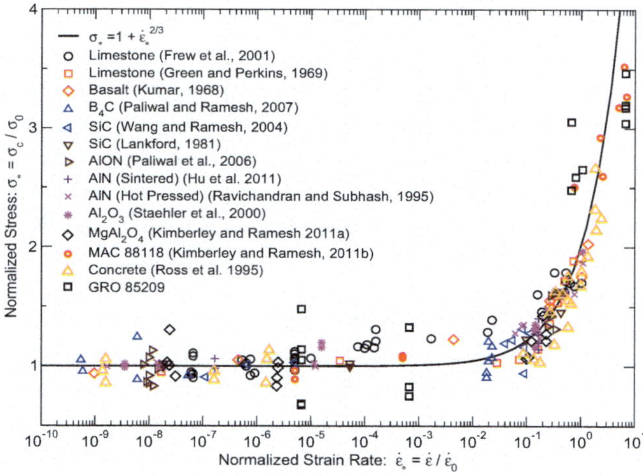

Fig. 5.2 Normalized experimental data for a wide range of brittle materials including rocks, engineering ceramics, and meteorite material. The normalized data compare well with the predicted model [17]

Building upon this wing crack model, Ravichandran and Subhash [16] presented a constitutive model applicable to brittle materials subjected to biaxial compressive loading. Brittle failure is based on a critical length for the wing cracks. This two-dimensional model was used to study material behaviours under uniaxial compressive constant strain rate loading. Based on their analysis, Ravichandran and Subhash suggested that the rate sensitivity exponent is a function of crack growth rate and the toughness of the material.

The high strain rate experimental data for various brittle solids are shown in Fig. 5.2. This figure presents data from both quasi-static and SHPB tests for a wide range of brittle solids, where the normalized flow stress is $\sigma^*(=\sigma_c/\sigma_0)$ and the normalized strain rate is $\dot{\varepsilon}_*(=\dot{\varepsilon}_*/\dot{\varepsilon}_0)$. σ_0 is the stress at a reference strain rate of $\dot{\varepsilon}_0$ and is usually taken as $1/s$ in most models. The wing crack analysis is an important theoretical study that supports the rate-dependent response of brittle solids as observed from the SHPB, rod-on-rod, plate-on-rod and plate impact experimental data on failure stress versus logarithmic strain rate shown in Chap. 4.

5.1.2 *Micro-plasticity in Ceramics*

Among the most significant contributions to ceramic research, Marsh [18] demonstrated that glass exhibits plastic flow under indentation testing. Analyses and theories from the 1960s to the 1980s often centred on plastic flows occurring in metallic or brittle specimens at local sites beneath an indenter during indentation tests. Lawn and Wilshaw [19] provided a comprehensive review titled "*Indentation fracture:*

principles and applications", discussing the evolution of inelastic strains in brittle specimens due to indentation by hard indenters of different tip shapes. Since elastic stress fields are well defined beneath an indenter, indentation tests lend themselves to fundamental analyses of the Griffith theory of fracture for elastic-brittle solids [20]. In their article, Lawn and Wilshaw delved into discussions on the so-called "*plastic*" flows in brittle solids and pressure-induced densification, considering intrinsic parameters like average flaw size, statistical distribution, surface strength, and flow velocity.

Hockey [1] reported networks of dislocations related to shear cracks in ceramics, indicating that indentation sites became favourable for shear crack nucleation due to a high triaxial stress state under compression. Similar observations of plasticity in brittle materials, based on transmission electron micrographs of dislocations within the plastic zone, have been reported [21]. The use of the term "*plasticity*" to describe inelastic deformation in ceramics or concrete stems from the extensive use of metal plasticity theories to elucidate inelastic strains in brittle solids due to micro-cracking [22–25].

In brittle solids, large-scale grain distortions are typically absent, and instead, dislocations and twinning-type defects are observed in confined samples recovered under ballistic impact loading conditions [1–5]. As illustrated in Fig. 5.3, dislocations and twins, micro-cracking, and pore collapse generate various forms of inelastic strains in brittle ceramics under shock and high strain rate loading conditions.

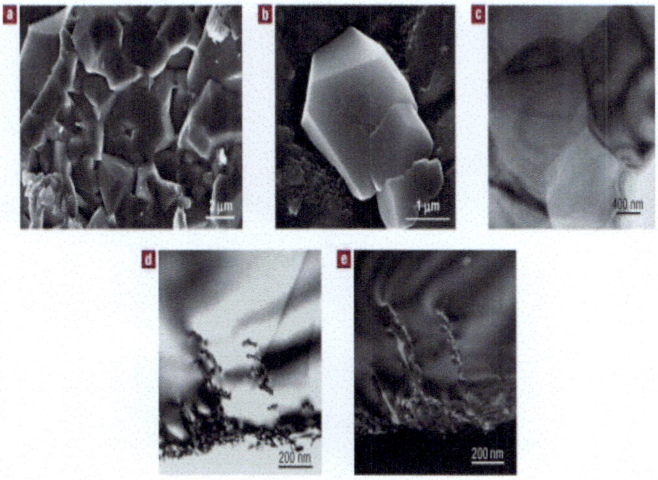

Fig. 5.3 **a** A SEM micrograph of the fragment surfaces taken over a large area. The feature of the fragment surface suggests that the failure is mainly along grain boundaries. **b** A grain with intact facets sticks out of the fracture surface. **c** TEM image taken from a large 6 GPa fragment without cracks. Crystal defects, such as dislocations and deformation twins, cannot be seen in (**d**) and (**e**), Bright-field (**d**) and dark-field (**e**) TEM images of dislocations and dislocation arrays emitted from a grain boundary. Partial dislocations and twins have not been seen in the 6 GPa fragments [26]

Extensive research, such as that conducted by Rice et al. [26], has evaluated the behaviour of various ceramic materials under quasi-static loading conditions, involving thorough micro-structural analyses as shown in Fig. 5.3. Brandt et al. [27] reported the presence of dislocations near micro-crack tips in permanently deformed ceramics, often referred to as "*plastic*". The term "*plastic*" is used in two ways: one denotes any observed non-linear, permanent deformation behaviour, and the other, when used in a micro-structural sense, implies plasticity is due to dislocation motions.

Staehler et al. [28] presented a study on micro-mechanisms of deformation in high-purity hot-pressed alumina. Under uniaxial stress conditions (such as in the SHPB tests), failure mechanisms were dominated by narrow micro shear bands followed by micro-crack nucleation. However, under uniaxial strain conditions (as in planar plate impact tests), deformation mechanisms were based on more dislocation activity along with extensive grain boundary cracking, as shown in Fig. 5.4.

The linear portion of the velocity profile between points **A** and **B** as shown in Fig. 5.5 is due to elastic deformation, while the curving (ramping) portion above point **B** indicates inelastic deformation, possibly due to microplasticity and/or micro-cracking. This section of the curve is often referred to as "*plastic*" in literature, describing the non-linear, loading path-dependent permanent deformations. Supporting evidence from micrographs of recovered ceramics from "*recovery*" plate impact tests reveals highly localized dislocations/twinning-based deformation due to microplasticity at the crack tips [29].

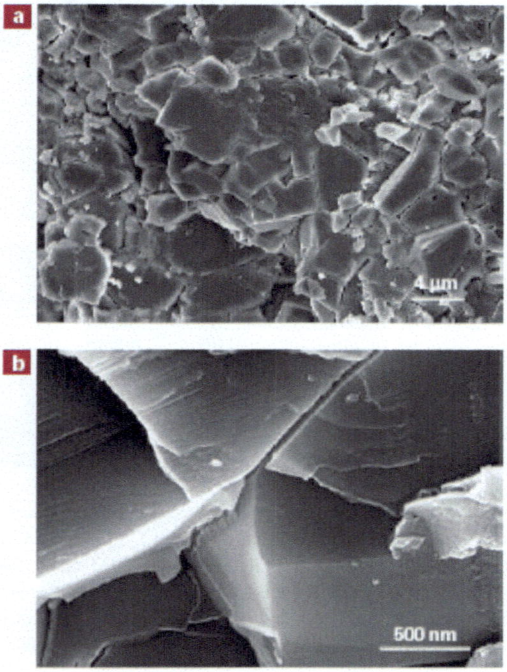

Fig. 5.4 **a** SEM micrograph of a fragment taken over a larger area **b** Cleavage steps on the fracture surface imaged at a high magnification. Both a and b suggest a cleavage failure mode of the alumina as the impact pressure is higher than the HEL [26]

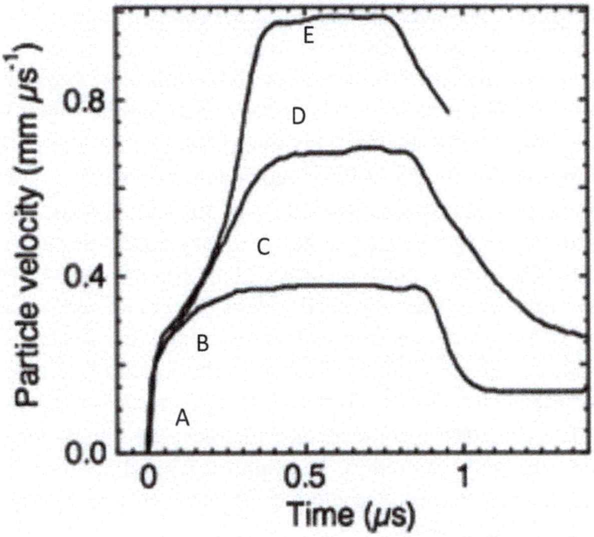

Fig. 5.5 Free surface particle velocity profiles for AD995 alumina from plate impact tests at three impact velocities. The "*plastic*" portions between **B-C, B-D**, and **B-E** are due to inelastic strains generated by shock propagation

Constitutive and failure model formulations require interpretations of the processes involved and the extent to which they dominate the inelastic response of the material. If inelastic deformation of brittle materials is primarily due to "*gross*" microplastic flow, the model formulation can be within the framework of plasticity theories. When inelastic deformation is mostly due to micro-cracking (nucleation and growth of micro-cracks), the models require incorporation of fracture mechanics-based laws and theories for realistic descriptions of inelastic strains under compression.

5.2 Constitutive Model Formulations

Rajendran and Cook [30] identified five distinct approaches for modelling brittle materials. They conducted a comprehensive review of the most pertinent models developed and utilized before the 1990s to characterize the response of brittle materials to shock and impact loading conditions. With the advancement of computational capabilities and sophisticated experimental and micro-structural techniques, a substantial number of physically based constitutive models were developed during the 1990s and the first decade of this century. For completeness, we will briefly outline the five modelling approaches suggested by Rajendran and Cook that still hold relevance today in the following paragraphs.

In the first approach, the material is assumed to be elastic, and stresses are derived from Hooke's law equations, as described in Sect. 2.5.3 of Chap. 2. Failure is then predicted using a generalized Griffith [20] and Margolin [31] fracture criteria. The second approach assumes that brittle materials under compressive loading behave as elastic-plastic solids. This approach disregards the details of crack growth and focuses on describing the effects of localized fracture on stress wave propagation. The concepts and equations parallel those derived for metal plasticity.

The third approach is grounded in fracture mechanics and microphysical theories, aiming to describe the deformation resulting from compressive failure processes. The fundamental idea of a microphysical model is to elucidate apparent inelastic behaviour while tracking microstructural evolutions under a given set of loading conditions. For ceramics, microcracking evolution is described using fracture mechanics theories governing crack growth. An example of a microphysical-based approach involves a statistical description of the number of cracks per unit volume as a function of position, crack size, and orientation. This approach combines Hooke's law-based elasticity equations with an effective moduli description linking microstructure to macroscopic material properties.

In the fourth approach, the theoretical formulation may not necessarily integrate fracture mechanics and continuum damage mechanics-based equations and theories. However, under this microstatistical approach, phenomenological concepts are derived from experimental observations, involving extensive microstructural analyses of recovered samples from plate impact and Split Hopkinson Pressure Bar (SHPB) tests. Curran et al. [3, 32, 33] proposed exponential distributions for defect nucleation and growth concerning stress or pressure to predict spallation-type tensile failures in both ductile and brittle solids. The deduction of a suitable statistical distribution of evolution is a crucial component of this approach, requiring extensive micro-structural analyses of specimens before and after testing.

The fifth approach considers constitutive and failure equations based on the theory of continuum damage mechanics. In this theory, the constitutive equation for stress or strain is derived from either Helmholtz or Gibbs free energy density functions. These constitutive relationships are subjected to the restrictions imposed by the second law of thermodynamics. Damage laws/equations are usually postulated based on a phenomenological approach. Sometimes, these laws can also be formulated from micro-statistical analyses of the actual fracture processes. In general, the modelling of brittle material behaviours often combines features from various approaches, making it challenging to categorize a model under a single approach. For instance, a model may follow a micro-statistical approach for nucleation laws while adopting a continuum damage approach for the stress–strain relationship. Therefore, in the following section, the various models described for brittle solids may not exclusively fall under only one of the five approaches. Instead, one model may incorporate concepts and equations from several approaches.

5.2.1 Mohr–Coulomb and Griffith Criteria Models

During the 1960s, due to a lack of advanced fracture mechanics theories and computational mechanics capable of capturing complex micro-mechanisms, many models employed a simplistic approach to calculate the compressive strength of brittle materials. One prevalent method during this period was the utilization of the Mohr–Coulomb law for predicting the compressive strength of brittle solids.

These envelopes shown in Fig. 5.6 can be experimentally determined for various brittle materials, especially rocks and soils. There exists a substantial body of literature on both Mohr–Coulomb and Drucker–Prager laws. Material behaviour is typically assumed to be elastic and is described by Hooke's law. When the principal stress lies within the envelope, the material is intact; when the stress state is outside, the material is considered failed.

The application of a Mohr–Coulomb failure criterion to ceramics is a straightforward modelling approach for materials with an extreme ratio of compressive to tensile strength. While this approach is practical in some limited applications, particularly when tensile failures are predominant, it proves inadequate for high-velocity impact problems where both compressive and tensile failures are significant. In two-dimensional calculations of ceramic impact, the effects of the Mohr–Coulomb model are easily visualized. However, this model's limitation lies in its nature as an instantaneous failure criterion, offering no information on the material's degradation from a solid state to fragmentation.

For tensile loading, a simple failure criterion involves the material fracturing when the applied pressure (P) falls below minimum value (Pmin). Under pure shear ($P = 0$), failure is assumed when the applied shear stress exceeds τ_0. While these instantaneous failure models have been historically used for limited engineering predictions, especially for soils, concrete, and rocks, the advent of computer algorithms and computing capabilities has led to the development of cumulative damage models. These models depict the degradation of stiffness and compressive strength before

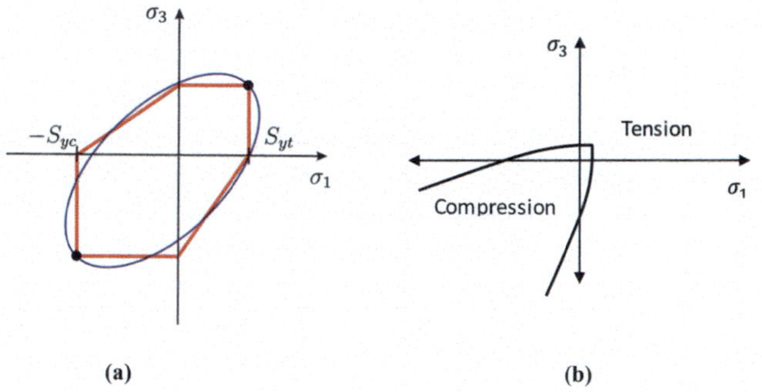

Fig. 5.6 **a** Mohr–Coulomb (red) and Drucker–Prager (blue) laws to describe failure of brittle solids. The axes are major and minor principal stresses. Yield stresses in tension and compression are S_{yt} and S_{yc}, respectively; **b** Fracture locus for biaxial stress state

complete material failure or catastrophic structural failure, showcasing impressive progress over the past two to three decades.

Griffith [20] proposed a fracture criterion for glass-like brittle solids, asserting that a flaw (crack) extends in an unbounded manner when the decrease in potential energy due to the creation of a new surface is equal to the increase in surface energy. Originally designed for static loading conditions, Griffith's criteria predict the failure of a crack-contained structure due to rapid crack extension. These criteria, utilized for single macro cracks, have found application in modelling dynamic fragmentation in brittle rocks, as demonstrated by Grady and Kipp [9]. Mescall and Tracy in an unpublished report [34], applied Griffith criterion to model the onset of fracture in ceramic under bi-axial compressive loads. This approach was applied to model the long rod penetration process into a confined aluminium oxide (Al_2O_3) ceramic target. The conditions for the onset of fracture are given by the following equations:

$$\begin{aligned} &\sigma_3 = \sigma_{cr} \quad \text{if} \quad (3\sigma_3 + \sigma_1) > 0 \\ &(\sigma_3 - \sigma_1)^2 - 8\sigma_{cr}(\sigma_3 + \sigma_1) = 0 \quad \text{if} \quad (3\sigma_3 + \sigma_1) < 0 \end{aligned} \tag{5.1}$$

where σ_1 and σ_3 are the applied compressive stresses. σ_{cr} is the uniaxial tensile strength. The failure envelope from these equations is shown in Fig. 5.6b. In this figure, the envelopes are based on the original Griffith [20] fracture criterion. Subsequently, McClintock and Walsh [35] modified the criterion to account for frictional effects in closed cracks, slightly enlarging the envelope in the third quadrant, although not shown in the figure. When one of the stress components is tensile, the ceramic is assumed to fail at relatively low stress values compared to when both principal stresses, σ_1 and σ_3 are compressive. This observation aligns with the inherent characteristic of brittle materials, which generally exhibit significantly higher compressive strength compared to tensile strength. The fracture criteria based on the Griffith criterion prove highly valuable in modelling brittle fracture in ceramics, particularly in computer code calculations. This methodology provides a robust framework for predicting the initiation of fracture in ceramics under complex stress states, such as those encountered during projectile penetration into confined ceramic targets.

The material is assumed to be elastic, making Hooke's law-based elasticity equations is sufficient to describe it. In computational analyses, the Griffith criterion was applied to map fracture zones in the target, qualitatively aligning with experimental observations. By encompassing both tensile and compressive fractures in the failure envelope, this approach allows for the identification of fracture regions based on failure modes. The use of fracture mechanics-based Griffith criteria in the 1980s marked progress in improving fracture models for ceramics.

Another simple model assumes material failure when the maximum principal stress exceeds a critical value. Alternatively, shear strength of brittle solids can be determined from quasi-static and dynamic experiments under various stress states, leading to a shear strength versus pressure plot as shown in Fig. 5.7.

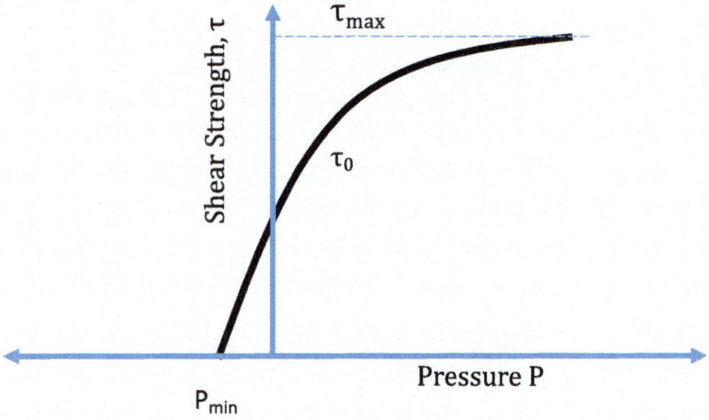

Fig. 5.7 A sketch of shear strength versus pressure plot for brittle materials

In general, constitutive models describing deformation and damage evolutions under compressive loading may not remain valid after the brittle solid is pulverized. The response of pulverized material to compressive loads, while under lateral confinement, requires a realistic constitutive description. For this purpose, many strength models employ the Mohr–Coulomb law. The strength of the pulverized material, Y_p is often modelled by the following relationship,

$$\tau_f = c + P tan\varphi \quad \text{or} \quad Y_P = \alpha + \beta P \tag{5.2}$$

where, P is the pressure, τ_f is the shear strength, c is apparent cohesion, φ is the angle of internal friction, and α and β can be considered as the model constants. High strain and shock data often guide the selection of suitable values for the pulverized material strength model.

5.2.2 *Empirical Damage Accumulation Models*

The Tuler and Butcher [36] introduced a time-dependent damage model to predict spall failure in metallic materials under a triaxial tensile loading state. Spall is assumed to occur when a time-dependent damage parameter reached a critical value. The Tuler–Butcher model can be expressed as,

$$D = \int_0^{t_f} (S - S_0)^\lambda \geq D_\lambda \tag{5.3}$$

where, S is the maximum principal stress or the axial stress in a plate impact testindexPlate impact test, S_0 is the threshold value of S and above which damage (D) accumulation occurs, λ is a model parameter, t_f is the time at failure when D (the value of the integral) is equal or above the critical value, D_λ. It's important to note that the material behaviour is assumed to be elastic, and the elastic modulus is not degraded due to the damage parameter, D. Though this model was applied to predict spallation in metals, this type of empirical cumulative damage model, which has no mechanistic or physical meaning, can also be used to model compressive failure. Although initially applied to predict spallation in metals, this type of empirical cumulative damage model, lacking mechanistic or physical meaning, can also be used to model compressive failure. However, it is possible to introduce D into the elastic equations to degrade the elastic moduli (or stiffness) as follows:

$$\sigma_{ij} = 2G(1 - D)e_{ij} + K(1 - D)e_v\delta_{ij} \tag{5.4}$$

where, G and K are shear and bulk moduli, respectively, and e_v and e_{ij} are volumetric strain and deviatoric strain. The strain can be expressed as, $\varepsilon_{ij} = e_{ij} + e_v\delta_{ij}$. This empirical stress–strain relationship with degraded moduli has been employed by Rajendran and Kroupa [37] to predict spall in silicon carbide when shocked above HEL. They used a strain rate dependent damage evolution law and reproduced the measured velocity profile in a plate impact test. We present more details on this phenomenological model in a later section.

The Tuler–Butcher type empirical damage models capture the effects of micro-cracks on the elastic modulus through a time-dependent damage parameter in a phenomenological sense, allowing the introduction of the effect of time on the failure process. This approach is an improvement over the instantaneous failure model where no damage parameter is involved. Simulations of impact and shock wave propagation problems using time and rate-independent models may exhibit serious mesh dependencies in the solutions. Substantial research has been conducted in the past to address these mesh effects, and these aspects will be discussed in Chap. 6 on "*Computational Codes*".

5.2.3 *Advanced Phenomenological Models*

5.2.3.1 Radial Return Correction

Many empirical model describe the compressive strength of brittle materials as a function of strain rate ($\dot{\varepsilon}$) and pressure (P). In computer code implementations, the "*inelastic*" strain is often calculated as plastic strain using conventional von Mises-based metal plasticity theories. The von Mises criterion states that plastic flow has occurred when the effective stress, σ_{eff} exceeds or equal to the value of Y for an isotropic and homogeneous solid.

$$\sigma_{eff} = \sqrt{\frac{3}{2}S_{ij}S_{ij}} = \sqrt{3J_2} \geq Y \tag{5.5}$$

where J_2 is defined as the second invariant of the deviatoric stress, S_{ij}. Note that, $\sigma_{ij} = S_{ij} + P\delta_{ij}$. The material strength Y is a function of strain rate, pressure, temperature, and damage. For brittle materials, we often ignore the effects of temperature on the strength under dynamic loading applications. Therefore, we can write:

$$Y = Y(\dot{\varepsilon}, P, D) \tag{5.6}$$

In computer code calculations, the strains (ε_{ij} and e_{ij}) are obtained at each time step. The deviatoric stresses are determined from elastic equations, $S_{ij}^{el} = 2Ge_{ij}$. The deviatoric strains, e_{ij} are considered elastic until the von Mises yield criterion is satisfied.

Therefore, the estimated S_{ij}^{el} using the Hooke's law need to be corrected to satisfy the von Mises yield criterion, which is, $\sqrt{\frac{3}{2}S_{ij}S_{ij}} = Y$, If we assume that the actual (or correct) S_{ij} is equal to $\eta\, S_{ij}^{el}$, then the correction factor, ε is calculated from $\sqrt{\frac{3}{2}\eta S_{ij}^{el}.\eta S_{ij}^{el}.} = Y$. Therefore η is obtained from the following equation,

$$\eta = \frac{Y}{\sqrt{\frac{3}{2}S_{ij}^{el}.S_{ij}^{el}.}};\ S_{ij} = \eta S_{ij}^{el} \tag{5.7}$$

This approach to correct S_{ij} for satisfying the von Mises criterion is often called as the "*Radial Return*" correction in the literature. In the code, the stress tensor is calculated by adding the deviatoric and volumetric parts as, $\sigma_{ij} = S_{ij} + P\delta_{ij}$. As the S_{ij} is obtained using the Y from the strength model in conjunction with von Mises yield criterion, the pressure P is determined from EOS for a given volumetric strain, e_v.

In numerical code simulations of dynamic problems, most empirical models calculate stresses using the radial return correction to satisfy the von Mises yield criterion, and the "*inelastic*" strains are obtained using plasticity theories. Most empirically based strength and constitutive models for cementitious materials follow plasticity theories in this manner.

5.2.3.2 Steinberg Model

Steinberg [38] proposed that all thermo-mechanical behaviours could be represented through macroscopic variables like strain, strain rate, temperature, and pressure. He adapted the essential aspects of his metal model to characterize ceramic material response under high strain rates and shock loading. The ceramic model posits that both yield strength and shear modulus vary with temperature, pressure, and strain rate. The governing equations are given by:

$$G(P,T) = G_0\left[1 + A\frac{P}{\eta^{1/3}} - B(\Delta T)\right]; \quad A = \frac{1}{G_0}\frac{dG}{dP} \quad \text{or} \quad B = \frac{1}{G_0}\frac{dG}{dT} \tag{5.8}$$

where η is the compressive strain, $\dot{\varepsilon}$ is effective strain rate, and G, P, and T are shear modulus, pressure, and temperature, respectively. G_0 is the shear modulus of the ceramic before shock loading. The expression for the yield (compressive) strength is:

$$Y = \left[D\dot{\varepsilon}^n + Y_A\right]\frac{G}{G_0}; \quad D = (3\rho_0 C_0 K_{1C}^2)^n \tag{5.9}$$

where ρ_0 is the initial density of the ceramic, C_0 is the longitudinal wave speed, K_{1C} is the fracture toughness, n is a model parameter which is assumed to be 1/3, and, according to Steinberg, Y_A is a material constant which can be sample dependent, as it is a function of purity, grain size, previous mechanical history, etc. Note that "D" is a not a damage variable. Steinberg used this scalar quantity to describe strength hardening. Y_A can be considered as the quasi-static compressive strength of ceramic. The initial shear modulus G_0 can be determined from ultrasonic and shear experiments. The strain is decomposed into elastic-viscoplastic components, and stress calculation in the computational implementation follows conventional plasticity theories.

Essentially, there are four model parameters: Y_A, A, B, and D which are estimated by adjusting to match the plate impact velocity profiles. Steinberg also incorporated a traditional void growth (spall) model to describe tensile cracking in ceramics, which requires three additional constants. It's important to note that this model is one of several phenomenological models employing empirical relationships to describe strength variation concerning strain rate, temperature, and pressure. However, it is not widely used for other types of brittle materials.

5.2.3.3 Johnson–Holmquist Model

Johnson and Holmquist [39, 40] developed a cumulative damage model as an extension of Drucker–Prager-type strength models, creating two versions: $JH1$ and $JH2$. These models were initially designed for describing projectile penetration into ceramic targets, but over time, they found applications in various fields, including tunnel blasting, rock fracturing, and oil drilling [41]. The later JHB model, introduced in 2005, utilized analytic functions to describe strength versus pressure curves, becoming the standard for brittle material behaviours under shock and impact loading. The $JH1$ model does not degrade strength due to damage until the solid is completely comminuted, and it doesn't account for pressure correction related to bulking in the equation of state. Therefore, the $JH1$ model has one strength versus pressure curve for the intact solid and another curve for the powdered solid. In contrast, the $JH2$ model continuously degrades strength with increasing damage (D), presenting a damage level-dependent third curve between intact and pulverized ceramics. The JH models are phenomenological, based on an elastic-viscoplastic

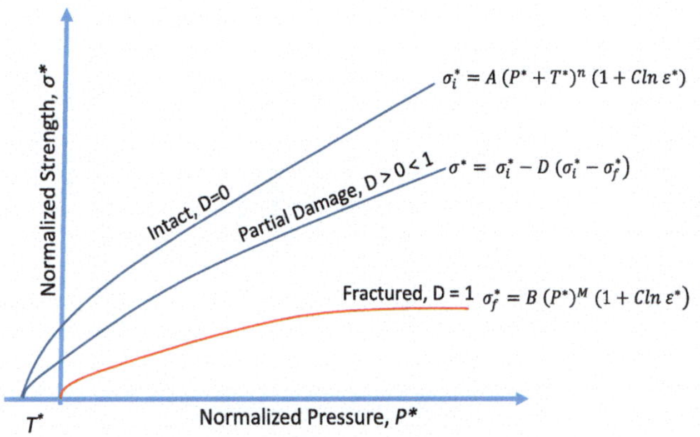

Fig. 5.8 Johnson–Holmquist ceramic model as depicted by a schematic of ceramic strength variation with respect to pressure, strain rate, and damage

(strain rate-dependent) approach, using metal plasticity theories for inelastic strain modelling. The $JH2$ model assumes plastic flow begins when the von Mises stress exceeds the quasi-static compressive strength (Y), determined by the von Mises yield criterion, $\sqrt{\frac{3}{2}S_{ij}S_{ij}} \geq Y$.

The $JH2$ model is shown through the plots in Fig. 5.8. Essentially, the strength or the equivalent stress increases entirely due to the pressure (hardening) and strain rate (hardening). A and N are pressure hardening constants and C is the strain rate sensitivity constant that appear in the equations in Fig. 5.8. The responses of solid, damaged, and fractured materials tend to be strain rate dependent. For simplicity, we can assume the rate sensitive parameter C is the same for these responses.

In Fig. 5.8, the non-dimensional strength σ^* is the ratio between material strength and the shock stress at the Hugoniot Elastic Limit (HEL). In a similar manner, the pressure P is normalized with P_{HEL}, the Hugoniot pressure at HEL. The normalized tensile strength is, $T^*(= T/P_{HEL})$. We discussed HEL as well as about how to obtain HEL from plate impact experimental data in Chaps. 3 and 4. The σ^* versus P^* curve (solid orange) represents the undamaged intact solid ($D = 0$) at a reference strain rate, $\dot{\varepsilon}^*$ of $1/s$ of $1/s$. As damage increases from $D = 0$ to $D = 1$, the strength continually decreases with D and therefore, the strength curve σ^* versus P^* for damaged solid is represented by the dashed line. When the damage process is complete ($D = 1$) under compression, the strength of the pulverized or fractured brittle solid is represented by the σ^* versus P^* curve in solid red. At some threshold pressure, the fractured material strength is saturated to $\sigma^*_{f\ max}$ and it is an important material model parameter which can be estimated or extracted from experiments.

As in the Johnson–Cook fracture model for metals, the damage (D) in brittle materials increases with increasing effective plastic (fracture) strain (ε^P_f) as shown in Fig. 5.9a. The plastic flow initiates when the computed von Mises stress exceeds σ^* for a given P^*. Using the radial correction method described in Sect. 5.2.3.1, the deviatoric stresses are modified so that the corrected von Mises stress follows (or lie

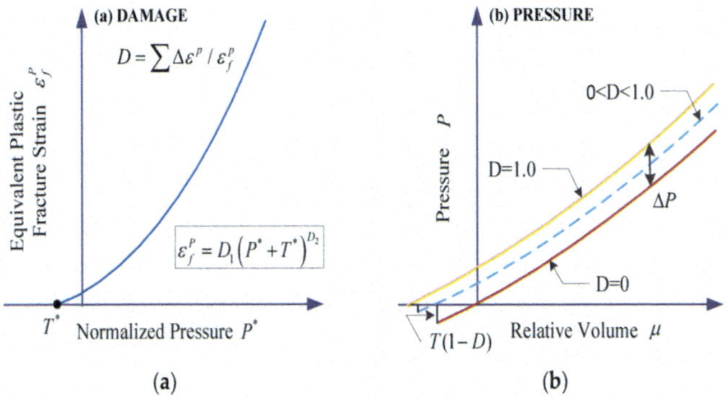

Fig. 5.9 The schematic of **a** damage evolution in **JH2** model, and **b** pressure correction due to bulking [39, 40]

on) the σ^* versus P^{**} curve in Fig. 5.8. It should be noted that this is a universal curve and therefore applicable to all stress states, such as uniaxial stress and uniaxial strain.

While the $JH2$ assumes a nonlinear description of the plastic strain with respect to pressure as in this figure, the $JH1$ assumes a linear variation between the fracture strain (ε_f^P) and pressure with a cap (saturated fracture strain). Johnson and Holmquist used the $JH1$ model in their modelling of a long rod penetration into a confined ceramic. It should be noted that unlike in the micro-mechanical models, the $JH1$ or $JH2$ fracture models do not degrade the elastic (bulk and shear) moduli due to damage.

In Fig. 5.9b, as damage evolves, pressure increases by "ΔP" to adjust for volume expansion under compression (bulking). Bulking is a characteristic response in rocks, especially during unloading under high strain rates. In 1976, Janach [42] addressed the role of bulking in rocks (such as the granite and limestone), under rapid compression. He used a one-dimensional heuristic or conceptual model to explain bulking in these materials. In later years, several micro-mechanical models based on the wing crack mechanism were developed to describe volume expansion under compression, both during loading and unloading.

The internal energy loss is converted to hydrostatic potential energy using an approximate equation for conservation of energy. Details of this equation are discussed by Holmquist and Johnson [39]. Tensile strength is low or insignificant for ceramics and it varies linearly with strain (μ) as illustrated in Fig. 5.9b.

Johnson and Holmquist applied the $JH1$ (or JHB) model to describe the penetration process of a long tungsten rod into a confined thick silicon carbide, as shown in Fig. 5.10. They used experimental data to determine intact strength, compressibility constants, and other parameters. The JH2 model has been employed in various loading and unloading conditions, illustrating its versatility. The determination of model constants can be challenging, but various experimental data and techniques, such as VISAR, time-resolved high-speed camera pictures, and stress and strain gauges, can be utilized for calibration.

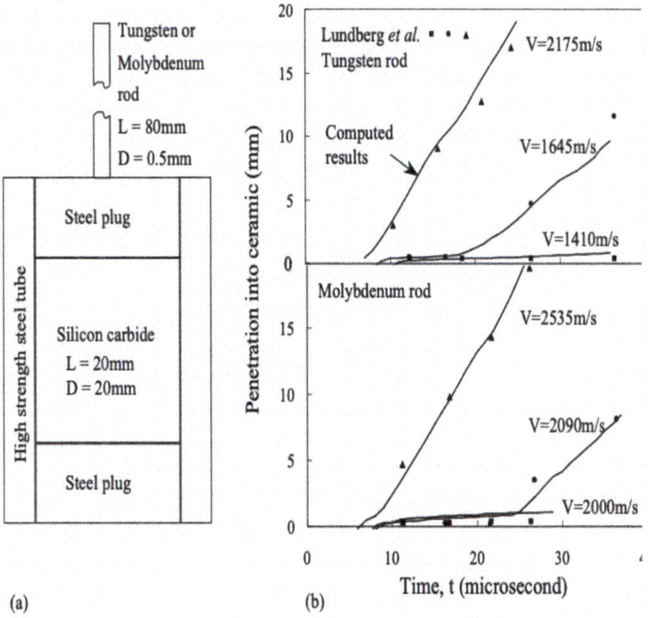

Fig. 5.10 Comparison of computational results and experimental results for tungsten and molybdenum rods impacting a confined silicon carbide target at various velocities. **a** Initial 2D geometry and **b** comparison of computed and experimental results

Johnson and Holmquist employed the plate impact data of Feng et al. [43], Grady and Moody [44], Dandekar and Bartkowski [45], and the uniaxial compression data of Pickup and Barker [46] to determine intact strength and compressibility constants. A very important characteristic of the Feng et al. [43] data is that both longitudinal and lateral stresses are provided, and this allows the complete stress state (both strength and pressure) to be determined. The slope constant for the failed material, ϕ, is provided by Klopp and Shockey [47].

Johnson and Holmquist [39] applied the **JH1** (or **JHB**) model to describe the penetration process of a long tungsten rod into a confined thick silicon carbide as shown in Fig. 5.10. They employed the plate impact data of Feng et al. [43], Grady and Moody [44], Dandekar and Bartkowski [45], and the uniaxial compression data of Pickup and Barker [46] to determine intact strength and compressibility constants. A very important characteristic of the Feng et al. data is that both longitudinal and lateral stresses are provided, and this allows the complete stress state (both strength and pressure) to be determined. The slope constant for the failed material ϕ, is provided by Klopp and Shockey [47]. In general, for brittle solids, this parameter (slope) characterizes the strength versus pressure variation of confined granular materials. Due to lack of experimental data, the maximum strength of the failed material, S^f_{max} cannot be determined but must instead be inferred by varying S^f_{max} until the computational depth of penetration of the projectile matches experimental data. The damage constant, ϕ is also obtained in a similar manner from the slope of equivalent strain versus pressure. The two experimental data points used to obtain these constants are the penetration of a tungsten rod into a silicon carbide target at 2175 m/s, and the duration of the dwell of a tungsten rod onto a ceramic target at 1645 m/s [48]. Holmquist and Johnson [39] provided a detailed approach to determine the **JH2** model parameters for silicon carbide. Additional details concerning the model and the constants are provided elsewhere [49].

In general, for brittle solids, this parameter (ϕ-slope) characterizes the $strength$ versus $pressure$ variation of confined granular materials. Due to lack of experimental data, the maximum strength of the failed material, S^f_{max} cannot be determined but must instead be inferred by varying S^f_{max} until the computational depth of penetration of the projectile matches experimental data. The damage constant, ϕ is also obtained in a similar manner from the slope of equivalent $strain$ versus $pressure$. The two experimental data points used to obtain these constants are the penetration of a tungsten rod into a silicon carbide target at 2175 m/s, and the duration of the dwell of a tungsten rod onto a ceramic target at 1645 m/s [48]. Holmquist and Johnson [39] provided a detailed approach to determine the $JH2$ model parameters for silicon carbide. Additional details concerning the model and the constants are provided elsewhere [49].

In general, the model parameters A and N can be determined from the locus of compressive strength data from uniaxial, biaxial, and triaxial quasi-static tests and the strain rate sensitivity parameter C from spilt Hopkinson bar tests. Johnson and Holmquist [39] illustrates the application of $JH2$ model for a range and type of loading and unloading conditions. In addition, they provide some procedures or guidelines for the determination of model constants.

Wang et al. [41] applied the $JH2$ model to describe rock failure under blast loading. The determination of various constants in JH models is non-trivial since some of the constants are difficult to determine from experiments, especially the saturated strength for the confined comminuted material, σ^*_{fmax} and the slope of the damage curve in Fig. 5.9a. These two constants can be calibrated by matching experimental data with hydrocode simulations of penetration experiments. Measured VISAR, time resolved high speed camera pictures, and stress and strain gauge data from different impact and shock experimental configurations can also be utilized as discussed in Chap. 4.

5.2.3.4 Modified TCK Model

One critical failure mode in brittle solids involves fragmentation due to micro-cracking under impact or explosive loading. Grady [50] proposed a model for dynamic fragmentation based on a stress intensity release energy balance criterion. The nominal fragment diameter d for a brittle solid is given by:

$$d = \left(\frac{\beta K_{Ic}}{\rho c\dot{\varepsilon}}\right)^n \tag{5.10}$$

where$\beta = \sqrt{20}$, K_{Ic} is the fracture toughness, ρ is is the material density, c is the uniaxial (stress) wave speed, and $\dot{\varepsilon}$ is the applied strain rate. This expression suggests that at low strain rates, the brittle materials fracture into large fragments, while at very high strain rates, complete pulverization occurs, especially when the shock stress exceeds the steady state value of HEL.

The energy balance equation proposed by Grady [50] initially omitted the consideration of elastic strain energy. Subsequently, Glenn and Chudnovsky [51] revised the fragmentation expression by incorporating elastic energy, resulting in a β value of $\sqrt{5}$. Building on this, Glenn et al. [52] introduced a unified fracture mechanics-based equation that encompasses both low and high strain rates while retaining solutions from Griffith and Grady. These fragmentation models play a crucial role in the formulation of damage models by Taylor et al. (TCK) [53], as well as Kipp and Grady [54].

To characterize the crack density, γ, under tension, the characteristic flaw size, a, is assumed to be directly proportional to the average fragment size. Thus, the expression for fragment size serves as the crack size descriptor in brittle constitutive and damage models.

In their modified version of the TCK model, Rajendran and Kroupa [37] postulated that under compression, brittle solids such as rocks, concrete, and ceramics exhibit strain rate-dependent compressive strength, while under tension, they fail (fragment) due to Mode I fracture. The modified model accommodates subsequent degradation of compressive strength due to tensile micro-cracking. Within this framework, the damage parameter, **D**, is associated with γ (crack density) and the degraded Poisson ratio $\bar{\nu}$ employing the expression derived by Budiansky and O'Connell [55]. Additionally, the average crack size, a, is determined by an expression derived by Kipp and Grady [54] for modelling fragmentation. Given that the number of microcracks in brittle solids follows a statistical distribution, the TCM model employs a Weibull description with respect to volumetric strain.

The TCK or modified TCK models prove highly valuable in computational modelling for describing the dynamic behaviours of rocks and concrete. As previously discussed, fragmentation models are employed to depict damage growth under tension and the subsequent degradation of compressive strength due to tensile damage.

The salient equations are summarized as follows:

$$\begin{aligned} &\text{Compression: } Y = Y_s(1 + B\log\dot{\varepsilon})(1 - D) \\ &\text{Tension: } \sigma_{ij} = 3K(1 - D)e_{kk}\delta_{ij} + 2G(1 - D)e_{ij} \end{aligned} \tag{5.11}$$

Damage is described by:

$$\begin{aligned} &D = \frac{16}{9}\left[\frac{1 - \bar{\nu}^2}{1 - 2\bar{\nu}}\right] C_d; \quad C_d = \beta N a^3; \quad a = \frac{1}{2}\left(\frac{\sqrt{20}K_{1C}}{\rho c\dot{\varepsilon}}\right) \\ &\bar{\nu} = \nu\left(1 - \frac{16}{9}C_d\right), \qquad N = k\varepsilon_v^m \end{aligned} \tag{5.12}$$

The number of cracks, N, that nucleate upon shock loading is assumed to follow Weibull distribution with respect to the volumetric strain, ε_v, k and m are model parameters. Y_s is the static compressive strength, B is the strain rate sensitivity parameter, K and G are bulk and shear moduli respectively, e_{ij} are deviatoric strains, σ_{ij} are total stresses, C_d is crack density, K_{IC} is static fracture toughness, $\bar{\nu}$ is degraded Poisson's ratio, β is a geometric factor, a is the crack (fragment) size, c is the sound speed, N is the number of flaws, ρ is material density, and $\dot{\varepsilon}$ is the strain rate. This model, implemented into the EPIC code [56], successfully replicated an experimentally measured shock profile from a plate impact test on silicon carbide.

Implementing this tensile failure model in commercial codes like ABAQUS and LSDYNA is relatively straightforward. For efficient solutions, the numerical implementation can utilize the 4th order Runge–Kutta method with an iterative scheme. However, for numerical stability and accuracy, Rajendran and Kroupa [37] opted for an iterative second-order implicit Runge–Kutta method.

The model parameters that matched the measured stress wave profile of AD85 (aluminium oxide) ceramic were: $Y_s = 2\,\text{GPa}$, $B = 0.175$, $K_{1c} = 3\,\text{MPa}$, $k = 10^{27}$, and $m = 6$. The shock stress at Hugoniot from plate impact experiments on AD85 is approximately 5.5 GPa or Y corresponding to HEL is about 4 GPa. The B parameter was estimated as 0.175 using the quasi-static strength and HEL. Notably, the model parameter values are not unique; calibration can be accomplished using other impact and high strain rate experimental data presented in Chap. 4. It is important to acknowledge the considerable variation for HEL and quasi-static compressive strength in open literature for brittle materials.

While the TCK model could be strengthened by incorporating a description of damage under compression, it treats damage as a scalar, and, in principle, splitting and faulting type failure modes cannot be modelled. The fragmentation-based models are mathematically simple, with only a few model parameters directly determined from modelling and simulation of plate impact tests. These strain rate-dependent TCK and Rajendran–Kroupa models are not difficult to implement into computer codes.

5.2.4 *Micro-mechanical Models*

During the 1980s, substantial research efforts were directed towards developing microstructure-driven models for brittle materials. Heterogeneous materials like concrete and rocks exhibit random distributions of defects and constituents. Despite this, concrete responses usually remain isotropic, while rocks may display anisotropic behaviour, depending on preferred planes.

Transparent glasses, being amorphous, lack short- or long-range order in their microstructures. In contrast, crystalline ceramics are predominantly isotropic and homogeneous. Therefore, a comprehensive understanding of a brittle solid's microstructure is crucial before attempting to model its constitutive response.

The complexity of defect generation and propagation under dynamic loading poses challenges in micro-mechanical models. Defining defects is inherently challenging, but depending on their geometry (e.g. dislocations, pores, cracks), various mechanisms come into play. In brittle materials, inelastic strains primarily arise from micro-cracking and pore collapse, though microplasticity and fragmentation are also possible. High-strength ceramics require the description of strains due to dislocation-type atomistic defects in constitutive modelling.

Microphysical theories rooted in fracture mechanics are employed to elucidate deformation during compressive failure processes. Microphysical models aim to describe apparent inelastic behaviour while tracking micro-structural evolutions under specific loading conditions. Models by Rajendran and Grove [57], Addessio and Johnson [58], and Espinosa et al. [59] follow this approach.

Micro-mechanical constitutive models have the following features or aspects:

1. The strain tensor is decomposed into elastic and irreversible (or permanent) plastic components, $\varepsilon_{ij} = \varepsilon^{e}_{ij} + \varepsilon^{p}_{ij}$.
2. The three-dimensional constitutive equation for the cracked solid is, $\sigma_{ij} = M_{ijkl}\varepsilon^{e}_{kl}$. The **M** (stiffness) and **C** (compliance) tensors are derived for cracked solids.There are many constitutive theories that analytically derive these tensors by decomposing them, $M_{ijkl} = D_{ijkl} + H_{ijkl}$. The elastic strain components can be decomposed further into strain in the matrix (or solid) material correspond to **D** and strains due to crack opening and sliding (**H**). Equations or expressions for the stiffness tensor **M** or the compliance tensor **C** can be obtained from the derivatives of either Gibbs free energy (GFE) function or Helmholtz free energy (HFE) function, respectively. The GFE function is expressed in terms of stress invariants while the HFE is in terms of strain (or stretch) invariants. These empirical functions will usually have the shear and bulk components of the solid in its initial state (or intact solid prior to the initiation and propagation of flaws). The rest of the terms in HFE could be related to micro-cracks and their orientations. Several examples of HFE are presented in the later sections of this Chapter.
3. The stiffness or compliance tensor is analytically derived from boundary value problems of a crack containing solid. The shape of the crack is usually assumed to be penny shaped. There are models that assume wing cracks emanating from a penny shaped flaw. In this approach, the stresses are averaged over the volume. The governing equations are derived by performing integration over both volume and crack surfaces. Margolin's model [60] is an example of this approach.
4. A dimensionless scalar damage parameter is defined, often as the crack density, $\gamma = \beta N a^3$, where, a is crack radius, N is number of micro-cracks per volume, and β is a dimensionless shape parameter to accommodate geometry of micro-cracks.
5. An evolution law with respect to time is required for the damage variable. This damage variable could be a function of several other variables, such as the strain rate, pressure, and any other state variable, like stress intensity factor, K_I or strain energy release rate, G_I, and an example is: $\dot{a} = n_1 C_R[1 - (G_{cr}/G_I)^{n_2}]$. The model constants are: n_1 and n_2. C_R (Raleigh wave speed) and G_{cr} (critical strain energy release rate) are material properties.

6. In the absence of evolution law for a, value for "*a*" need to be determined from appropriate expressions or laws associated with the crack evolution.
7. A fracture mechanics or physically based failure criterion, such as $\gamma = \gamma_{cr}$ is required to predict pulverization or fragmentation of the brittle solid.
8. Mohr–Coulomb or Drucker–Prager type strength model is often employed to describe the flow behaviour of granular or pulverized materials that are under inertial or lateral confinement.
9. If the response of a brittle material is considered anisotropic, then the inelastic behaviour due to crack orientations requires a "*multi-plane*" models.

In general, it is difficult to obtain readily available values for the model parameters. For brittle materials, material properties like the elastic constants, shock speed, and fracture toughness also enter into the constitutive/damage models. Those parameters which cannot be measured directly from impact and shock experiments can be "*tuned*" or "*calibrated*" by comparing the "*in-situ*" high speed photos, strain and stress gauge data, X-ray pictures, VISAR signal with hydrocode simulations.

Micromechanical models, while mathematically complex and user-unfriendly, can be implemented into any hydrocode, as discussed in Chap. 6. Analytical derivations related to boundary value problems often necessitate solutions to complex algebraic or coupled partial difference equations. However, these models explicitly capture defect evolutions due to various mechanisms, such as micro-crack opening and sliding, micropore collapse, voids growth, and microplasticity.

Until 1988, shock-wave propagation-based finite element/difference numerical codes lacked material models capable of describing the inelastic behaviours of brittle ceramics under shock and high strain rate loading conditions. Due to ceramics' brittle nature, the damage incurred during loading often involves the nucleation and growth of micro-cracks, eventually coalescing to form macro cracks leading to failure. Therefore, a fracture mechanics-based approach to model micro-crack growth in ceramics seems plausible, providing a sound theoretical basis for analysing the cracking process.

In the 1990s, significant progress was made in developing advanced models that utilized fracture mechanics concepts and principles to describe brittle material behaviours under both compressive and tensile loading conditions. While numerous damage models address the quasi-static behaviours of rocks, it appears that with appropriate modifications, these models can be extended to include the dynamic behaviour of ceramics. The following subsections briefly identify and present salient features of different advanced models, mostly developed for high-strength ceramics but applicable to describing the shock and high strain rate behaviours of concrete and rocks. Two important research monographs, Krajcinovic [61] and Nemat Nasser and Hori [62], provide significant fundamental results on the failure of brittle materials due to the growth of inhomogeneities and defects in the virgin solid. These monographs offer mathematical and physical insights into the effects of ensembles of micro-cracks on the macroscopic compliance of a brittle solid.

In essence, the overall or average strain and stress relation in the cracked solid [62, 63] can be described by,

$$\bar{\varepsilon}_{ij} = C^0_{ijkl}\bar{\sigma}_{kl} + \alpha_{kl} \tag{5.13}$$

where

$$\alpha_{ij} = \frac{1}{2V}\int_{S_c}(\eta_i V_j + \eta_j v_i)\,dS$$

The additional strain, α_{ij}, in the solid is derivable from a knowledge of the displacement field on the surface of each crack and it is often proportional to a dimensionless parameter, crack density, $\gamma = Na^3$ where, N is the number of cracks per unit volume and a is average micro-crack size. The effective compliance tensor, $\bar{C}_{ijkl}$ can be then written as,

$$\bar{C}_{ijkl} = C^0_{ijkl} + \frac{\partial \alpha_{ij}}{\partial \bar{\sigma}_{kl}} \tag{5.14}$$

Several micromechanical models strive to derive analytical expressions for the damaged moduli using the BVP approach under different loading conditions.

The crux of developing constitutive models for brittle materials lies in determining expressions for the strain induced by the presence of flaws, micro-cracks, or pores in the solid under various loading conditions. In this context, we present several microphysics constitutive models from the 1980s known for their ability to realistically describe the effects of micro-structural damage evolution in brittle materials. These models have laid the foundation for the development of newer models in subsequent years. If implemented in today's commercial and research codes, these fundamental models can provide realistic solutions for structural responses to blast and impact loading. For the benefit of our readers, we delve more elaborately into two models in this section: the Rajendran–Grove model and a hyperelastic continuum damage model.

In the subsequent sections, we provide summaries of various complex micromechanical models found in the open literature, presenting essential equations necessary for developing user-defined subroutines in commercial and research hydrocodes. Additionally, we include a recent brittle failure model designed for glass-reinforced plastics. This inclusion aims to demonstrate the effectiveness of a hydrocode finite element (HFE) based approach in modelling both the brittle matrix (epoxy) and the hyperelastic fibrous material, such as E2 or S2 glass fibres.

5.2.4.1 Margolin Model [31, 63, 64]

Many fundamental processes modelled under static loading conditions can be effectively extended to describe the damage processes in rocks, akin to brittle solids, under dynamic loading. This approach has been employed by Margolin [31] to model the dynamic behaviour of rocks subjected to tension and compression. His bedded crack model [63, 64] is grounded in the decomposition of the strain tensor. This decomposition involves strain in the uncracked matrix or solid material due to applied far-field stresses and additional strains resulting from the displacements of surfaces of penny-shaped cracks distributed throughout the body.

Margolin derived the effective modulus, expressed in terms of a compliance tensor, using analytical expressions that account for crack opening and sliding under both tensile and compressive loading. The model also incorporates the effects of strain rate, pressure, and damage on compressive strength. The sliding mode arises from frictional forces across the crack faces under compression. For non-interacting, penny-shaped micro-cracks of various sizes and random orientations, and in the case where all three principal stresses are positive, Margolin [63] provided the following expression for the elements of an isotropic compliance tensor **C**:

$$C_{ijkl} = c_1 \delta_{ik} \delta_{jl} + c_2 \delta_{il} \delta_{jk} + c_3 \delta_{ij} \delta_{kl} \tag{5.15}$$

where

$$c_1 = B_0 + \frac{1}{4G}, \qquad c_2 = D_0 + \frac{1}{4G}, \qquad \text{and} \qquad c_3 = A_0 - \frac{\nu}{2(1+\nu)G} \tag{5.16}$$

In the above equations, δ is the Kronecker delta, G and ν are the shear modulus and Poisson's ratio, respectively, of the undamaged material, and A_0, B_0, and D_0 are parameters whose values depend on the principal stress state and the extent of damage. Before damage initiates, $A_0 = B_0 = D_0 = 0$.

For stress states in which all three principal stresses are compressive, Margolin proposed that

$$A_0 = -(1+\nu)\gamma^*, \qquad B_0 = 4(1+\nu)\gamma^* \quad \text{and} \quad D_0 = A_0$$

where

$$\gamma^* = \frac{16\gamma}{45E} \quad \text{and} \quad \gamma = N_0^* a^3 \tag{5.17}$$

In the above equations, γ is the micro-crack density (a dimensionless measure of damage), E is the Young's modulus of the undamaged material, N_0^* (a model constant) is the number of micro-cracks per unit volume, and a is the maximum crack length in the exponential distribution of micro-cracks. Upon substituting Eq. (5.17) into Eq. (5.16) and inverting the compliance tensor **C**, we obtain the following expressions for the shear and bulk moduli ($\bar{G}_c$ and $\bar{K}_c$, respectively) of the damaged ceramic material under compression, which is given by:

$$\bar{G}_c = \frac{2G}{(2+3B_0G)}, \qquad \bar{K}_c = \frac{2G(1+\nu)}{3(1-2\nu)} \tag{5.18}$$

The above equations result in degradation of the shear modulus, but not of the bulk modulus, because under compression only crack movement of the closed micro-cracks (modes II and III) is permitted.

Margolin [31] described crack growth initiation through a generalized Griffith criterion and assumed micro-cracks to grow at a fraction of the Rayleigh wave speed, $\dot{a} = \delta C_R$. The finite element simulations of explosive fracture of oil shale using the bedded crack model showed reasonable agreement with data from field experiments [64].

In summary (see Table 5.1), the strengths of this model are:

- Three-dimensional constitutive equations
- Crack growth under modes I, II, & III
- Micro-mechanical in terms of crack growth under all three fracture modes
- Simple dynamic crack growth law

Table 5.1 Specific features of the Margolin model [63, 64]

Constitutive Relationship
• Three-dimensional stress/strain state $\bar{\varepsilon}_{ij} = C^0_{ijkl}\bar{\sigma}_{ij} + \alpha_{ij}$
• Not derived from any Helmholz or Gibbs free energy function, but from a boundary value problem of a cracked body
• General solution is crack orientation and crack size dependent
Definition of Damage (D)
• Statistically prescribed distribution of bedded penny shaped cracks
• D is a scalar and derived for random orientation of the micro-cracks
• Damage evolution under: (a) all principal stresses positive, (b) all negative, and (3) two have opposite sign to the third
• $D = \int \beta N(a) a^3 \, da$
where, β is crack geometry parameter
a is the micro-crack size
$N(a)$ is the statistical crack distribution
Micro-Crack Distribution
• $N(a) = N_0 e^{-\frac{a}{a_0}}$; a_0 is characteristic crack size
Micro-Crack Growth Laws
• Mode 2 (shear mode) crack growth
• Generalized Griffith criterion used to initiate crack growth
• Crack extension rate is constant and equal to a fraction of the Rayleigh wave speed
Failure Criterion
• $D = D_{cr}$
• The material element failed when the relative volume of the micro-cracks reached a critical value
Other Mechanisms
• None

- Model has been used to describe shock and cratering experiments
- Strain-rate dependent damage calculation under both compression and tension.

5.2.4.2 Horii and Nemat-Nasser Model [13, 14, 65]

Horii and Nemat-Nasser [13] derived expressions for the elastic moduli of a cracked solid based on the micro-mechanisms of rock failure, specifically axial splitting and shear failure, under plane stress conditions. Under axial compression, tension cracks nucleate at the tips of preexisting model flaws (see Fig. 5.1), grow with increasing compression, and become parallel to the direction of the maximum far-field compression. When lateral compression is introduced, crack growth stabilizes and halts at some finite crack length. This fundamental mechanism of axial splitting is observed in uniaxially compressed rock or, more generally, in any brittle solids. The authors estimated the "*ultimate strength*" and the orientation of the overall fault plane with confining pressure using the wing crack model.

While the wing crack model is not a constitutive model, we include some salient features of this two-dimensional analytically tractable model that describes shear failure of rocks based on the "*wing" crack* mechanism. The model assumes that each flaw can slip in its own plane, transmitting a shear stress, $\sigma_{xy} = \tau_c + \mu\, \sigma_y$, where τ_c is the cohesive stress, σ_y is the compressive stress on the flaw surface, and μ is the friction coefficient. The σ_y is the projection of axial compression (σ_1) perpendicular to the flaw. Horii and Nemat-Nasser performed all their calculations with zero cohesive stress and $\mu = 0.4$. The 2D model used several parameters associated with the fixed initial flaw size ($2c$), space between adjacent flaws (d), the wing crack length (l), and the inclination or angle of the flaw (Ø) with respect to the axial compression, and the ratio between the axial and lateral compression. The primary aim of this 2D plane stress model was to establish an analytical model explaining splitting and faulting types of rock failures. Experiments on plates of Columbia resin CR39 containing preexisting flaws supported the analytical results qualitatively.

In this section, we provide a further summary of the elastic stress-strain relationship for a micro-crack-damaged brittle solid, as developed by Horii and Nemat-Nasser [14]. Their study focuses on a solid with randomly distributed penny-shaped flaws, and analytical relationships for elastic moduli are derived. Notably, this model departs from the use of Helmholtz free energy or Gibbs free energy expressions. Instead, strains are calculated based on the modelling of Budiansky and O'Connell [55] for a cracked elastic solid.

Within this model, damage is represented as a single scalar, independent of the orientation of micro-cracks. No attempt is made to describe the microstatistics of the crack nucleation process, and the growth of micro-cracks is not considered. The primary objective of this model is to compute the elastic moduli of cracked bodies, assuming the average number of cracks per volume and the average crack size are known. The strain is then decomposed into elastic strain in the matrix and inelastic strain resulting from micro-cracking. This approach offers a simplified yet effective representation of the complex behaviour of micro-crack-damaged brittle solids.

The stress–strain is given by,

$$\varepsilon_{ij} = \bar{D}_{ij}\,\sigma_{ij} \tag{5.19}$$

The overall compliance, $\bar{D}_{ij}$ can be expressed as,

$$\bar{D}_{ijkl} = D_{ijkl} + H_{ijkl} \tag{5.20}$$

The "*elastic*" compliance tensor of the matrix material is **D** and the inelastic compliance is **H**. The strain due to crack opening is written as,

$$\varepsilon_{ij}^{c} = H_{ijkl}\,\sigma_{kl}\,, \qquad H_{ijkl} = \sum K_{ijkl} \tag{5.21}$$

where summation is taken over all cracks in a unit volume.

For an unbounded penny shaped crack, components of the non-zero elements of the compliance tensor **K** is derived as,

$$K'_{3333} = \frac{16\left(1-\upsilon^2\right)}{3E}a^3; \qquad K'_{1313} = K'_{2323}\frac{8\left(1-\upsilon^2\right)}{3E(2-\upsilon)}a^3 \tag{5.22}$$

and all other components of K_{ijkl} are zero. The damage parameter is defined as, $d = Na^3$.

The components of the "*inelastic*" compliance tensor **H** are given by,

$$\begin{aligned}
H_{1111} = H_{2222} = H_{3333} &= \frac{16\,(10-3\bar{\upsilon})\left(1-\bar{\upsilon}^2\right)}{45\,(2-\bar{\upsilon})\,E}d\\
H_{1122} = H_{2233} = H_{3311} &= \frac{16\,\bar{\upsilon}\left(1-\bar{\upsilon}^2\right)}{45\,(2-\bar{\upsilon})\,E}d\\
H_{1212} = H_{2323} = H_{3131} &= \frac{16\left(1-\bar{\upsilon}^2\right)(5-\bar{\upsilon})}{45\,(2-\bar{\upsilon})\,E}d\\
H_{ijkl} = H_{jikl} = H_{ijlk} = H_{klij}; &\quad \text{otherwise } H_{ijkl} = 0.
\end{aligned} \tag{5.23}$$

Using these equations, Horii and Nemat-Nasser derived expressions for the degraded Young's modulus and damage variable d as,

$$\begin{aligned}
\frac{\bar{E}}{E} &= 1 - \frac{16\left(1-\bar{\upsilon}^2\right)(10-3\bar{\upsilon})}{45\,(2-\bar{\upsilon})}d\\
d &= \frac{45(2-\bar{\upsilon})(\upsilon-\bar{\upsilon})}{16(1-\bar{\upsilon}^2)(10\upsilon-\bar{\upsilon}(1+3\upsilon))}
\end{aligned} \tag{5.24}$$

The degraded bulk modulus was determined using the expressions of Budiansky and O'Connell [55] for penny-shaped cracks. The resulting degraded elastic moduli, influenced by frictional sliding, were then applied to investigate the loading path effect on the overall response of a micro-crack-containing brittle solids. In the specific case where randomly distributed penny-shaped micro-cracks are all open, the two-dimensional analysis utilized the solution provided by Budiansky and O'Connell. The study also explored the effects of crack closure and load-induced anisotropy using the degraded moduli equations. However, the micro-mechanical constitutive model developed by Budiansky and O'Connell, as applied by Horii and Nemat-Nasser [65], did not consider damage growth and dynamic loading for brittle solids.

Subsequently, Horii and Nemat-Nasser devised a failure model to simulate experimentally observed failure modes under axial compression, both with and without lateral confining pressures. This model successfully predicts splitting and faulting type failure modes. While the model focuses solely on the final failure process, its complexity arises from the stress intensity factor calculations for the tensile cracks that nucleate and grow at pre-existing but closed microflaws. Although limited to describing the final failure, the model can predict the brittle-to-ductile transition of brittle materials. The stress–strain relationship from Eqs. (5.19)–(5.21) can be implemented in hydrocodes, enabling the modelling of compressive failure under shock and impact loading.

In summary (see Table 5.2), the strengths of this model are:

- Proven capability in predicting modes 1 (opening), 2 (sliding), and 3 (tearing)
- Includes the effects of confining pressure
- Ability to model splitting and faulting modes of failure
- Describes brittle to ductile transition.

5.2.4.3 Ilankamban and Krajcinovic Model [66]

Ilankamban and Krajcinovic [66] introduced a mathematically intense constitutive model describing the nonlinear behaviour of progressively deteriorating brittle solids under static loading. Some specifics of this model are briefly outlined in Table 5.3. Initially, Krajcinovic and Fanella [67] developed a concrete model based on microstructure geometry, encompassing coarse aggregate and micro-cracks within general continuum mechanics. They decomposed strain into two components: one from the isotropic, homogeneous cement paste-aggregate matrix response and the other from crack surface displacements. The Helmholtz Free Energy in this model is a function of damage and strain components.

From a phenomenological perspective, the energy associated with a crack is assumed to be related to tensile and shear strain rather than stress. The Helmholtz Free Energy (HFE) is expressed using isotropic invariants of ε and $[(\omega^i, N^i), i = 1, 2, ...n][(\omega^i,\ N^i),\ i = 1, 2, \ldots, n]$.

However, from a phenomenological point of view, it is assumed that the energy associated with a crack is related to tensile strain and shear strain rather than tensile

Table 5.2 Specific features of the Horii and Nemat-Nasser model [14]

Constitutive Relationship

- Three-dimensional for an Elastic Solid Containing Open Penny-Shaped
- Frictional sliding of micro-cracks contributes to anisotropic moduli in two-dimensional under both tension and compression
- Includes crack interaction effects
- $\varepsilon_{ij} = (S_{ijkl} + H_{ijkl})\sigma_{kl}$; **S** is the elastic compliance of the matrix; **H** is compliance due to inelastic strains of cracked material

Definition of Damage (D)

- $D = \bar{N}\bar{a}^3$
- $\bar{N}$ is the average number of cracks ($\bar{N}$) per volume and $\bar{a}$ is the average crack size

Micro-Crack Distribution

- Size distribution of cracks and crack orientations are not considered

$\bar{N}$ is a constant and independent of the crack size. Random distribution of cracks is assumed

Micro-Crack Growth Laws

- Abrupt failure is assumed when the stress intensity factor, $K_I = K_{Ic}$ under crack opening mode

Failure Criterion

- Critical ratio of average crack size to crack spacing (STATIC)

Other Mechanisms

- None

Table 5.3 Ilankamban and Krajcinovic damage model for rocks

Material parameters	Unit	Quartzite	Norite
λ	MPa	1.36×10^{10}	6.36×10^{10}
μ	MPa	2.333	0.667
C_1/λ		−84.5	−35.0
C_2/λ		−5.91	−2.10
$\mathscr{R}$		1.9×10^2	7.80×10^{-7}
p		0.15	0.15

stress and shear stress. The Helmholtz free energy (HFE) can therefore be written using the isotropic invariants of ε and $[(\omega^i, N^i), i = 1, 2, ...n][(\omega^i,\ N^i),\ i = 1, 2, \ldots., n]$.

$$\rho\Psi = \left(\frac{\lambda}{2}\right)\varepsilon_{KK}\varepsilon_{LL} + \mu\varepsilon_{KL}\varepsilon_{KL} + C_1\ \Sigma_i\ \omega^i\ [< N^i_K\varepsilon_{KL}N^i_L >< N^i_M\varepsilon_{MN}N^i_N >]$$

$$+ C_2 \sum_i \omega^i\ [< N^i_K\varepsilon_{KL}\varepsilon_{LM}N^i_M - (N^i{}_K\varepsilon_{KL}N^i_L)(N^i{}_M\varepsilon_{MN}N^i_N)] \tag{5.25}$$

where λ and μ are Lame's constants, C_1 and C_2 are material parameters associated with cracks. The value of the term inside the "*Macaulay's bracket*" is equal to one when it is positive, otherwise zero. The first two terms are the elastic response of the intact solid or solid in its virgin state. The first term is the bulk response and the second term is the shear response. Since N_I is a unit vector, the quantities within the first and second square brackets in the above equation are squares of the normal (tensile) and shear strain on the plane with the normal N_i, respectively. The stress components when derived from the HFE expression become a function of micro-crack density in different orientations.

In this model, the energy associated to the compressive strain over the faces of the cracks is considered zero. The stress–strain and thermodynamic force $R^i - \varepsilon_{KL}$ relationships are given by,

$$\sigma_{IJ} = \rho \frac{\partial \Psi}{\partial \varepsilon_{IJ}} = K_{IJKL}\, \varepsilon_{KL} \tag{5.26}$$

$$R^i = -\rho \frac{\partial \Psi}{\partial \omega^i} \tag{5.27}$$

Since stress components are derived from this function, the model allows damage to degrade the elastic moduli. By applying the symmetry requirement on stress and strain tensors and considering an isotropic distribution of micro-cracks ($\omega^1 = \omega^2 = \ldots \omega$), the stiffness tensor $\mathbf{K}$ for the case of hydrostatic tension reduces to the following form,

$$\begin{aligned} K_{IJKL} = &\left[\lambda + \frac{4\pi}{15}\,\omega\,(C_1 - C_2)\right]\delta_{IJ}\delta_{KL} + \\ &\left[\mu + \frac{4\pi}{15}\,\omega\,(C_1 + 1.5\,C_2)\right](\delta_{IK}\delta_{JL} + \delta_{IL}\delta_{JK}) \end{aligned} \tag{5.28}$$

<u>*Under uniaxial compression*, anisotropic deformation develops due to the compressive strain experienced by micro-cracks</u> at certain angles with respect to the loading direction. The deformation becomes transversely isotropic, and therefore, Ilankamban and Krajcinovic [66] derived the following expression for determining incremental stress components as,

$$\delta\sigma_{IJ} = K_{IJKL}\,\delta\varepsilon_{KL} + \Sigma_i\, A^i_{IJ}\delta\omega^i \tag{5.29}$$

where

$$\begin{aligned} A^i_{IJ} = &-2C_1 < N^i_K \varepsilon_{KL} N^i_L > N^i_I N^i_J - C_2[N^i_K \varepsilon_{KI} N^i_J + N^i_K \varepsilon_{KJ} N^i_I \\ &-2\,(N^i_K \varepsilon_{KL} N^i_L)\, N^i_I N^i_J] \end{aligned} \tag{5.30}$$

This equation suggests that the total stress increment is due to incremental elastic strain ($\delta\varepsilon_{KL}$) of the intact solid and the damage-induced stress. The second term captures the contribution of all damages ($\delta\omega^i$).

The complete incremental stress–strain relation is given by,

$$\delta\sigma_{IJ} = (K_{IJKL} + \bar{K}_{IJKL})\,\delta\varepsilon_{KL};, \quad \bar{K}_{IJKL} = \Sigma_i A^i_{IJ} H^i_{KL}$$

$$H^i_{KL} = -c^i \left(\frac{p}{\lambda}\right)\left(\frac{R^i}{\lambda}\right)^{p-1} A^i_{KL} \tag{5.31}$$

The incremental crack density can be obtained from $\delta\omega^i = H^i_{KL}\delta\varepsilon_{KL}$. The crack growth function, g^i is given by,

$$g^i = (\frac{R^i}{\lambda})^p - \left(\mathscr{R}_0 + \omega^i\right) = 0 \tag{5.32}$$

The $c^i = 1$ when $g^i = 0$ or $\frac{\partial g^i}{\partial R^i}\,\delta R^i \geq 0$; otherwise $c^i = 0$. Note that the "p" and $\mathscr{R}_0$ are model parameters. $\mathscr{R}_0$ defines the onset of micro-cracking in a plane. Therefore, the constitutive equations under tension is given by Eqs. (5.26)–(5.29), and compression is given by Eq. (5.31), respectively. The damage in terms of crack density can be calculated from Eq. (5.32).

Ilankamban and Krajcinovic performed computations to show the effect of micro-crack-induced damage on the stress–strain curve (or strength) under both uniaxial and biaxial compression for two rock materials: Witwatersrand Quartzite, and Norite. For example, Fig. 5.11 for Norite shows the development of nonlinearity in the stress–strain curve due to damage approximately just above an axial compressive strain of 1.6×10^{-4}. However, the volumetric strain under compression starts decreasing much earlier at about 0.6×10^{-4} indicating crack opening (or dilation) due to crack surface displacements (tensile or shear strain).

The experimental stress–strain data match the model results as shown in Fig. 5.11. Although this model does not explicitly model wing crack growth, it captures the effects of this mechanism on volume expansion under compression. We can call this effect splitting due to volume expansion under uniaxial compressive stress [13, 14]. The strength (axial stress) also increases with confining pressure for the three rock materials. The model demonstrates the observed trends in micro-crack distributions. Near failure, micro-crack growth accelerates rapidly over a large range of orientations, leading to faulting.

Given the significant data on rocks, cement, and concrete from split Hopkinson pressure bar tests indicating strong strain rate effects on strength, it is possible to modify the Ilankamban and Krajcinovic model by introducing rate-dependent dynamic crack growth laws. It is also possible to introduce crack density (γ) into the model and modify the model to include crack coalescence through a critical failure criterion, such as $\gamma = \gamma_c$.

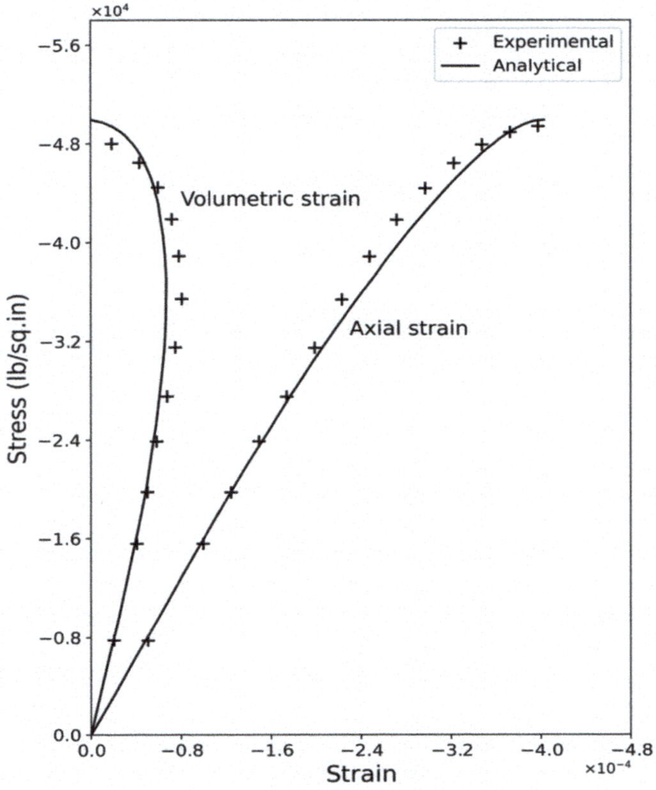

Fig. 5.11 Comparison of Ilankamban and Krajcinovic damage model with quasi-static uniaxial compression experiment data for Norite Rock

In summary (Table 5.4), the advantages of this model are

- Modelling dilatancy
- Modelling anisotropic damage
- Includes confining pressure effects
- Ability to describe faulting and splitting under compression
- User friendly and relatively easy implementation into any hydrocode
- Computationally efficient due to the direct derivation of elastic stiffness moduli and a simple damage evolution law.

5.2.4.4 Costin–Stone Multi-orientation Micro-crack Model [68, 69]

The Costin and Stone model presents a comprehensive approach considering different microcrack orientations and anisotropic microcrack growth in brittle rock-type

Table 5.4 Specific features of the Ilankamban and Krajcinovic model [66]

Constitutive Relationship

- Continuum damage mechanics / Helmholtz Free Energy
- Three-dimensional stress state

Definition of Damage (D)

- No microcrack interaction
- Proportional to vector space of micro-crack densities
- Both under tension and compression

Microcrack Distribution

- Vector space of micro-crack density

Micro-Crack Growth Laws

- Damage surface which separates the elastic domain from the inelastic domain
- Micro-crack density empirically related to its conjugate thermodynamic force

Failure Criterion

- Not provided

Other Mechanisms

- None

materials. Key features of this model are summarized in Table 5.5. Damage (D) is defined as a vector with components in 13 different orientations for two-dimensional problems and 144 directions for three-dimensional problems. Constitutive equations are derived from the Helmholtz Free Energy (HFE) function:

$$\rho\Psi = \left(\frac{\lambda}{2}\right)\varepsilon_{KK}\varepsilon_{LL} + \mu\varepsilon_{KL}\varepsilon_{KL} + C_1\,D_K^{\alpha}\,\varepsilon_{KL}\,D_L^{\beta}\,\varepsilon_{MM} + C_2\,D_K^{\alpha}\,\varepsilon_{KL}\,\varepsilon_{LM}D_M^{\beta} \tag{5.33}$$

where ρ is the density and C_1 and C_2 are scalar-valued material parameters. The α and β are different damage types, such as micro-cracks and micropores. The Costin derived the stress–strain relation from this function as.

$$\begin{aligned} \varepsilon_{ij} &= S_{ijkl}\sigma_{kl} \\ S_{ijkl} &= \frac{1+\nu}{E}\delta_{ik}\delta_{jl} - \frac{\nu}{E}\delta_{ij}\delta_{kl} + C_1(D_i D_l\delta_{jk} + D_j D_k\delta_{il}) + C_2 D_i D_j\delta_{kl} \end{aligned} \tag{5.34}$$

The inelastic strain due to micro-cracking is given by:

$$\varepsilon_{ij}^{c} = [C_1(D_i D_l\delta_{jk} + D_j D_k\delta_{il}) + C_2 D_i D_j\delta_{kl}]\sigma_{kl} \tag{5.35}$$

The model parameters C_1 and C_2 are zero when the indices are not the same ($i \neq j$). The moduli are computed and degraded using the compliance matrix **S** which is a function of the components of damage vector **D**.

Table 5.5 Specific features of the Costin and Stone model [68, 69]

Constitutive Equations

- Continuum damage mechanics (Helmholtz Free Energy)
- Degraded modulus of an elastic solid
- Both under tension and compression
- Pressure dependent
- Includes crack interaction effects

Definition of Damage (D)

- Damage is assumed to be a vector
- Crack opening - Mode 1
- Penny-shaped cracks with radius a
- $D_i = \frac{1}{a_0} \int_\Omega e_i(\vec{a} - \vec{a}_0)\, H(K_I)\, d\Omega$

where a and a_0 are current and initial micro-crack sizes, respectively
$H(K_I) = 0;\ K_I < K_{Ic},\ H(K_I) = 1;\ K_I > K_{Ic}$

Micro-Crack Distribution

- ($\vec{a}_0 = a_0\vec{n}$), where a_0 is the initial flaw size as a function of various orientations; n is the unit normal perpendicular to the crack size

Micro-Crack Growth Laws

- $K_I = K_{Ic}$ (static crack growth); The estimation of K_I is empirically formulated from the results of Horii and Nemat-Nasser (1985)
- Micro-crack density empirically related to its conjugate thermodynamic force

Failure Criterion

- Micro-crack size reaches a critical value with respect to the micro-crack spacing

Other Mechanisms

- None

$$D_i = \frac{1}{a_0} \int_\Omega n_i(a - a_0)\, H\, |K_I\, (n)|\, d\Omega \tag{5.36}$$

where a and a_0 are current and initial micro-crack sizes, respectively. Note that "n_i" is the components of unit normal vector $\vec{n}$ that is perpendicular to the crack face. The heavy side function, $H(K_I) = 0$; when $K_I < K_{Ic}$ and, $H(K_I) == 1$; when $K_I > K_{Ic}$. K_{Ic} is the fracture toughness of the brittle material.

Costin and Stone [69] applied their fracture mechanics and continuum damage-based approach by modelling the deformation and failure behaviours of rocks under quasi-static compressive loading. Their model captures the effects of nucleation, growth, and coalescence of micro-cracks on material degradation. In their numerical implementation, both the magnitude and direction of micro-crack growth in each element are computed to model anisotropy due to crack growth in preferred directions and the stress history dependence of micro-crack growth. As loading increases, micro-crack interactions lead to coalescence and subsequent material failure or softening. Several sample calculations modelled the softening portion of the stress–strain curves for Westerly granite (Fig. 5.12). The multi-plane Costin–Stone model can

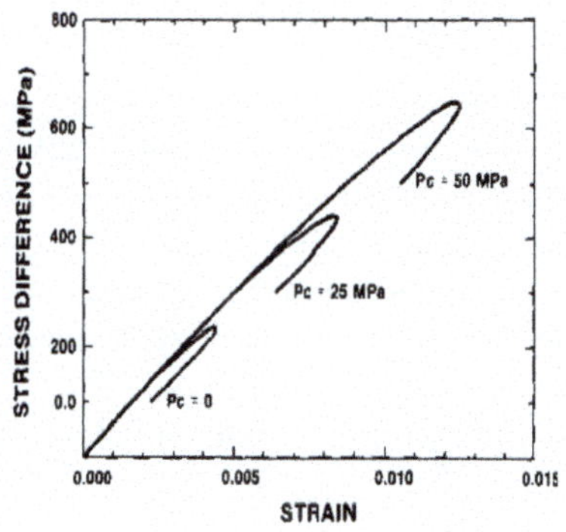

Fig. 5.12 Costin and Stone model simulation of triaxial tests on Westerly granite at different pressures (P_c). Stress difference is $\sigma_{11} - P_c$

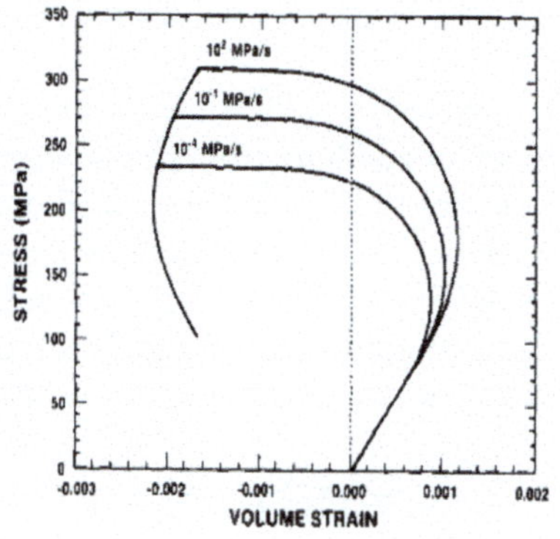

Fig. 5.13 Model simulation of uniaxial compression of Westerly granite at different stress rates. The plots show volumetric responses for three different stress rates. The positive x-axis is compressive strain and negative is tensile strain

simulate splitting and faulting failure modes due to its ability to model anisotropic damage.

A crucial aspect of any continuum damage mechanics (CDM) or phenomenological constitutive model for brittle materials is its ability to capture dilation or volume expansion under axial compression with and without lateral confinement. In Fig. 5.13, the Costin–Stone model simulates stress–strain responses for Westerly granite at different stress rates, demonstrating this feature. As axial strain increases linearly with axial compressive stress, damage initiates at around 0.1% strain, causing volume expansion and eventually leading to negative (tensile) strain under compression. The failure envelope is also depicted in the second quadrant. The model realistically captures the effects of pressure and loading rate on strength and ductility (strain to failure) under uniaxial compression in brittle materials like rocks.

In summary (Table 5.5), the strengths of this model are:

- Applicable to quasi-static loading conditions
- Demonstrated model capability to describe stress–strain curves
- Dilatancy under compressive loading is modelled
- Describes failure under compression
- Includes confining pressure effects
- Models anisotropic effects due to micro-crack growth
- Few model parameters
- Methods are available to obtain model parameters.

5.2.4.5 Addessio–Johnson Ceramic Model [58]

This model by Addessio and Johnson [58] focuses on determining inelastic strains in ceramics resulting from penny-shaped micro-crack growth under tension and compression. In this approach, inelastic strains are obtained by integrating individual crack strains over a material volume, considering all crack sizes and orientations. By making assumptions about crack size and orientation, simplified expressions for inelastic strains are derived. In 1990, Addessio and Johnson proposed relationships for the deviatoric parts of the inelastic strain components (e^c_{ij}):

$$e^c_{ij} = \beta^e \bar{c}^3 S_{ij}, \ \beta^e = 2(5-\nu)\beta \ \text{(compression)}; \ \beta^e = 6\beta \ \text{(tension)}$$
$$\text{where} \ \beta = \frac{64\pi}{15}\left(\frac{1-\nu}{2-\nu}\right)\frac{N_0}{G} \tag{5.37}$$

where S_{ij} are the stress deviators, $\bar{c}$ is the sound wave speed, ν is Poisson's ratio, and N_0 is the number of flaws. In the deviatoric elastic stress–strain relationship, the shear modulus (G) is degraded using the following expression:

$$\hat{G} = \frac{G}{1 + \delta G \beta^e \bar{c}^3} \tag{5.38}$$

where δ is a material model parameter that was arbitrarily introduced into the above expression. Addessio and Johnson obtained crack growth criteria by considering an energy balance on isolated cracks. The crack growth rate is described by:

$$\dot{\bar{c}} = a\dot{c}_{max} \tanh(d_s) \tag{5.39}$$

where $\dot{c}_{max}$ is assumed to be the shear wave speed, "a" is a factor that will reduce the crack speed, and d_s is a measure of the distance the state of stress exceeds the damage surface. The main model parameters are $\bar{c}$, N_0, τ_0, μ, δ, and a. In addition to these parameters, the model also requires one or two other parameters.

This seemingly simple model was revisited by Addessio and his colleagues [70] in 2006 to describe brittle failure under dynamic loading through stress state-dependent failure envelopes. They developed a rate-dependent, continuum damage mechanics (CDM) based model to describe the dynamic response of brittle materials containing an isotropic distribution of cracks. The damage surface $F(\sigma, \bar{c}) = 0$ takes on one of four different forms, depending on the stress state (the signs and relative magnitudes of the principal stresses). The four forms include tension (F^0), compression (F^c), combined opening and shear (F^{0s}), and pure shear (F^{ps}). When the stress state lies within the surface no damage is developed as in any failure envelope for a brittle solid. When the stress state hits the surface, micro-cracks are activated and for stress states outside the surface, damage accumulates as cracks propagate dynamically.

Zuo et al. [70] derived separate expressions for these four failure surfaces and considered the following three stress states:

$$\sigma_1 \geq \sigma_2 \geq \sigma_3 \geq 0 \ (pure - tension);$$
$$0 \geq \sigma_1 \geq \sigma_2 \geq \sigma_3 \ (pure - compression);$$
$$\sigma_1 > and \ \sigma_3 < 0 \ (tension - compression).$$

The failure envelope shifts according to the value of coefficient of friction (for cracks under compression or pure shear) especially when the principal stresses are compressive.

The new constitutive model decomposes the total strain rate into three parts: crack free matrix strain rate ($\dot{\varepsilon}_m$), an elastic crack (damage) strain rate $\dot{\varepsilon}_c^d$, and an inelastic part of the crack strain rate $\dot{\varepsilon}_c^{gr}$. The constitutive equation is written in tensorial variables as.

$$\dot{\sigma} = (\mathbf{C}_m + \mathbf{D}(\bar{c}))^{-1} (\dot{\varepsilon} - \dot{\varepsilon}_c^{gr}) \tag{5.40}$$

$$\dot{\varepsilon}_c^{gr} = (3\dot{\bar{c}}/\bar{c}) \, \mathbf{D}(\bar{c}) \, \sigma \tag{5.41}$$

$$\dot{\bar{c}} = \dot{c}_{max} \left(1 - \frac{1}{1 + < F(\sigma, \bar{c} >}\right) \tag{5.42}$$

The details of the D and F functions can be found in Ref. [69]. This model also can simulate the effects of strain or stress rates on the axial compressive stress–strain curves (Fig. 5.14) as Costin–Stone model (see Fig. 5.13).

5.2.4.6 Riuo–Cottenot–Boussuge Tensile Damage Model [71]

This model [71] assumes elastic deformation in ceramics below the Huguenot Elastic Limit, HEL and plastic deformation when shock amplitudes exceed HEL. Damage initiation occurs when the principal stress exceeds a threshold stress σ_{th}. Damage initiates in planes perpendicular to principal stress directions when all principal stresses are tensile and exceed this threshold. Therefore, damage can initiate and propagate when $\sigma_i \geq \sigma_{cr}$. The initial value of σ_{cr} is σ_{th} and this threshold stress reduces according to: $\sigma_{cr} = (1 - d_i)\sigma_{th}$. "$d_i$" is the ratio of total crack surface of penny-shaped cracks over the total solid surface, and its expression is:

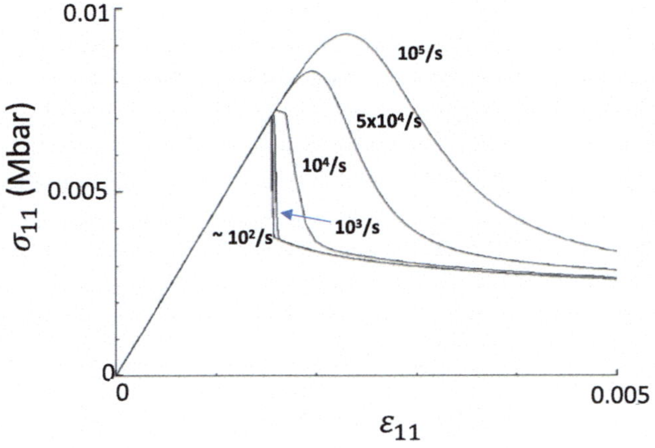

Fig. 5.14 The stress–strain curves for various strain rates under axial compression [70]

$$d_i = N_s \pi \left(\frac{a_i}{2}\right)^2 \tag{5.43}$$

where N_s and the initial micro-crack size ai are model parameters. An evolution law for crack growth can be derived from Eq. (5.43), assuming N_s as a model constant and $\dot{a}_i = V_f$. If N_s changes with volumetric strain (as in the TCK model, see Eq. (5.12)), then the evolution law for crack growth is coupled to the evolution law for the number of cracks. In the Riuo–Cottenot–Boussuge model, crack extension or growth is assumed to be at a constant speed, V_f, and damage increases from 0 to 1. These assumptions impose a bound on the maximum crack size and are valid only when the initiation criterion and the time rate of change of the principal stress are positive. Stresses degrade according to $(1 - d_i)\sigma_e$, where σ_e is the principal elastic stress. A Mohr–Columb law indexMohr–Coulomb model describes the strength of the pulverized ceramic.

This model was applied to simulate a three-dimensional configuration involving the impact of a steel cylinder (20 mm long, 11 mm in diameter) on a silicon carbide beam (20 mm thick). In experiments, stress measurements were taken using a stress gauge embedded between the back surface of the beam and a thin steel plate, and fracture patterns were captured with a high-speed camera. The model was validated through three-dimensional simulations of this configuration, with computational results qualitatively agreeing with observed crack patterns. The model also predicted stress gauge measurements well, though an elastic analysis without damage also reasonably matched the experiments.

5.2.4.7 Espinosa's Multi-plane Model [59]

Espinosa's multi-plane model assumes that micro-cracking can occur on a discrete number of orientations. Nine orientations were selected at intervals of 45° along three mutually perpendicular planes. Inelastic strain is entirely due to the opening/sliding of penny-shaped micro-cracks oriented normal to these nine directions. The average inelastic strains are given by:

$$\varepsilon_{ij}^{c} = \sum_{k=1}^{9} N^{(k)} S^{(k)} \frac{1}{2} \left(\bar{b}_i^{(k)} n_j^{(k)} + n_i^{(k)} \bar{b}_j^{(k)} \right) \tag{5.44}$$

The superscript k represents the orientation, N is the number of flaws per unit volume, S denotes the surface of the micro-crack, $\mathbf{n}$ is the corresponding unit normal, and $\bar{\mathbf{b}}$ is the average displacement jump vector across the surface S. The $\bar{\mathbf{b}}$ has been analytically derived for normal tractions under both tension and compression. The corresponding expressions are:

$$\bar{b}_i^{(k)} = \frac{16(1-\nu^2)}{3E(2-\nu)} a^{(k)} \left(2\sigma_{ij} n_j^{(k)} - \nu \sigma_{jl} n_j^{(k)} n_l^{(k)} n_i^{(k)} \right) \tag{5.45}$$

$$\text{and} \quad \bar{b}_i^{(k)} = \frac{32(1-\nu^2)}{3E(2-\nu)} a^{(k)} f_i^{(k)} \tag{5.46}$$

where $a^{(k)}$ is the crack radius of the penny-shaped micro-cracks on orientation k and $f^{(k)}$ is the effective shear traction vector on orientation k. The micro-crack growth law is very similar to the one that Rajendran [72] and Rajendran and Grove [73], using the stress intensity factor instead of the strain energy release rate. The two crack growth constants, n_1 and n_2; these two constants can take on different values for tension and compression as discussed in the following section.

5.2.4.8 Rajendran–Grove Continuum Damage Model [57, 72, 73]

The evolution of damage in brittle materials under compression is primarily associated with micro-cracking and pore collapse. When subjected to high triaxial stresses (pressure), concrete and rocks tend to fracture through splitting and faulting failure modes. These micro-level mechanisms differ from those observed in metallic materials, as they are not dependent on defect nucleation caused by dislocations or twins. However, in polycrystalline ceramics, the existence of micro plasticity under high pressure and high strain rates has been extensively documented. Micro-structural studies conducted on dynamically deformed specimens or targets offer evidence of the presence of defects, such as dislocations and twins.

The Rajendran–Grove (RG) ceramic failure model formulation decomposes the strain due to elastic, elastic cracking, pore collapse, and plastic flows. The model assumes preexisting randomly distributed penny shaped flaws and presence of pores. The plastic flow and pore collapse occur when subjected to shock above the Hugoniot elastic limit (HEL). The model utilizes the analytical expressions from the Margolin model to describe degradation of elastic moduli in both compression and tension due to micro-cracking. RG model assumes pulverization of the brittle solid at a critical crack density.

The total strain is divided into elastic and plastic components. Elastic strains encompass those within the intact matrix material and strains resulting from crack opening/sliding, calculated by subtracting plastic strains from total strains. To characterize the ceramic material's response to plastic deformation under high-compression pressure, a strain rate-dependent strength relationship is applied. The model employs the conventional radial return approach (refer to Sect. 5.2.3.1), commonly utilized in implementing viscoplasticity models in codes, to determine deviatoric stresses. Stress components are derived by adding the deviatoric stress to the pressure calculated from the equation of state equation. Unloading and reloading paths in the damaged material due to micro-cracking adhere to the degraded elastic modulus (secant modulus), enabling full recovery of strains.

The elastic stress-strain relationship for the damaged aggregate material is expressed as:

$$\dot{\sigma}_{ij} = M_{ijkl}\left(\dot{\varepsilon}_{kl} - \dot{\varepsilon}^{p}_{kl}\right) \tag{5.47}$$

where $\dot{\sigma}_{ij}$ is the total stress rate, $\dot{\varepsilon}_{kl}$ is the total strain rate, $\dot{\varepsilon}^{p}_{kl}$ is the plastic strain rate due to viscoplastic flow and pore collapse, the strain rate difference $(\dot{\varepsilon}_{kl} - \dot{\varepsilon}^{p}_{kl})$ is the elastic strain rate. The components of the stiffness tensor M are determined by analytically inverting the compliance tensor. To describe the elastic strain due to crack sliding, the RG model used the Margolin's derivation [63] for the compliance tensor, C_{ijkl}, which is the inverse of M_{ijkl} (Sect. 5.2.4.1).

During tensile loading, the RG model assumes rapid degradation of both bulk and shear moduli due to micro-crack opening (Mode I). Based on the damaged stiffness solution proposed by Budiansky and O'Connell [55], expressions for the shear and bulk moduli ($\bar{G}_t$ and $\bar{K}_t$, respectively) under tension are provided.

$$\bar{G}_t = G\left\{1 - \left(\frac{32}{45}\right)\frac{(1-\bar{\nu}_t)(5-\bar{\nu}_t)}{(2-\bar{\nu}_t)}\gamma\right\} \tag{5.48}$$

$$\text{and} \quad \bar{K}_t = \frac{2\bar{G}_t(1+\bar{\nu}_t)}{3(1-2\bar{\nu}_t)} \tag{5.49}$$

$$\text{where} \quad \bar{\nu}_t = \nu\left(1 - \frac{16}{9}\gamma\right) \tag{5.50}$$

The equations above, which are valid only for tension, impose a limit on the micro-crack density γ to 9/16. At this critical value, the ceramic material experiences a complete loss of stiffness ($\bar{K}_t = \bar{G}_t = \bar{\nu}_t = 0$) and subsequently becomes unable to carry any tensile load. Micro-crack growth/extension (indicated by an increase in 'a') leads to an increase in micro-crack density, $\gamma = Na^3$, resulting in the relaxation of stresses in the cracked ceramic material. The crack growth (damage evolution) law, derived from a fracture mechanics-based relationship for a single crack propagating under dynamic loading conditions, is described as follows:

$$\dot{a} = n_1^{\pm} C_R \left[1 - \left(\frac{G_{cr}}{G_I^{\pm}} \right)^{n_2^{\pm}} \right] \tag{5.51}$$

In the above equation, model constants, n_1^+, n_2^+, and n_2^- are arbitrarily assumed to be equal to 1.0, while n_1^- is generally assumed to be equal to 0.1. These four constants are used to limit the micro-crack growth/extension rates. C_R is the Rayleigh wave speed, G_{cr} is the critical strain energy release rate for micro-cracking, and $G_I^{\pm}$ are the applied strain energy release rates. The "+" superscript corresponds to crack opening under tension (Mode I), while the "−" superscript relates to micro-crack extension under compression (Modes II/III). Micro-cracks nucleate and extend when the stress state satisfies a generalized Griffith criterion [20] developed by Margolin [31] and Dienes [74]. For this criterion, G_{cr}, G_I^+, and G_I^- are calculated as:

$$G_{cr} = \frac{K_{IC}^2 (1 - \nu^2)}{E} \tag{5.52}$$

$$G_I^+ = \frac{4(1 - \bar{\nu}^2) a}{\pi \bar{E}} \left[\sigma_{kk}^2 + \frac{2(\sigma_{ik}^2 + \sigma_{jk}^2)}{(2 - \bar{\nu})} \right], \quad i \neq j \neq k \tag{5.53}$$

$$G_I^- = \frac{8(1 - \bar{\nu}^2) a}{\pi \bar{E}(2 - \bar{\nu})} \left[\sqrt{\sigma_{ik}^2 + \sigma_{jk}^2} + \mu \sigma_{kk} \right]^2, \quad i \neq j \neq k \tag{5.54}$$

In the above equations, K_{IC} (a model constant) is the fracture toughness of the "*undamaged*" material; however $\bar{E}$ and $\bar{\nu}$ are the degraded Young's modulus and Poisson's ratio, respectively, of the "*damaged*" *(cracked)* material, and μ is the dynamic friction coefficient. Note that G_{cr} is computed from the undamaged (virgin) material properties. Micro-crack opening occurs when G_I^+ exceeds G_{cr}, or micro-crack extension occurs when G_I^- exceeds G_{cr}.

Under high triaxial tensile stress loading conditions, brittle materials often undergo two distinct damage processes: (1) debonding of hard carbide and oxide particles from the matrix, and (2) non-spherical pore (planar crack) growth at triple point grain boundaries. To account for the effects of damage from these processes, the RG model incorporates a critical stress-based spall criterion, in addition to the Griffith criterion.

This spall criterion assumes initiation and growth of damage when all three principal stresses are tensile, and the maximum principal stress exceeds a critical spall threshold stress, σ_s (a model constant). In this case, the crack growth rate, $\dot{a}$ is assumed to be proportional to the Rayleigh wave speed, C_R. The spall damage criterion is only applied to tensile stress states that fail to satisfy the generalized Griffith criterion for crack opening/extension. Consequently, the micro-crack density γ accumulates continuously due to either micro-cracking or spall damage.

The RG model defines the material to be in a comminuted (pulverized) state when the micro-crack density (γ) exceeds γ_p during compressive loading. Typically, γ_p is set to 0.75, assuming that pulverization occurs when micro-cracks coalesce. Before pulverization, the compressive strength Y of the matrix (void-free) ceramic material is described by the following strain rate-dependent relationship:

$$Y = A(1 + C \log \dot{\varepsilon}^p_{eff}) \tag{5.55}$$

Here, A is the quasi-static maximum strength, C is the strain rate sensitivity parameter, and $\dot{\varepsilon}^p_{eff}$ is the normalized (dimensionless) equivalent plastic strain rate; A and C are model constants. The model assumes that pore collapse may occur during compressive loading above the HEL due to local microscopic plastic flow in the matrix material surrounding the pores. The pore collapse strain components are derived from Gurson's pressure-dependent yield surface [75]:

$$3J_2 - Y_m^2 \left[1 + f^2 - 2f \cosh\left(\frac{3P}{2Y_m}\right)\right] = 0 \tag{5.56}$$

Here, J_2 is the second invariant of the deviatoric stress in the porous aggregate material, Y_m is the effective stress in the matrix (void-free) material, P is the compressive pressure in the porous aggregate material, and f is the porosity or void volume fraction. Note that in the absence of porosity (i.e. when $f = 0$), the equation reduces to the von Mises yield condition. If pore collapse occurs, the effective shear and bulk moduli of the damaged aggregate material are defined using a modified form of Mackenzie's relationship [76] and Johnson [77].

$$G_{eff} = \bar{G}\left[\frac{((1 - f)\left[\frac{6K+12G}{9K+8G}\right] f}{(1 - f_0)\left[\frac{6K+12G}{9K+8G}\right] f_0}\right] \tag{5.57}$$

$$K_{eff} = \bar{K}\left[\frac{(1 - f)\left(1 + \left[\frac{3K}{4G}\right] f_0\right)}{(1 - f)\left(1 + \left[\frac{3K}{4G}\right] f_0\right)}\right] \tag{5.58}$$

where

$$f = 1 - (1 - f_0)e^{-\varepsilon^p_v}$$

In the expressions above, G_{eff} and K_{eff} are the effective shear and bulk moduli (respectively) of the aggregate material, $\bar{G}$ is the degraded shear modulus of the cracked matrix material (either $\bar{G}_c$ or $\bar{G}_t$), $\bar{K}$ is the degraded bulk modulus of the cracked matrix material (either $\bar{K}_c$ or $\bar{K}_t$), f_0 (a model constant) is the initial porosity, $\varepsilon^p_v(= \varepsilon^p_{11} + \varepsilon^p_{22} + \varepsilon^p_{33})$ the plastic volumetric strain due to pore collapse, f is the porosity, and G and K are the shear and bulk moduli (respectively) of the virgin material.

The RG model utilizes the following modified Mie–Gruneisen relationship to compute the pressure P in the aggregate material before pulverization:

$$P = \begin{cases} \left(K_{eff}/K\right)\left(b_1\eta + b_2\eta^2 + b_3\eta^3\right)(1 - 0.5\Gamma\eta) + \Gamma E_s(1+\eta), \eta \geq 0 \\ \left(K_{eff}/K\right)(b_1\eta)(1 - 0.5\Gamma\eta) + \Gamma E_s(1+\eta), \eta < 0 \end{cases} \tag{5.59}$$

Note that the volumetric strain η is given by, "$\eta = \rho/\rho_0$", where ρ and ρ_0 are the current and initial densities, respectively. Γ is the Gruneisen coefficient, b_1, b_2, and b_3 are empirical constants used for a cubic fit to the Hugoniot curve for the virgin material, E_s is the internal energy per initial volume, K is the bulk modulus of the virgin material, and K_{eff} is the effective bulk modulus of the aggregate material (see Eq. 5.58).

As mentioned earlier that when the crack density γ reaches a critical value of 0.75, the model assumes pulverization of the brittle solid under compressive loading. At this stage of the material state, the RG model assumes that pore collapse and crack growth/extension no longer occur (i.e. $\dot{f} = \dot{a} = 0$). Additionally, the model assumes that the pulverized material is unable to carry a tensile load, resulting in $\sigma_{ij} = P = 0$ in tension.

However, in compression, the stresses and pressure are computed using the effective shear and bulk moduli (G_{eff} and K_{eff}) corresponding to the values of γ and f at the time of pulverization. The pressure P and strength Y of the comminuted material are described by:

$$P = \begin{Bmatrix} -K_{eff}\varepsilon_v^e, \varepsilon_v^e < 0 \\ 0, \varepsilon_v^e \geq 0 \end{Bmatrix} \tag{5.60}$$

$$Y = \begin{Bmatrix} \min\left(\beta_p P, Y_p^{\max}\right), P > 0 \\ 0, P \leq 0 \end{Bmatrix} \tag{5.61}$$

In the above equation, β_p is the dynamic friction coefficient for the comminuted material. ε_v^e is the elastic volumetric strain, and Y_p^{max} is the model constant that sets the upper limit on the compressive strength of the pulverized ceramic material. Since experimental data for the fractured strength is generally either unavailable or difficult to determine, β_p is usually set to "1" and Y_p^{max} is calibrated to match the measured penetration depths from projectile penetration experiments. Seven RG ceramic model constants require calibration with experimental data: strain rate sensitivity parameter (C), initial crack size (a_0), micro-crack number density (N_0^*), dynamic friction coefficient (μ), coefficient to limit the mode II/III crack extension rate (n_1^-), critical spall stress (σ_s), and the maximum compressive strength of the pulverized material (Y_p^{max}). We employed the following set of model constants to describe the dynamic response of AD995 ceramic in this study: $\rho_0 = 3.89\,\text{g/cm}^3$, $G = 156\,\text{GPa}$, $K = 231\,\text{GPa}$, $b_1 = 231\,\text{GPa}$, $b_2 = -160\,\text{GPa}$, $b_3 = 2774\,\text{GPa}$, $\Gamma = 2.3$, $A = 2.3\,\text{GPa}$, $C = 0.2$, $K_{IC} = 3\,\text{MPa}\sqrt{\text{m}}$, $f_0 = 0.0$, $a_0 = 1.5 \times 10^{-6}$, m, $N_0^* = 2.0 \times 10^{+11}$, m^{-3}, $\mu = 0.60$, $n_1^- = 0.1$, $\sigma_s = 0.5\,\text{GPa}$, $\gamma_p = 0.75$, $\beta_p = 1.0$, and $Y_p^{max} = 4.5\,\text{GPa}$.

The RG model was implemented into the later versions of the EPIC finite element code [56], and simulations of these impact configurations were performed. The EPIC is a well-established three-dimensional hydrocode initially developed in the early 1970s to characterize the response of solid materials to dynamic impact loading. To maintain the stability of the explicit finite element solution, an iterative scheme based on a second-order diagonally implicit Runge–Kutta method [78], in conjunction with the conventional radial return correction, was employed in the RG model solution algorithm. The importance of a robust "*sub-cycling time step*" based numerical implementation of constitutive and damage models is discussed in Chap. 6 on Computer Modelling in this book.

Over the years, this code has undergone continuous development, integrating various advanced algorithms specifically designed to model projectile penetration into armour plates. Chapter 6 provides illustrative examples of the most complex impact failure problems successfully solved using this explicit Lagrangian finite element code. The RG model has been implemented in the 3-Dimensional version of the EPIC Lagrangian code, CTH—Eulerian code [79], and in the ABAQUS—Lagrangian code [80].

Rajendran and Grove [73] further verified the generality of the RG model constants for 99.5% pure aluminium oxide (AD995) through computer simulations of the following four impact configurations:

- Plate Impact
- Rod-on-Rod Impact
- Graded-Density-Plate-on-Rod Impact
- Projectile Penetration.

Details of these experimental configurations are provided in Chap. 4 of this book. Initial values for the model parameters were obtained from the high strain rate SHPB tests and through the calibration of VISAR or Stress gauge data from the plate impact experimental data.

In the following section, simulations using the EPIC code are presented to demonstrate the capabilities of the RG model, as applied to four impact experimental configurations

Planar Plate Impact

Rajendran and Grove initially calibrated the model constants using EPIC's one-dimensional (1D) strain option through simulations of two AD995 plate impact tests:

Low Velocity Test: This test involved a flyer thickness of 4 mm, a target thickness of 8 mm, and an impact velocity of 83 m/s, conducted by Dandekar and Bartkowski [45]. In this "*below HEL*" test, the stress in the window (or backup plate) was measured using stress gauges.

High Velocity Test: Grady and Moody [44] reported a test with a flyer thickness of 5 mm, a target thickness of 10 mm, and an impact velocity of 1943 m/s. The free surface particle velocity in this "*above HEL*" test was measured using the VISAR.

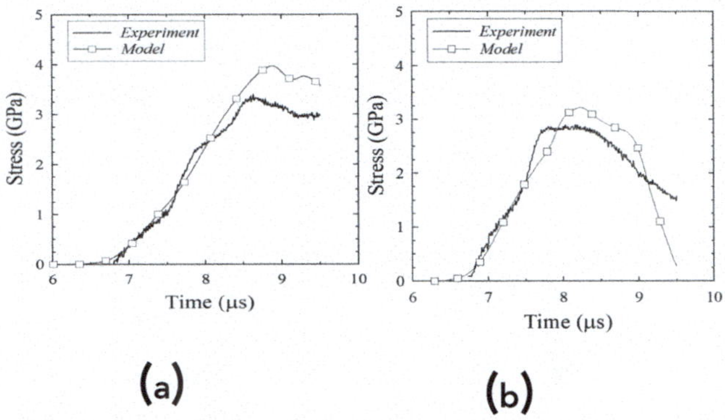

Fig. 5.15 Comparison of computational results with rod-on-rod impact data

This was reported in Chap. 4 (Fig. 4.25) and it illustrates that the computed spall signals (profiles beyond point S) align with the measured profiles. Spall strength below HEL is non-zero as evident from the presence of a spall signal in Fig. 4.25a, whereas the spall strength is zero above the HEL as depicted in Fig. 4.25b. For an interpretation of spall strength in ceramics or any other brittle solids above and below the HEL stress, readers can refer to Chap. 4.

Rod-on-Rod Impact

EPIC's two-dimensional (2D) axisymmetric geometry option was utilized to simulate the rod-on-rod impact configuration at two different impact velocities: 175 and 278 m/s. The AD995 striker rod was 50 mm long and 12.5 mm in diameter with an L/D aspect ratio of 4, while the AD995 target rod was 100 mm long and 12.5 mm in diameter ($L/D = 8$). Rajendran [81] compared the model-predicted stress histories with experimental measurements in Fig. 5.15. Although the model does not precisely match the data, the peak stress levels closely approximate the measurements.

Graded-Density Plate-on-Rod Impact

Two-dimensional axisymmetric simulations were conducted for the graded-density plate-on-rod impact configuration, considering both unsleeved (bare) and sleeved AD995 ceramic target rods. The impact velocities for both cases were approximately 300 m/s. The flyer plate was represented as a layered circular disk with a diameter of 50 mm and a thickness of 22 mm. A continuous finite element grid was employed for the graded flyer plate to simulate a "*perfect*" bond between adjacent layers of material.

The target was modelled as a solid rod with a diameter of 19 mm and a length of 74 mm. In contrast, the steel sleeve was modelled as a hollow rod with an inner diameter of 19 mm, an outer diameter of 38 mm, and a length of 74 mm. Frictionless sliding was allowed between the inner surface of the sleeve and the outer surface of the ceramic rod. Figure 5.16 presents a comparison between the model-predicted

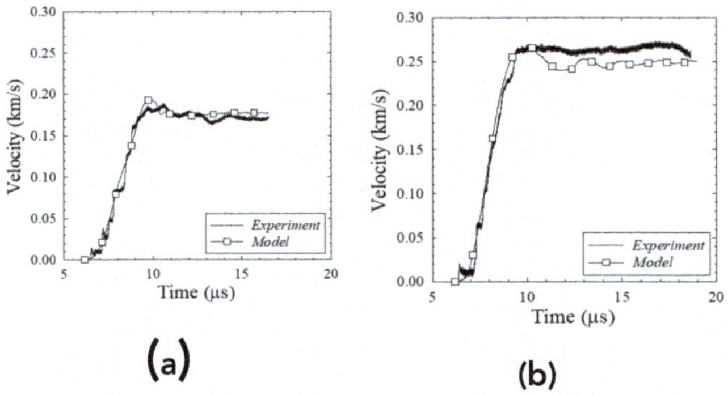

Fig. 5.16 Comparison of computational results with graded-density plate-on-rod impact data. **a** impact velocity = 300 m/s; **b** impact velocity = 321 m/s

velocity histories and those measured experimentally by Chhabildas et al. [82]. The figure demonstrates the model's excellent performance in predicting both the peak velocity levels and the constant-velocity behaviour, attributed to the spallation of the ceramic rod near the free end.

Projectile Penetration

In the two-dimensional axisymmetric simulations of the projectile penetration configuration, the target was modelled as an AD995 ceramic disk with a diameter of 152.4 mm. This ceramic disk was backed by a thick steel cylinder with a diameter of 203.2 mm and a thickness of 127 mm. Additionally, the ceramic disk was radially confined by a steel ring fixed to the surface of the steel cylinder. The steel ring had an inner diameter of 152.4 mm, an outer diameter of 203.2 mm, and a thickness of 50.8 mm. The projectile, in this case, was represented as a tungsten rod with a diameter of 7.87 mm and a length of 78.7 mm, impacting the target with a velocity of 1500 m/s.

Simulations of this configuration were performed for seven different thicknesses (ranging from 10.2 to 50.8 cm) of AD995 ceramic disks. The penetration process was modelled through EPIC's erosion algorithm (using an erosion strain of 150%). Figure 5.17 compares the measured [83] and computed residual depths of penetration (DOP) into the backup steel block versus the areal densities (mass per unit area) of the ceramic disks. The straight line in this figure represents a linear least-squares fit to the experimental data, indicating consistency between the model-predicted depths of penetration and the experimental measurements.

The RG ceramic failure model's governing equations and a set of model constants for AD995 ceramic were described in detail by Grove and Rajendran [84]. These constants were used in finite element simulations for a benchmark suite of experimental impact configurations, demonstrating the model's generality and predictive capability across a wide range of impact loading conditions.

For completeness, Table 5.6 summarizes the model features and key properties of various ceramic materials. These properties, experimentally measured and averaged

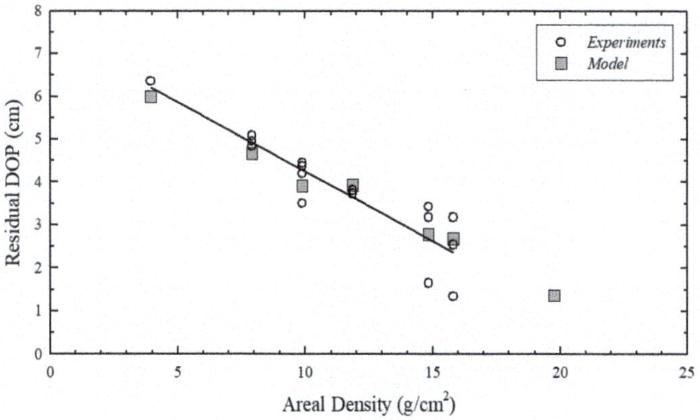

Fig. 5.17 Comparison of computational results with projectile penetration data

over several plate impact test data, provide insight into the shock properties of ceramic materials and their dependence on factors such as density, microstructures, porosity, and ceramic processing techniques.

Most of the data in Table 5.6 came from [85]. It should be noted that the ceramic material shock properties depend on several factors, such as density, microstructures, porosity, ceramic processing (hot sintering vs. cold sintering), grain morphology, etc. However, the properties in this table are experimentally measured and averaged over several plate impact test data. Also in the above table, the data on A luminous Nitride comes from [86, 87], and the single edge V-notch beam specimen test data (column corresponding to Silicon carbide) are taken from [88]. For the case of Zirconium dioxide (the last row in the above table), dental ceramic has a value of 5 $\text{MPa}\sqrt{\text{m}}$. Other Zirconium Oxides show values around 2 $\text{MPa}\sqrt{\text{m}}$.

For completeness, the model features are summarized in Table 5.7. The key aspects of the RG model include its ability to describe:

(a) Three-dimensional constitutive equations
(b) Pore collapse or expansion
(c) Micro plasticity based permanent deformation
(d) Elastic strain of the intact solid
(e) Elastic strain of the cracked solid under tension, compression, and shear
(f) Strain rate dependent or dynamic crack growth
(g) Scalar damage variable (crack density) evolution
(h) Damage nucleation, growth, and coalescence or pulverization
(i) Degradation of strength and stiffness under all modes.

Table 5.6 Ceramic properties

Material	Density (g/cm^3)	Poisson's Ratio	K (GPa)	G (GPa)	Wave Speed CL(km/sec)	Fracture Toughness, KIC (MPa m1/2)	HEL (GPa)
Tungsten Carbide WC	14.9	0.20–0.22	315	217	6.4	8.5	4–6
Silicon nitride, Si_3N	3.2	0.26	216	123	10.9	6.8	~ 9.24
Aluminium oxide Al_2O_3	3.9	0.23–0.25	253	152	10.7	4.5	~10.8
Titanium diboride TiB_2	4.52	0.05–0.11	246	259	11.3	5.0	14–17
Aluminium nitride AlN	3.25	0.24	201–204	131	10.79	3.1 [86, 87]	~9.66
Boron carbide B_4C	2.51	0.16–0.17	230–233	200	14.08	2.5–3.2	~18.2
Silicon carbide SiC	3.15	0.16–0.18	213–226	180–191	12.2	4.9* [88]	15–19
Silicon dioxide SiO_2	2.63	0.08	36	42	5.9	0.58–0.78	7–9
Zirconium dioxide ZrO_2	6.03	0.30–0.31	176–194	83	7.0	2.0–5.0@	13–16

Most of the data in this table came from: Grady, D., "Dynamic Properties of Ceramic Materials," Sandia Report: SAND94-3266 • UC-704 Unlimited Release, February 1996 and Grady, D., "Fundamentals and dynamic failure," Chapter 1 in "Physics of Shock and Impact," Copyright © IOP Publishing Ltd 2017, Online ISBN: 978-0-7503-1254-7. It should be noted that the ceramic material shock properties depend on several factors, such as density, microstructures, porosity, ceramic processing (hot sintering vs. cold sintering), grain morphology, etc. However, the properties in this table are experimentally measured and averaged over several plate impact test data. *Single edge V-notch beam specimen test data. The fracture toughness values varied widely between 2.5 to 6.4 for different grades of Silicon carbides [88]. @Zirconium oxide as a dental ceramic has a value of $5MPa\sqrt{m}$. Other zirconium oxides show values around $2MPa\sqrt{m}$

5.2.4.9 Bhat–Rosakis–Sammis (BRS) Wing Crack Model [89]

The BRS model [89], a recently developed micro-mechanical model, exemplifies mathematical complexity attributed to its three-dimensional and nonplanar crack geometry, specifically the wing crack configuration. Constructing constitutive (stress–strain) relationships necessitates rigorous mathematical or analytical solutions, involving solving boundary value problems through an iterative numerical approach.

Horii and Nemat-Nasser [14] initially modelled the inelastic strains in brittle materials caused by the nucleation and growth of "*wing cracks*", emanating from pre-existing flaws, under two-dimensional stress/strain constitutive equations. Their work demonstrated the model's efficacy in simulating the effects of lateral confinement on failure modes, such as the splitting and faulting mode.

Table 5.7 Specific features of the Rajendran and Grove model [57, 72, 73, 84]

Elastic-Viscoplastic-Microcraking Damage Moduli

- Continuum damage using Helmholtz or Gibbs Free Energy
- Both tension and compression
- Three-dimensional stress state; Strain rate and pressure dependent strength model
- Isotropic

Definition of Damage

- Scalar
- Elements of the moduli are degraded by a scalar damage parameter, γ

Micro-Crack Distribution

- No explicit formulation; however, a distribution function is used in Margolin's formulation for degraded moduli

Damage Growth Law

- An empirical growth law based on dynamic crack propagation equation:

$\dot{a} = n_1^{\pm} C_R \left[1 - \left(\frac{G_{cr}}{G_I}^{\pm}\right)\right]$; $\gamma = Na^3$, N is number of microcracks; a is current crack radius. A comminuted (pulverized) state is assumed when the micro-crack density γ exceeds γ_p (a model constant) during compressive loading. The tensile strength becomes zero

Failure Criterion

- The moduli go to zero as damage reaches its critical value

Other Mechanisms

- Porous yield function to describe rate dependent pore collapse
- Mohr–Coulomb strength model for comminuted materials under compression loading

While the Horii and Nemat-Nasser approach aptly captures the effects of lateral compressive stress on constitutive response under quasi-static loading, addressing rate effects during dynamic crack growth necessitates calculating the dynamic stress intensity factor for predicting the failure of brittle solids under various loading rates. Bhat et al. [89] expanded upon this by developing a three-dimensional constitutive model, known as the BRS model, based on the "*wing crack*" mechanism.

In 2017, Kari Kolari [90] further contributed to this field by presenting a three-dimensional continuum model centred on the concept of "*wing crack growth*" in confined granular brittle materials.

The schematic in Fig. 5.18a illustrates potential pre-existing flaws assumed to be penny-shaped micro-cracks situated between grain boundaries. When a brittle material is subjected to axial and lateral compression, the sliding of penny-shaped cracks (see Fig. 5.18b generates wedge-like forces on the flaw surfaces. This fundamental failure process in brittle solids is extended to a generalized three-dimensional solution for the wing crack geometry. Wing cracks are assumed to emanate from the tips of pre-existing flaws at some angle dependent on the confinement stress levels. Bhat et al. [89] applied this model to describe the behaviour of a high-quality white marble known as Dionysus Pentelicon marble (or rock). Stress–strain curves for various strain rates were determined using their model for this marble, and the calculated peak stress successfully matched experimental measurements.

The BRS constitutive model serves as an extension of models developed and reported by Ashby and Sammis [91] and later by Deshpande and Evans [92]. The governing and salient equations describing the growth of wing cracks are derived using dynamic fracture mechanics and the Gibbs free energy function under isothermal conditions. Dilatancy resulting from wing crack opening and shear deformation generates inelastic strains, leading to a reduction in bulk and shear modulus. The BRS model captures the micro-mechanics of fracture through the evolution of stress intensity factors of wing cracks and a crack density-based damage parameter, D.

Damage: As described earlier in Sect. 5.2.4, the expression for the scalar damage follows, $\gamma = \beta N a^3$ where N is the fixed number of preexisting flaws, per unit volume, a is the flaw radius, and β is a geometry parameter. The BRS model assumes the initial crack density related to damage as, $D_0 = \frac{4}{3}\pi N(\alpha a)^3$. The projection ($\alpha$) of the flaw "$\alpha a$" in the direction of the maximum principal compression stress is the effective size of the flaw that contributes to damage. The modelling of wing crack-like geometry involves two length parameters: the pre-existing flaw size (a) and the length of the emanating wing cracks (l). The BRS model considers an effective "$overall$" length of a representative flaw/crack system as the sum of "$l + \alpha a$", where l is the length of wing crack, and $2a$ is the size of flaw, as shown in Fig. 5.18.

Since the degradation of elastic tensor (or elastic stiffness) is due to damage evolution (D), the constitutive response (stress vs strain curves) of brittle materials is strongly influenced by D. The rate form for the damage is given by,

$$\frac{dD}{dt} = \left(\frac{3D^{2/3}D_0^{1/3}}{\alpha a}\right)\frac{dl}{dt} \tag{5.62}$$

Since the flaw size a is assumed to be constant, the evolution of damage is then described by the extension of the wing crack, in other words, the current length of the

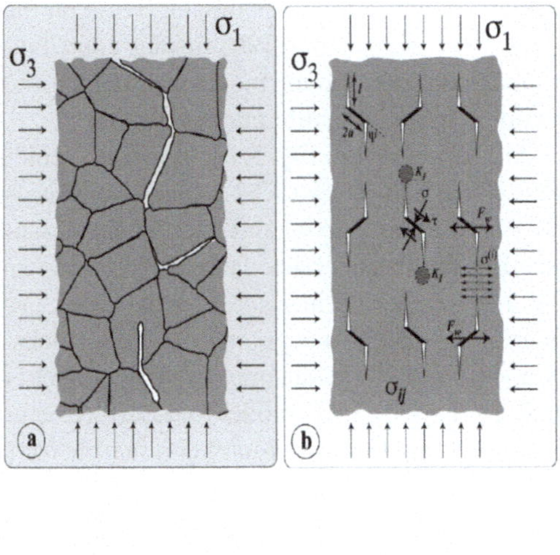

Fig. 5.18 The schematic of penny-shaped flaws and wing cracks geometry in the micro-mechanical damage mechanics model by Ashby and Sammis [91]

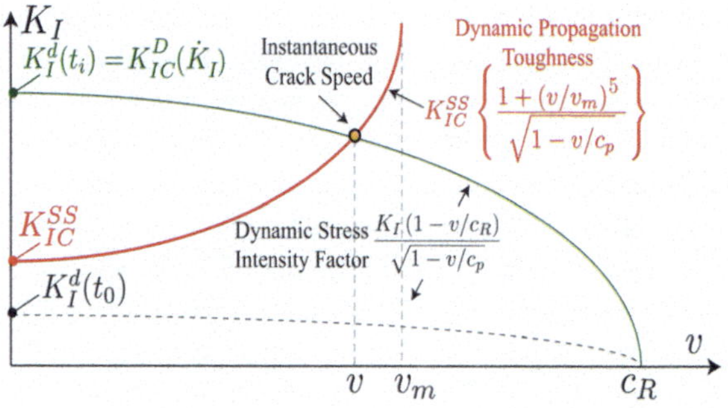

Fig. 5.19 Graphical representation of the solution to Eq. (5.63)

l. An implicit evolution law for l or the crack speed, dl/dt is given by the following equation:

$$\frac{K_I\,(1 - v/c_R)}{\sqrt{1 - v/c_p}} = K_{IC}^{SS}\left\{\frac{1 + (v/v_m)^5}{\sqrt{1 - v/c_p}}\right\} \tag{5.63}$$

where, $v(= dl/dt)$ is the crack speed, c_R and c_p are *Rayleigh* and *Dilatational* wave speeds, respectively, v_m is the limiting(or branching) speed, K_I and is the quasi-static stress intensity factor, and K_{IC}^{SS} is the quasi-static limit of the dynamic initiation toughness. While the parameters, v_m, c_R, c_p, and K_{IC}^{SS} are material model parameters or constants, v and K_I are applied loading controlled state variables. Additionally, note that the expression on the left-hand side represents the dynamic fracture toughness, K_I^d. The BRS model provides appropriate equations that are required to calculate the stress intensity factor, K_I. For a given K_I, the crack speed v is determined from the above expression. We can then calculate the dynamic stress intensity factor, K_I^d using the values for v and K_I.

The graphical solution of Eq. (5.63) to determine the current value for the wing crack (l) is shown in Fig. 5.19.

This fracture mechanics-driven model necessitates governing equations for stress intensity factors that approximately capture the crack tip conditions of both the pre-existing penny-shaped flaws and the emanating wing cracks. The wing cracks are assumed to coalesce when D approaches 1, leading to the stress intensity parameter (K_I) becoming unbounded.

The constitutive equations are derived from the Gibbs free energy function for cracked (or damaged) solids. The change in the free energy (dW) due to the generation of new surfaces, in other words, due to local incremental extensions of micro-cracks is described by the energy function given below.

$$dW = \int_{\Gamma} \left\{\left[G(\sigma, S) - 2\gamma_s\right] ds\right\} d\Gamma$$

where, G is the elastic energy release rate, S represents the current damaged state of the material, γ_s is the surface energy, Γ is the locus of all crack fronts in the damaged solid. The derivatives of this energy integral with respect to stress will provide an expression for the inelastic strains due to micro-crack extension. An analytical expression, that is based on linear elastic fracture mechanics, is readily available for the Gibbs energy function in terms of stress intensity factors. The BRS model provides all salient constitutive equations for three regimes: (I) isotropic linear elastic solid response, (II) inelastic (crack opening) response due to wedging forces generated due to the frictional sliding of the pre-existing flaws under globally compressive loading, and (III) inelastic response (strains) due to crack opening under global tensile loading. For these three loading conditions, constitutive equations are derived using the Gibbs energy function, given by

$$W(\sigma, S) = W^e(\sigma) + \frac{1-\nu^2}{E}\int_{\Gamma}\left[K_I^2(\sigma, S) + K_{II}^2(\sigma, S) + \frac{K_{III}^2(\sigma, S)}{(1-\upsilon)}\right] ds\, d\Gamma \tag{5.64}$$

where "K' are the stress intensity factors for different modes of crack propagation, The BKS model includes expressions for the three stress intensity factors, σ is the stress tensor, υ is Poisson's ratio, and **E** is Young's modulus. After manipulating and approximating all salient equations for the stress intensity factors, the stress–strain equations for these three loading regimes are derived as,

Regime I

$$\sigma_{ij}(\varepsilon, D) = 2\mu\varepsilon_{ij} - \frac{2\mu v}{(1-2v)}\varepsilon\delta_{ij} \tag{5.65}$$

Note that $\varepsilon = \varepsilon_{11} + \varepsilon_{22} + \varepsilon_{33}$ and μ is the shear modulus.

Regime II

$$\begin{aligned}\sigma_{ij}(\varepsilon, D) = 2\mu &\left\{\left(1 + \frac{2\bar{A}_1\bar{B}_1\varepsilon}{\gamma} + 2\bar{B}_1^2\right)\varepsilon_{ij}\right) \\ &+\left(\frac{v}{(1-2v)} - \frac{2\bar{A}_1\bar{B}_1\varepsilon}{3\gamma} - \frac{2\bar{B}_1^2}{3} + \bar{A}_1^2\right)\varepsilon\delta_{ij} + (\bar{A}_1\bar{B}_1)\gamma\delta_{ij}\right\}\end{aligned} \tag{5.66}$$

where

$$\gamma = \sqrt{2e_{ij}e_{ij}}\ and\ e_{ij} = e_{ij} - e_v\delta_{ij}/3, \text{ where } e_v = \text{volumetric strain} \tag{5.67}$$

$$\bar{A}_1(D) = \frac{1}{2}(A_1a_1 + B_1b_1); \tag{5.68}$$

$$\bar{B}_1(D) = \frac{1}{2}(A_1a_1 + B_1b_2) \tag{5.69}$$

$$a_1 = \frac{1}{\Gamma}\left(1 + \frac{B_1^2}{2}\right); b_1 = -\frac{1}{\Gamma}\left(\frac{A_1B_1}{2}\right); b_2 = \frac{1}{\Gamma}\left[\frac{A_1^2}{2} + \frac{3(1-2\upsilon)}{2(1+\upsilon)}\right] \tag{5.70}$$

$$\Gamma = \left[\frac{3(1-2\upsilon)}{2(1+\upsilon)} + \frac{3(1-2\upsilon)B_1^2}{4(1+\upsilon)} + \frac{A_1^2}{2}\right] \tag{5.71}$$

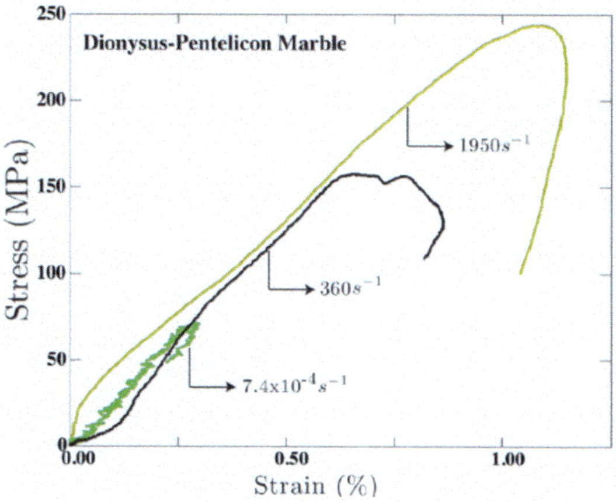

Fig. 5.20 The experimental stress–strain curves for the Marble at three different strain rates

Regime III

$$\sigma_{ij}(e_v, D) = 2\mu\left(1 + 2\bar{E}_1^2\right)\varepsilon_{ij} + \left(\frac{\upsilon}{(1-2\upsilon)} - \frac{2\bar{E}_1^2}{3} + \bar{C}_1^2\right) e_v\delta_{ij} \tag{5.72}$$

where

$$\bar{C}_1\,(D) = \frac{C_1}{\frac{3(1-2\upsilon)}{(1+\upsilon)} + C_1^2}\ ;\ \bar{E}_1\,(D) = \frac{E_1}{2 + E_1^2}; \tag{5.73}$$

$$C_1 = C\sqrt{\frac{\pi D_0}{\alpha^3(1+\nu)}},\quad E_1 = E\sqrt{\frac{\pi D_0}{\alpha^3(1+\nu)}} \tag{5.74}$$

In most micro-mechanical models, the constitutive equations for three-dimensional wing cracks incorporate the crack density-based effective scalar damage, denoted as D. This damage parameter is derived through the iterative numerical solution of the coupled Eqs. (5.62) and (5.63). The governing equations were solved to generate stress–strain curves for different loading rates under uniaxial compressive stress, and these results were then compared with experimental data for Dionysus Pentelicon marble. Quasi-static data were acquired using a conventional load frame apparatus, while the high strain rate response was determined through the split Hopkinson pressure bar technique.

Figure 5.20 shows the experimental stress–strain curves for marble at three distinct strain rates. The stress or strength increases significantly from approximately 70 MPa at a quasi-static strain rate to about 240 MPa at a strain rate of 1950/s, highlighting substantial strain rate effects on the compressive strength of marble. To simulate

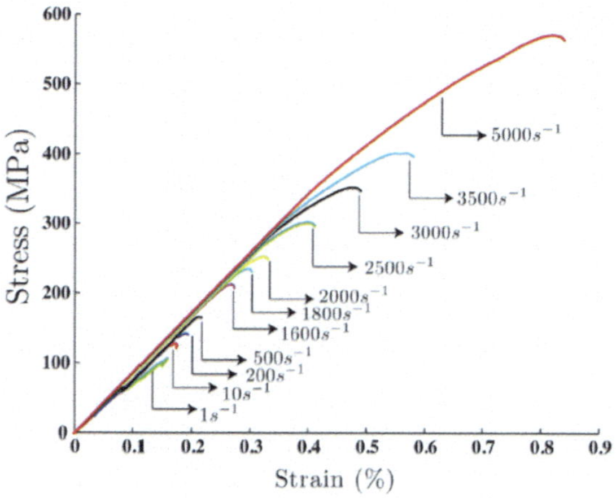

Fig. 5.21 Numerically obtained stress–strain curves at different strain rates for Dionysus Pentelicon marble based on the BRS constitutive and damage models

these effects, the three-dimensional wing crack-based BRS constitutive equations were integrated into the commercial finite element code ABAQUS-Explicit.

The damage evolution (Eq. (5.62)) is calculated through a 5th order Runge–Kutta scheme with adaptive step-size control reported by Press et al. [93]. Note that the solution of Eq. (5.63) provides a value for the instantaneous wing crack tip speed, v at any given time step. Using this value in Eq. (5.62), the current damage rate is obtained for further use in the constitutive equations for different regimes. The crack-tip speed-dependent equations capture strain rate effects on fracture toughness and compressive strength (or stress–strain curves). The BRS model has a few material parameters that were set to: initial penny crack ($a = 1$ mm), friction coefficient between the faces of the penny crack ($f = 0.7$), initial damage ($D_0 = 0.1$), and the crack branching speed as ($v_m = 0.35 \times C_R$). Numerical simulations were performed using the ABAQUS code to demonstrate BRS model capabilities to show strain rate effects on the maximum strength under axial compressive loading conditions.

Figure 5.21 displays numerically obtained stress–strain curves at different strain rates for Dionysus Pentelicon marble based on the BRS constitutive and damage models. The simulated curves reveal a pronounced strain rate sensitivity of marble and other brittle solids where fracture results from wing crack nucleation and propagation under both quasi-static and dynamic loading conditions. Given that the model considers both pressure and strain rate effects on fracture behaviours, it is applicable for predicting both splitting and faulting types of global failure modes often observed in brittle solids. This model was employed in simulating dynamic earthquake rupture propagating along an interface that involves damaged rock. Table 5.8 summarizes the salient features of this advanced and mathematically intense micro-mechanical model.

Table 5.8 Specific features of the Bhat–Rosakis–Sammis model [89]

Elastic-Microcraking Damage Moduli
• Continuum damage using Gibbs Free Energy
• Both tension and compression (Mode I and Mode II dominant fracture)
• Three-dimensional stress state; Strain rate and pressure dependent strength model; but results focus only on the strain rate effects
• Isotropic elastic solid containing penny shaped cracks
Definition of Damage
• Scalar
• Compliance tensor elements are degraded by a scalar damage parameter D, which is based on crack volume density
Micro-Crack Distribution
• No explicit formulation; Nucleation and growth of wing cracks
Damage Growth Law
• Opening of both wing and penny shaped micro-cracks. The crack growth rate is given by: $\frac{dD}{dt} = \left(\frac{3\ D^{2/3} D_0^{1/3}}{\alpha\ a}\right) v$; where, $v = \frac{dl}{dt}$; $D = \frac{4}{3}\pi N(l + \alpha a)^3$; $D = D_0$ prior to wing nucleation or when l is zero. N is number of existing penny shape cracksin a unit volume with a radius, a and l is the wing crack length. Initially l is zero until nucleation due to axial compressive loading
$\frac{K_I(1-v/c_R)}{\sqrt{1-v/c_p}} = K_{IC}^{SS}\left\{\frac{1+(v/v_m)^5}{\sqrt{1-v/c_p}}\right\}$
This nonlinear equation is solved to obtain the crack speed v
Note that the stress intensity factor, K_I is a function of applied stresses, and K_{IC}^{SS} is the quasi-static initiation toughness (material property)
v_m is the limiting crack speed or branching speed of v
Failure Criterion
• As D approaches 1, K_I becomes unbounded representing the limit when neighbouring wing cracks coalesce
Other Mechanisms
• No pore collapse or micro-plasticity are considered for the rock material considered in the modelling

5.2.4.10 Hyperelastic Model [94]

In this section, we present a hyperelastic model based on continuum damage mechanics (CDM) [94], designed to capture the response of materials exhibiting both brittle and elastometric (rubber-like) behaviour under three-dimensional loading conditions. This hyperelastic model is particularly applicable to fibre-reinforced plastics (FRP), with a specific focus on laminates composed of layered fabrics. Readers are encouraged to consult Chap. 2, where a comprehensive collection of hyperelastic constitutive equations is presented. This chapter offers an extensive overview of various forms available in textbooks and open literature, providing valuable con-

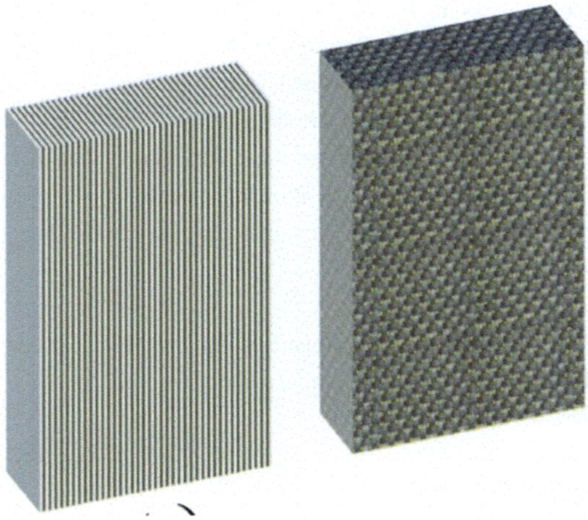

Fig. 5.22 Schematic of layered unidirectional and woven GRP composite. Unidirectional

text and additional resources for those interested in delving deeper into hyperelastic modelling.

While we do not aim to encompass all types of layered composite systems, our emphasis is on laminates constructed from layered fabrics. These fabrics consist of fibres bound together by a matrix material, such as epoxy or polyester resin. The architectural arrangement of the fabric is influenced by layering angles, including 0°/90°, both in woven and unidirectional laminates, illustrated in Fig. 5.22 (Brazil et al. [95]). It's noteworthy that the matrix material often exhibits brittle behaviour, while the polymer fibres display a hyperelastic response characterized by substantial stretching under tensile loading, especially along the fibre directions.

The CDM approach is relatively straightforward, involving the linear degradation of bulk and shear moduli in a generalized Hooke's law through a damage parameter, D. An illustrative example of this approach is the modified TCK model proposed by Rajendran and Kroupa [37], as expressed in Eq. (5.75):

$$\sigma_{ij} = 3K(1 - D)e_{kk}\delta_{ij} + 2G(1 - D)e_{ij} \tag{5.75}$$

The damage evolution in the TCK model, detailed in Sect. 5.2.3.4, relies on a strain rate and fracture toughness-dependent fragmentation model developed by Kipp and Grady [51]. It is important to note that this model is applicable only for "*small strains*" and it serves well in describing the matrix brittle response. However, the total stretch of the Fibre Reinforced Plastic (FRP) also encompasses the hyperelastic response of the fibre systems.

In the case of Glass Fiber Reinforced Plastics (GRP), Scott et al. [96] recently employed a strain-based damage model developed by Barham et al. [94] to simulate shock wave propagation. This hyperelastic constitutive model, derived from a Helmholtz Free Energy (HFE) function for cross-fibre composites by Holzapfel [97], incorporates damage evolution using the continuum damage mechanics approach, akin to the TCK model. This hyperelastic damage model is designed to account for energy partitioning in different fibre orientations and the isotropic polyester (or epoxy) matrix.

The HFE function, ψ, is taken to be of the form,

$$\psi = \frac{G}{2}\left(\bar{I}_1 - 3\right) + \frac{K}{\beta^2}\left(\beta lnJ + J^{-\beta} - 1\right) + \frac{k_1}{2}(I_4 - 1)^2$$
$$\frac{k_2}{2}(I_6 - 1)^2 \tag{5.76}$$

where, $\bar{I}_1 = I_1\ J^{-2/3}$ is the volume preserving part of the first invariant of the Right Cauchy Green tensor, $\mathbf{C}$ and $I_1 = tr(\mathbf{C})$. $J = det\mathbf{F} = (det\mathbf{C})^{1/2}$ where $\mathbf{F}$ is the deformation tensor. $\mathbf{C}$ is defined by $\mathbf{C} = \mathbf{F}^{\mathbf{T}}\mathbf{F}$ for an isotropic material such as the polyester matrix. More descriptions for these tensors can be found in Chap. 2.

The second term on the right-hand side of the equation signifies the Ogden model for compressible hyperelasticity, addressing volume changes within the material. In this term, K represents the bulk modulus of the matrix material, β is the Ogden parameter, and J is the Jacobian of the deformation gradient tensor.

The third and fourth terms on the right-hand side account for fibre deformation under tension or compression and denote the free energy of the fibres in two orthogonal directions. The fourth invariant ($I_4 = \mathbf{M_1}.\mathbf{CM_1}$) and sixth invariant ($I_6 = \mathbf{M_2}.\mathbf{CM_2}$)) are directionally dependent strain (or stretch) terms, where M_1 and M_2 represent the reference fibre directions. The parameters k_1 and k_2 correspond to the moduli of the two fibre systems. In the case where the Glass Fibre Reinforced Plastic (GRP) has only one fibre system (e.g. S-2 glass), $k_1 = k_2$. This distinction helps tailor the model to the specific fibre architecture of the material under consideration.

The resulting stress–strain relationship is derived by differentiating the HFE function ψ, with respect to the strains, B_{ij}; the 2nd Piola-Kirchhoff (**S**) and Cauchy (σ) stresses are defined by,

$$\mathbf{S} = 2\frac{\partial\psi}{\partial\mathbf{C}};\quad \sigma = \frac{1}{J}\,\mathbf{FSF}^T \tag{5.77}$$

The following constitutive equation is written in terms of the Cauchy stress, σ.

$$\sigma = \frac{G}{J}\left(\bar{\mathbf{B}} - \frac{tr\left(\bar{\mathbf{B}}\right)}{3}\mathbf{I}\right) + \frac{K}{J\beta}\left(1 - J^{-\beta}\right)\mathbf{I} +$$
$$\frac{k_1}{J}\left(\mathbf{m_1}.\mathbf{m_1} - 1\right)\mathbf{m_1}\bigotimes\mathbf{m_1} + \frac{k_2}{J}\left(\mathbf{m_2}.\mathbf{m_2} - 1\right)\mathbf{m_2}\bigotimes\mathbf{m_2} +$$
$$\frac{k_s}{J}\left(\mathbf{m_{s1}}.\mathbf{m_{s1}} - 1\right)\mathbf{m_{s1}}\bigotimes\mathbf{m_{s1}} + \frac{k_s}{J}\left(\mathbf{m_{s2}}.\mathbf{m_{s2}} - 1\right)\mathbf{m_{s2}}\bigotimes\mathbf{m_{s2}} \tag{5.78}$$

where $\mathbf{m_1} = \mathbf{FM_1}$ and $\mathbf{m_2} = \mathbf{FM_2}$ are the current fibre directions; $B = \mathbf{FF^T}$ and $\bar{\mathbf{B}} = \mathbf{B}J^{-2/3}$. Note that $\bar{\mathbf{B}}$ is the volume preserving part of the left Cauchy Green tensor. The variables in "*bold*" are either a vector or a tensor quantity. The symbol $\bigotimes$ in the above equation signifies the Kronecker product of two matrices. The first term of Eq. (5.78) with the shear modulus, G, represents the deviatoric (or shear) part of the deformation gradient tensor. The second term, with the bulk modulus K, is the volumetric part of the deformation gradient. The other terms describe the stretch in the direction of the fibre systems, where m_i and m_{si} (with $i = 1, 2$) representing the directions of the two-fibre systems. For a GRP with only one fibre system, the cross terms between the two different fibre systems (i.e. the k_s terms) drop out in Eq. (5.78).

Using Eq. (5.78), Barham et al. [94] proposed the following hyperelastic continuum damage mechanics (HECDM) based constitutive relationship given by,

$$\begin{aligned}\sigma(F) = & \left(1 - d^M\right)\frac{G}{J}\left(\bar{B} - \frac{tr\left(\bar{B}\right)}{3}I\right) + \left(1 - d^M\right)\frac{K}{J\beta}\left(1 - J^{-\beta}\right)I + d^M\frac{K^D}{\nu\beta}\left(1 - \nu^{\beta}\right)I \\ & + \left(1 - d^{f_1}\right)\frac{k_1}{J}\left(m_1.m_1 - 1\right)m_1 \bigotimes m_1 \\ & \left(1 - d^{f_2}\right)\frac{k_2}{J}\left(m_2.m_2 - 1\right)m_2 \bigotimes m_2 \\ & + \left(1 - d^{f_s}\right)\frac{k_s}{J}\left(m_{s1}.m_{s1} - 1\right)m_{s1} \bigotimes m_{s1} \\ & + \left(1 - d^{f_s}\right)\frac{k_s}{J}\left(m_{s2}.m_{s2} - 1\right)m_{s2} \bigotimes m_{s2}\end{aligned} \tag{5.79}$$

Note that the terms $(\mathbf{m_1.m_1} - 1)$ and $(\mathbf{m_2.m_2} - 1)$ in Eq. (5.78) represent fibre strains in the 1 and 2 directions, respectively. Stiffness degradation is described through the terms with $(1 - d)$. For instance, the damaged matrix shear modulus, G_{eff}, is defined as $G_{eff} = \left(1 - d^M\right)G$, where G is the intact shear modulus of the matrix material and the model parameter d^M defines the matrix damage. When d^M reaches one, the load carrying capacity of the pulverized matrix under compression is described by the volumetric strain part containing the degraded bulk modulus, K^D. Note that ν is the volumetric strain that appears in the 3rd term of the right-hand side of Eq. (5.79).

Damage nucleation and growth can be described for different modes of damage: matrix cracking under shear, matrix expansion under compression and tension, delamination, fibre buckling (or fibre debonding from the matrix), and fibre breakage under tension.

Damage Initiation and Evolution Models

In the model, the damage growth law for all modes of failure follows the relationship:

$$d = \begin{cases} 0, \text{ if } R < 1 \\ 1 - \exp\left(\frac{-(R-1)}{\iota}\right), \text{ if } R \geq 1 \end{cases} \tag{5.80}$$

When the damage criterion is satisfied for a specific damage mode (i.e. $R \geq 1$), the initiation of damage accumulation occurs in the Glass Fibre Reinforced Plastic (GRP). It's important to emphasize that R serves as the damage nucleation parameter and varies for different modes. The exponential form of the damage evolutionary equation plays a crucial role in controlling stress relaxation rates upon the initiation of damage ($R \geq 1$). If the damage parameter d reaches a value of 1 instantaneously, it results in abrupt fracture rather than a gradual accumulation of damage.

Additionally, the variable ι influences the slope of the softening portion of the stress-strain curve. For instance, when the value of ι is approximately 0.1, the unloading modulus (E_f) is roughly 10 times the Young's Modulus, indicating sudden or abrupt failure.

The exponential damage growth model defines different R values for each failure mode. The five primary failure modes are Fibre Breakage in Tension (R^{FiT}), Fibre Buckling in Compression (R^{FiC}), Matrix Cracking under Shear (R^{MS}), Matrix Cracking under Compression due to Bulking (R^{ME}), and Delamination of the Fibres and Matrix (R^{D}). Each of these five R parameters is expressed by an equation, such as Eq. (5.81) given below. It's essential to note that the failure criterion stipulates that damage initiation occurs only when R^{FiT} and R^{FiC} (with $i = 1, 2$) exceed 1. In other words, when the applied equivalent strain, either under tension or compression, surpasses a critical value, the corresponding R becomes greater than 1, denoted as:

$$R^{FiT/C} = \frac{\varepsilon^{i}}{\varepsilon^{FiT/C}} \tag{5.81}$$

where $\varepsilon^{FiT/C}$ is the maximum strain (a model constant) at which stress softening begins due to incipient damage and ε^{i} is the applied (stretch) strain in the direction of the fibre. When the applied fibre strain ε^{i} is positive, the damage is described by the R^{FiT} parameter, and when it is negative R^{FiC} models the damage process under compression. The maximum strain can be experimentally determined from the ratio of the maximum stress (or fibre strength) and the fibre Young's modulus.

When the fibre bundle is under compression, two types of damage can occur. The variable R^{FiC} models the fibre buckling and fibre debonding. Equation (5.82) defines this parameter as the ratio of applied fibre compressive strain and the critical strain at damage initiation.

$$R^{FiC} = \begin{cases} \frac{\varepsilon^{i}}{\varepsilon^{FiC}}, & \text{damage in compression} \\ 0, & \text{buckle in compression} \end{cases} \tag{5.82}$$

In the plate impact simulation, the buckling mode is assumed to be absent due to the three-dimensional compressive stress state.

The HECDM model introduces parameters R^{MS} and R^{ME} to account for the matrix damage modes capable of degrading both the shear and bulk moduli. The R^{MS} parameter is defined as,

$$R^{MS} = \frac{\varepsilon^{VM}}{\varepsilon^{MS}} \tag{5.83}$$

where, ε^{VM} is the von Mises strain (or equivalent strain) and ε^{MS} is the critical shear strain (a strain rate dependent model constant) to failure.

Bulk Modulus Degradation Due to Bulking

In Sect. 5.1.1, we explored the fracture mechanisms associated with the "*Wing Crack*" phenomenon in brittle materials under combined axial and lateral compression loading. Studies conducted by Chen and Ravichandran [98], as well as Yuan et al. [99], focused on the dynamic compressive behaviour of brittle materials, specifically soft ceramic Macor and Westerly Granite, respectively. To validate the wing crack-type mechanisms, their experimental research centred on the lateral confinement effects on brittle fracture. These investigations significantly contribute to our understanding of how brittle materials respond under dynamic compressive loading conditions and play a crucial role in validating theoretical models based on experimental observations.

Therefore, it is noteworthy that a key aspect of brittle fracture involves crack opening (Mode I) under overall compressive loading conditions. In the HECDM model, the volume expansion under compression or the bulk damage criterion is expressed through the following empirical equation:

Therefore, the volume expansion under compression or the bulk damage criterion is expressed through the following empirical equation:

$$R^{ME} = \frac{J-1}{J_e-1} \tag{5.84}$$

where J represents *Jacobian* which is less than 1 under compression and has a limiting value of zero. It's crucial to note that this expression is applicable for both tension ($J > 1$) and compression ($J < 1$). J_e is the critical relative volume expansion of the matrix under tensile damage. This is a user-controlled parameter and controls degradation of the bulk modulus with respect to the applied volumetric strain. In this model, the bulk damage evolution was employed to describe "*bulking*" or volume expansion of the matrix due to damage accumulation under compressive impact loading by degrading the bulk modulus.

Recently, Fraser [100] used the following alternative expression for R^{ME}. The bulk damage criterion is defined as:

$$\mathrm{R^{ME}} = \begin{cases} \alpha(1-\mathrm{J})\mathrm{e}^{\frac{\mathrm{J}}{\mathrm{J_e^c}}} & ,\ \mathrm{J} \le 1 \\ \frac{\mathrm{J}-1}{\mathrm{J_e^t}-1} & ,\ \mathrm{J} > 1 \end{cases} \tag{5.85}$$

where α is an arbitrary empirical parameter, $\mathrm{J_e}^{\mathrm{c}}$ is the critical matrix relative volume expansion strain under compression (often called as bulking), and $\mathrm{J_e}^{\mathrm{t}}$ is the critical matrix relative volume expansion strain in tension. α, $\mathrm{J_e}^{\mathrm{c}}$, and $\mathrm{J_e}^{\mathrm{t}}$ are user-defined variables dictating the extent of bulk modulus degradation with respect to applied volumetric strain.

The rationale or motivation behind this empirical relationship is to incorporate the effects of the ratio between axial compression (σ_1) and lateral compression (σ_2) on bulking. The upper piecewise function in Eq. (5.85) describing compressive damage evolution captures the physical "*bulking*" effect when a material undergoes shock loading. This equation limits the degradation of the bulk modulus as the lateral compression reaches an arbitrary threshold level and allows R^{ME} to decrease below 1 to prevent further bulking.

The degraded bulk and shear moduli of the GRP are given by the two following equations, respectively:

$$K = (1 - d^M)K^{undamaged} + d^M K^{damaged} \tag{5.86}$$

$$G = (1 - d^M)G^{undamaged} \tag{5.87}$$

where $K^{damaged}$ is some arbitrarily-set value for the residual bulk modulus of a pulverized GRP. Physically, the term $K^{damaged}$ term is justified by the fact that shock waves can still propagate through a pulverized GRP when under compression. If under tension, however, the pulverized GRP disintegrates and $K^{damaged}$ becomes zero. In modelling and simulation, this user defined parameter can be calibrated or adjusted to improve matching between experimental data, such as the depth of penetration into a thick brittle solid and simulated results.

The parameter R^D models ply delamination and depends on the through-thickness normal and shear stresses. Applying Tsai-Hill criterion at the lamina level, a smooth elliptical failure envelope representing delamination failure, can be expressed as,

$$R^D = \frac{\tilde{\sigma}_{33}^2}{\left(t^d\right)^2} + \frac{\tilde{\sigma}_{13}^2 + \tilde{\sigma}_{23}^2}{\left(s^d\right)^2} \tag{5.88}$$

where the 3-3 direction is through the thickness of the GRP (perpendicular to the 1 and 2 directions in which the fibres are oriented). This combined tension/shear delamination criterion is applied only when the volumetric strain is positive (tension).

Fiber damage criteria R^{fit} and R^{fic} represent fibre breakage in tension and compression respectively. These criteria apply for both fibre systems, denoted by $i = 1, 2$. Unlike the matrix damage, the tensile failure of fibres occurs abruptly without any damage accumulation. In the model, as soon as R^{fit} exceeds 1, in other words once the applied fibre strain in the "ith" fibre, ε_i exceeds the critical tensile strain, ε^{fit}, fibres fail under tension. The fibre damage in tension criterion is:

$$R^{fit} = \frac{\varepsilon_i}{\varepsilon^{fit}}; \qquad \varepsilon_i = \mathbf{m}_i \bullet \mathbf{m}_i - 1 \tag{5.89}$$

When fibres in compression surpass their elastic compressive strain limit, either buckling or breakage ensues depending on if there is existing matrix or fibre damage. Buckling will only occur if both the matrix and fibre are undamaged. During buckling, the fibre and matrix de-bond, transferring a greater portion of the compressive load onto the matrix. However, since the buckled fibre has not failed, it still possesses a fraction of its original stiffness and contributes to the GRP strength in the fibre direction; Eq. (5.90) reflects this:

$$\mathrm{k_i}^{\mathrm{buckled}} = \eta \mathrm{k_i}^{\mathrm{undamaged}} \tag{5.90}$$

where η is the buckling constant. In the HECDM model, the fraction of residual strength (η) of the buckled fibre is arbitrarily chosen between 0 and 1. Unfortunately, there is no experimental value exists for such modelling parameter (or constant). Therefore, η can be calibrated to match some measured shock or impact data, such as the VISAR measured velocity profile or depth of penetration in a ballistic test. We mentioned in the introduction chapter (Chap. 1) about the need for new creative impact experiments to directly estimate model constants such as the residual strength due to fibre breakage.

Fibre breakage in compression occurs when the matrix surrounding the fibre is damaged. In this scenario, the surrounding matrix provides less support, meaning the fibres shoulder a greater compressive load. Like in tension, when fibre breakage occurs, all fibre stiffness is lost. The compression damage fibre criterion is:

$$\mathrm{R}^{\mathrm{fic}} = \frac{\varepsilon_{\mathrm{i}}}{\varepsilon^{\mathrm{fic}}} \tag{5.91}$$

where $\varepsilon^{\mathrm{fic}}$ is the critical compressive strain in fibre system 'i'. The damage term governing the contribution of fibre strengths is summarized by Eq. (5.92):

$$\mathrm{d}^{\mathrm{f}} = \begin{cases} 1, & \mathrm{R}^{\mathrm{fit}} > 1 \\ 1-\eta, & \mathrm{R}^{\mathrm{fic}} > 1 \text{ and } \mathrm{d}^{\mathrm{M}} = 0 \\ 1, & \mathrm{R}^{\mathrm{fic}} > 1 \text{ and } \mathrm{d}^{\mathrm{M}} > 0 \end{cases} \tag{5.92}$$

The fibre damage term and fibre stiffness are related by Eq. (5.93):

$$k_i = (1 - d_i^f) k_i^{undamaged} \tag{5.93}$$

The fibre failure under compression is also assumed to be instantaneous without any damage accumulation.

Most fibre-reinforced composites exhibit strain rate-dependent behaviour. Therefore, it is beneficial to introduce strain rate effects on damage accumulation through damage mode constants, such as the $\varepsilon^{\mathrm{MS}}$ and α. A simple logarithmic expression that is often used in strength models is assumed to govern any viscous (strain rate) effects in GRP. The corresponding equation is,

$$\mathrm{var_{current}} = \mathrm{var}\left(1 + \mathrm{c_1} \ln \frac{\dot{\varepsilon}_{\mathrm{eq}}}{\dot{\varepsilon}_{\mathrm{ref}}}\right) \tag{5.94}$$

where "*var*" is some selected variable, such as the ε^{MS}, $\mathrm{c_1}$ is some arbitrary strain rate sensitivity constant, $\dot{\varepsilon}_{\mathrm{eq}}$ is the equivalent strain rate, and $\dot{\varepsilon}_{\mathrm{ref}}$ is the reference strain rate which is in most models assumed to be unity. For instance, the strain rate dependent critical strain ε^{MS} to initiate matrix damage under shear can be modelled using Eq. (5.83); the "*var*" in this case is ε^{MS} in Eq. (5.94). Note that this model

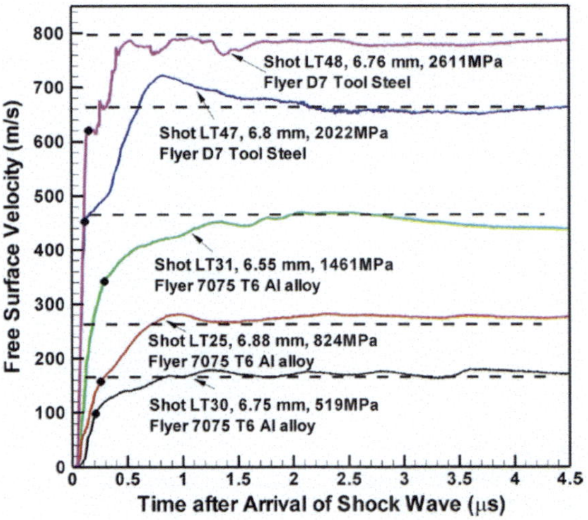

Fig. 5.23 VISAR data of free surface particle velocity. The GRP thickness varies between 6.55 and 6.88 mm, velocity varies between 112 m/s (LT 30) and 418 m/s (LT 48). Flyers are 25.4 mm thick steel and aluminium. The solid dot on the profiles indicates the velocity level at which "*elastic–elastic cracking (EEC)*" is assumed to occur

allows different values for c_1 for different mode of failures in GRP. When c_1 is zero, the damage mode is considered to be strain rate insensitive and the damage model constants remain the same for all strain rates.

Model Validation

The CDM-based hyperelastic model validation used the plate impact experimental data on GRP targets. Detailed description of the plate impact configuration is provided in Chap. 4. Tsai et al. [101] obtained data for a 6.8 mm GRP target with 10 plies which was impacted by a 25.4 mm thick steel or aluminium at different velocities. The free surface particle velocities were determined by the VISAR probe. Tsai et al. [101] results are shown below.

At this juncture, it's crucial to grasp the distinction between "HEL" and "EEC" (Elastic–Elastic Cracking). While the HEL in metals and ceramics arises from the "$elastic - plastic$" transition, the HEL-like solid dots depicted in Fig. 5.24 stem from elastic micro-cracking of the brittle matrix material. Scott et al. [96], have provided a detailed discussion on this aspect concerning GRP under shock loading conditions. This differentiation highlights that the observed HEL in certain materials, particularly those with brittle matrices like GRP, may be attributed to micro-cracking phenomena rather than a conventional elastic–plastic transition seen in metals and ceramics.

Fraser [100] performed ABAQUS [80] simulations for plate impact tests #LT30 and #LT48 in Fig. 5.23 to understand how the damage model parameters influence the computed wave profiles, which are time histories of particle velocity at the back surface of the GRP target (refer to Chap. 4). A parametric sensitivity study is crucial to determine suitable values for the model parameters that best match the experimental velocity profiles. Fraser focused on three HECDM model parameters: the Ogden parameter β, crack growth parameter ι, and strain rate sensitivity parameter

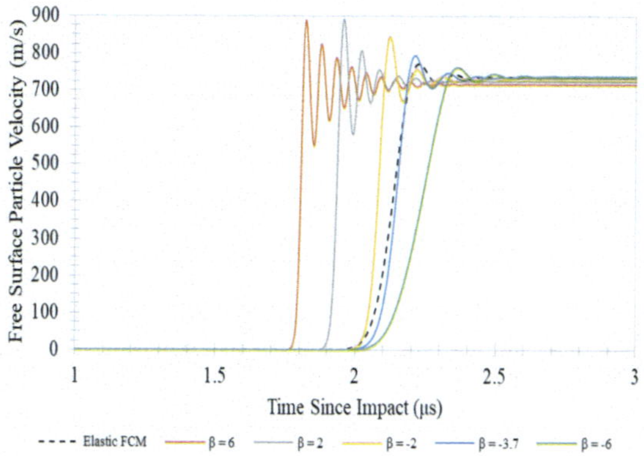

Fig. 5.24 Beta effect on undamaged hyperelastic model [100]

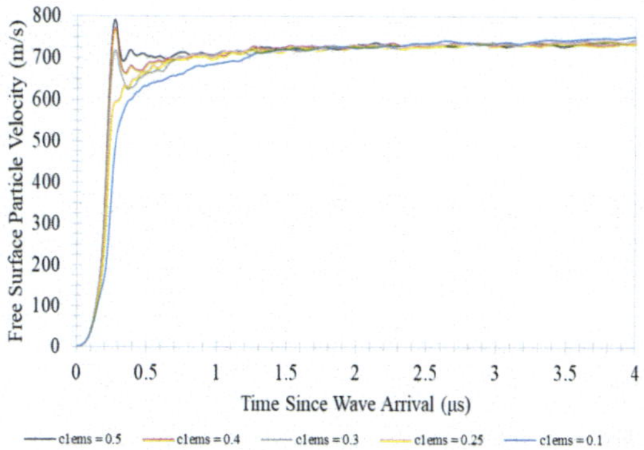

Fig. 5.25 Matrix shear damage strain rate effect on velocity profiles [100]

c_1. Understanding how each parameter modifies the wave profile accelerates and enables the tuning or calibration of the parameters.

Figure 5.24 depicts the effect of β on the velocity profiles by comparing with the elastic undamaged hyperelastic model. Fraser [100] obtained the "*elastic FCM*" solution using ABAQUS. The value of the Ogden parameter β influences the volumetric strain (2nd term in Eq. (5.78)), the shock wave speed (see Chaps. 3 and 4) changes with this parameter. This is because the bulk modulus varies for different values of β and wave speed (C_L) is given by $((K + 4G/3)/\rho)$. Therefore, based on this sensitivity study, a suitable value for β can be assigned.

Figure 5.25 illustrates the effects of the strain rate sensitive parameter $c_{1,\,\mathrm{ems}}$ on the velocity profiles and the matrix shear damage criterion. The c_1 is the strain rate sensitivity parameter in Eq. (5.94). For matrix cracking (ε^{MS}) as per Eq. (5.83), c_1 is labeled as "$c1ems$" in this figure. The nonlinear portion above the EEC transition point is significantly influenced by the strain rate sensitivity. The rise time to reach the peak velocity becomes shorter with increasing rate dependency. However, the shock wave arrival time is obviously not influenced by the different levels of rate sensitivity as one would expect. As c_1 increases, the strain, ε^{MS} to initiate shear damage also increases. Therefore, the applied equivalent strain (see Eq. (5.83)) has

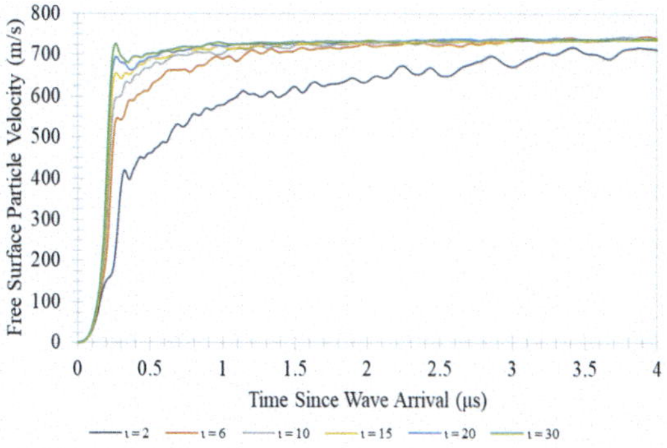

Fig. 5.26 "ι" dampening effect on particle velocity profiles [100]

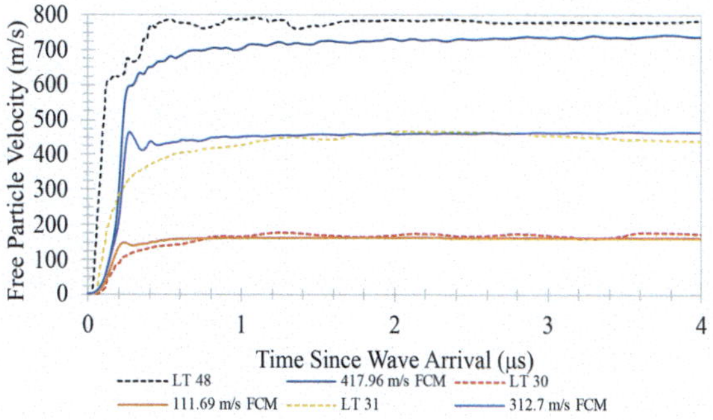

Fig. 5.27 Model prediction for free surface particle velocity at low, intermediate, and high velocity plate impact experiments [100]. The FCM denotes simulation results obtained by Fraser using Abaqus

to exceed ε^{MS} in order to initiate damage ($R^{MS} > 1$). In Fig. 5.25, EEC occurs at higher velocities for larger c_1 due to increasing applied strain.

The influence of damage rate parameter ι on the particle velocity profiles is displayed in Fig. 5.26. The slope of the particle velocity profile in this figure prior to reaching its peak is significantly influenced by the ι value. An observation of the damage rate Eq. (5.83) reveals that the damage rate is higher for a lower value of ι. The EEC is also significantly influenced by the damage growth rate. The velocity profile difference between $\iota = 2$ and $\iota = 6$ is much greater than the difference between $\iota = 20$ and $\iota = 30$. This behaviour is due to ι appearing in the denominator of Eq. (5.83). Based on the matching between experimentally measured profile and Abaqus simulations for a wide range of ι values from 2 through 40, a value of 10 was selected by Fraser [100]. It should be noted that this value is contingent on the assigned values for the other user-defined parameters.

To assess the hyperelastic damage model's ability to match measured wave profiles (VISAR data), Fraser compared computed particle velocity profiles from three specific plate impact tests, labelled as #LT48, #LT31, and #LT30 in Fig. 5.23. Fraser selected a best set of values for the model parameters to maximize the matching between HECDM and VISAR data at different impact velocities in Fig. 5.27.

Given the complexity of the GRP architecture, and the inherent variability in VISAR data from test to test, the model-generated velocity profiles exhibited a reasonable degree of agreement with the experimental data. In all tests, the ABAQUS-

Table 5.9 Parameters used in ABAQUS simulations [100]

Elastic constants	GPa	GRPproperties		HECDM constants			
$C_{11} = C_{22}$	31.55	G(GPa)	4.63	C_{1ems}	0.25	tcrit	0.057
C_{33}	20.12	K(GPa)	13.24	J_e^c	0.04	ccrit	0.00395
$C_{44} = C_{55}$	4.63	β	-3.7	J_e^t	2	η	0.8
C_{66}	4.94	k1(GPa)	86.9	ι	10	α	$1E-9$
C_{12}	15.86	k2(GPa)	86.9	ε^{MS}	0.0215	C_{1alpha}	0.001
$C_{13} = C_{23}$	9.75			ε^{fit}	0.057	Critical tension fiber strain	
				ε^{fic}	0.00395	Critical compression fiber strain	

simulated peak velocities agreed with the elastic Hugoniot established by the Rankine–Hugoniot conditions (see Chap. 3) and also showed excellent agreement with the peak velocity predictions by the ALE3D code [102] by Scott et al. [96].

During shock compression in plate impact simulations, fibre damage is not anticipated due to the prevailing one-dimensional strain state in the GRP target. However, in cases of multi-axial loading during projectile penetration into GRP, the HECDM model is capable of modelling damage due to fibre breakage, fibre buckling and ply delamination. Unfortunately, this model has not been validated using depth of penetration (DOP) data for GRP targets [103]. Crucial parameters influencing energy dissipation due to fibre failure modes include the fibre stiffness parameters $\mathbf{k_1}$ and $\mathbf{k_2}$. Achieving agreement between experimentally measured DOP with hydrocode simulations indexHydrocodesnecessitates the calibration of damage parameters associated with fibre damage (Table 5.9).

Note that C_{1ems} and C_{1alpha} are strain rate sensitive parameters for matrix cracking and bulking, respectively.

Table 5.10 summarizes the various salient features of this HECDM model.

The subroutines for VUMAT in ABAQUS developed and validated by Fraser [100] are given in Appendix A at the end of the book.

This chapter has presented an array of constitutive, damage, and failure models suitable for application under shock and impact loading conditions. When these damage models are correctly implemented in any type of hydrocode, whether Lagrangian or Arbitrary Lagrangian Eulerian, it is reasonable to anticipate that simulated results from different hydrocodes will closely align with dynamic experimental data. This aspect is further explored and elaborated upon in the subsequent Chap. 6.

Table 5.10 Specific features of the hyperelastic model [94, 96, 100]

Hyperelastic

- Phenomenological Helmholtz Free Energy expression:

$$\psi = \frac{G}{2}\left(\bar{I}_1 - 3\right) + \frac{K}{\beta^2}\left(\beta ln J + J^{-\beta} - 1\right) + \frac{k_1}{2}\left(I_4 - 1\right)^2 + \frac{k_2}{2}\left(I_6 - 1\right)^2$$

with several invariants of the stretch tensor.

- Three-dimensional stress state with special terms to incorporate the effects of different fibre systems in the derived constitutive model
- Anisotropic deformation/isotropic damage

Definition of Damage

- Continuum damage mechanics based.
- Scalar under different stress states, both tension, compression, and shear
- Damage variables are equivalent strain and volumetric strain based
- Different modes of damage; matrix cracking, dilatation under compression, and fibre breaking.

Micro-Crack Distribution

- No explicit formulation; implicitly follows volume expansion under compression.

Damage Nucleation

- When strain-based damage parameter, R exceeds 1, damage nucleates; There are four different expressions for R based on the mode of failure.

Damage Growth Law

- Empirical exponential growth law: $d = \begin{cases} 0 & \text{if } R < 1 \\ 1 - \exp\left(\frac{-(R-1)}{\iota}\right), & \text{if } R \geq 1 \end{cases}$

Failure Criterion

- The moduli go to zero as damage, d reaches the value of one.

Other Mechanisms

- None

5.3 Summary

Many constitutive equations utilized in advanced finite element codes find application in the design analyses of real-life scenarios involving brittle materials subjected to shock and impact loading conditions. The implementation of constitutive and damage models, once integrated into hydro codes, and the determination of model parameters either directly from experimental measurements or through calibrations to align with measured stress or velocity signals facilitate the simulation of complex multi-axial problems, contributing to the design of impact-resistant structural systems.

For instance, models like the "$Rajendran-Grove$" model, based on penny-shaped micro-cracks and microplasticity, and the advanced phenomenological "$Johnson-Holmquist$" model for ceramics, have proven successful in the analysis of vehicular and body armour design. These models accurately predicted the depth of penetration of tungsten long rod projectiles in confined thick ceramic targets [104]. Similarly, in understanding strain rate-dependent damage propagation in media hosting earthquake faults, the BRS model was employed in dynamic earthquake rupture applications.

In situations where highly diagnostic laboratory experiments are impractical or cost-prohibitive for addressing real-life design questions, physics-based or advanced phenomenological models often provide solutions through high-fidelity massively parallel computing. Given the highly nonlinear response of brittle materials under high-velocity impact or explosions, coupled with geometric nonlinearity, the design methodology necessitates the use of high-performance computational modelling.

Realistic constitutive, damage, and failure models in advanced finite element codes are crucial for designing ballistic and blast-resistant structures. It's essential to recognize that solutions from hydrocodes can be influenced by mesh or particle resolution, time increments, and various numerical artefacts. With the advancements in computing capabilities, the development of advanced numerical algorithms in codes, and realistic material models, we can now make realistic predictions, minimizing the need for prohibitively expensive quarter- or full-scale impact tests at a wide range of velocities with advanced diagnostics.

The implementation of any material constitutive model in commercial hydrocodes often involves iterative numerical schemes, such as the 4th and 2nd order Runge–Kutta methods. Readers of this book may identify CDM-based models in the open literature to describe not only brittle materials but also brittle fibre/epoxy-based laminates and implement them in hydrocodes through user-defined subroutines. This is frequently done as the available models in the codes may not be suitable for certain impact scenarios. Chapter 6 addresses some of the issues involved in the implementation in commercial codes.

References

1. Hockey BJ (1971) Plastic deformation of Aluminium Oxide by Indentation & Abrasion. J Amer Ceramic Soc 54(5):223–231
2. Shockey DA, Marchand AH, Skaggs SR, Cort GE, Burkett MW, Parker R (1990) Failure phenomenology of confined ceramics targets and impacting rods. Int J Impact Eng 9:263–275
3. Marla TB, Chen HW, Howitt DG, Kelsey PV, Korth GE, Williamson RL (1988) Dislocation microstructures in explosively deformed hard materials. Mater Sci Eng, A A105(106):293–298
4. Espinosa HD, Raiser G, Clifton RJ, Ortiz M (1992) Experimental observations and numerical modelling of inelasticity in dynamically loaded ceramics. J Hard Mater 3:285–313
5. Lankford J, Predebon WW, Staehler JM, Subhash G, Pletka BJ, Anderson CE (1998) The role of plasticity as a limiting factor in the compressive failure of high strength ceramics. Mech Mater 29:205–218
6. Kennedy JM, Cook N (1987) Crack models for the failure of rocks in compression. In: Desai CS, Krempl EK, Kiousis PD, Kundu T (eds) Constitutive laws for engineering materials theory and applications. Elsevier Sciences Publications, p 879

7. Shockey DA, Curran DR, Seaman L, Rosenberg JT, Petersen CF (1974) Fragmentation of rock under dynamic loads. Int J Rock Mech Mining Sci & Geomech Abstr II:303–317
8. Costin LS, Holcomb DJ (1983) Continuum model of inelastically deformed brittle rock based on the mechanics of micro-cracks, SAND-82-1633C; CONF-830104-1. Sandia National Labs, Albuquerque, NM
9. Grady DE, Kipp ME (1980) The Micro-mechanics of Impact fracture of rocks. Int J Rock Mech Mining Sci & Geomech Abstr 7:147
10. Nemat-Nasser S, Horii H (1982) Compression induced non-planar crack extension with application to splitting, exfoliation, and rockburst. J Geo Phys Res 87:6805–6821
11. Nemat-Nasser S, Deng H (1994) Strain-rate effect on brittle failure in compression. Acta Metallurgica Materialia 42(3):1013–1024
12. Nemat-Nasser S, Sarva S (2002) Micro-mechanisms of compression failure. In: McCauley J, Crowson A, Gooch W, Rajendran AM, Bless SJ, Logan KV, Normandia M, Wax S (eds) Ceramic armour materials by design, Ceramic Transactions, vol 134, pp 403–419
13. Horii H, Nemat-Nasser S (1985) Compression-induced micro-crack growth in brittle solids: axial splitting and shear failure. J Geophys Res 40(B4)
14. Horii H, Nemat-Nasser S (1983) Overall moduli of solids with micro-cracks: load-induced anisotropy. J Mech Phys Solids 31(2)
15. Ning J, Ren H, Li P (2008) Mechanical behaviours and damage constitutive model of ceramics under shock compression. Acta Mech Sin 24:305–315
16. Ravichandran G, Subhash G (1995) A micro-mechanical model for high strain behavior of ceramics. Int J Solids Struct 32(17):2627–2646
17. Kimberley J, Hu G, Ramesh KT (2011) A scaled model describing the rate-dependent compressive failure of brittle materials. In: Proulx T (ed) Dynamic behavior of materials, vol 1. Proceedings of the society for experimental mechanics series. Springer, New York
18. Marsh DM (1963) Flow and fracture in glass. In: Drucker DC, Gilman JJ (eds) Fracture of solids. Wiley, New York, 143–155
19. Lawn BR, Wilshaw TR (1975) Review: indentation fracture: principles and applications. J Mater Sci 10:1049–1081
20. Griffith AA (1920) The phenomena of rupture and flow in solids. Philosoph Trans R Soc Lond 221:163–198
21. Ohr SM, Kobayashi S (1980) In-situ observation of crack propagation by transmission electron microscopy. J Metals 32(5):35–38
22. Chen AC, Chen WF (1985) Constitutive relations for concrete. J Eng Mech Div ASCE 101(EM4):465–481
23. Takahashi Y (1983) Elastic-plastic constitutive modelling of concrete. Argonne National Laboratory Report, ANL-83-23
24. Dragon A, Mroz Z (1979) A continuum model for plastic-brittle behavior of rock and concrete. Int J Eng Sci 17:121–137
25. Bazant ZP, Kim S (1979) Plastic-fracturing theory for concrete. J Eng Mech Div ASCE 105(EM3):407–428
26. Chen MW, McCauley JW, Dandekar DP, Bourne NK (2006) Dynamic plasticity and failure of high-purity alumina under shock loading. Nat Mater 5:614–618
27. Brandt RC, Hasselman DPH, Lange FF (1983) Fracture mechanics of ceramics. Plenum Press, NY
28. Staehler JM, Predebon WW, Pletka BJ, Subhash G (2000) Micro-mechanisms of deformation in high-purity hot pressed alumina. Mater Sci Eng A 291(1–2):37–45
29. Chen MW, McCauley JW, Dandekar DP, Bourne NK (2006) Dynamic plasticity and failure of high-purity alumina under shock loading. Nat Mate 5:614–618
30. Rajendran AM, Cook WH (1988) A comprehensive review of modelling of impact damage in ceramics. Air Force Report, AFATL-TR-88-143, Air Force Armament Laboratory, Eglin Air Force Base, Florida 32542-5434
31. Margolin LG (1984) A generalized griffith criterion for crack propagation. Eng Fract Mech 19(3):539–543

32. Curran DR, Shockey DA, Seaman L (1973) Dynamic failure in solids. J Appl Phys 44:4025–4038
33. Curran DR, Seaman L, Shockey DA (1987) Dynamic failure of solids. Phys Rep 147:253–388
34. Mescall JF, Tracey CA (1986) Improved modelling of fracture in ceramic armours. Technical Report of Army Materials Technology Laboratory, Watertown, MA
35. McClintock FA, Walsh JB (1962) Friction on griffith cracks. In: Proceedings of the 4th U.S. national congress of applied mechanics, Berkeley, CA, pp 1015–1021
36. Tuler FR, Butcher BM (1968) A criterion for the time dependence of dynamic fracture. Int J FractMech 4:431–437
37. Rajendran AM, Kroupa JL (1989) Impact damage model for ceramic materials. J Appl Phys 8(66):3560–3565
38. Steinberg DJ (1992) Computer studies of the dynamic strength of ceramics. Shock compression of condensed matter. Elsevier Science, pp 447–450
39. Johnson GR, Holmquist TJ (1992) A computational constitutive model for brittle materials subjected to large strains, high strain rates and high pressures, shock wave and high-strain-rate phenomena in materials. In: Meyers M, Murr L, Staudhammer K . CRC Press, Boca Raton, FL, USA, pp 1075–1081
40. Johnson G, Holmquist TJ (2005) Characterization and evaluation of silicon carbide for high-velocity impact. J Appl Phys 97:093502
41. Wang J, Yin Y, Luo C (2018) Johnson-Holmquist-II(JH-2) Constitutive Model for Rock Materials: Parameter Determination and Application in Tunnel Smooth Blasting. Appl Sci 8(9):675
42. Janack W (1976) The role of bulking in brittle failure of rocks under rapid compression. Int J Rock Mech Mining Sci Geomechan Abstar 13:77–186
43. Feng R, Gupta YM, Wong MKW (1997) Dynamic analysis of the response of lateral Piezoresistance Gauges in shocked ceramics. J Appl Phys 82:2845–2854
44. Grady DE, Moody RL (1996) Shock compression profiles in ceramics, Sandia Report No. SAND96-0551, Sandia National Laboratory, Albuquerque, NM
45. Dandekar DP, Bartkowski P (1994) Shock response of AD995 alumina. High-pressure science and technology, Part 1. In: Schmidt SC, Shaner JW, Samara GA, Ross M. AIP Press, New York, pp 733–736
46. Pickup IM, Barker AK (2000) Deviatoric strength of silicon carbide subject to shock, shock compression of condensed matter. In: Furnish MD, Chhabildas LC, Hixson RS (eds). American Institute of Physics
47. Klopp RW, Shockey DA (1991) The strength behavior of granulated silicon carbide at high strain rates and confining pressure. J Appl Phys 70:7318–7326
48. Lundberg P, Renstrom R, Lundberg B (2000) Impact of metallic projectiles on ceramic targets: transition between interface defeat and penetration. Int J Impact Eng 24:259
49. Timothy J, Holmquist, Johnson GR (2002) Response of Silicon Carbide to High Velocity Impact. J Appl Phys 91(9):5858–5866
50. Grady DE (1982) Local inertia effects in dynamic fragmentation. J Appl Phys 53:322–325
51. Glenn LA, Chudnovsky A (1986) Strain energy effects on dynamic fragmentation. J Appl Phys 59(4):1379–1380
52. Glenn LA, Gommerstadt BY, Chudnovsky A (1986) A fracture mechanics model of fragmentation. J Appl Phys 60(3):1224–1226
53. Taylor LM, Chen EP, Kuszmaul JS (1986) Micro-crack-induced damage accumulation in brittle rock under dynamic loading. Comput Methods Appl Mech Eng 55:301–320
54. Kipp ME, Grady DE (1985) Dynamic fracture growth and interaction in one dimension. J Mech Phys Solids 33:399–415
55. Budiansky B, O'Connell RJ (1976) Elastic moduli of a cracked solid. Int J Solids Struct 12:81–97
56. Johnson GR, Stryk RA (1986) User instructions for the EPIC-2 code. Air Force Report, AFATL-TR-86-51

57. Rajendran AM, Grove DJ (1996) Modelling the shock response of silicon carbide, boron carbide, and titanium diboride. Int J Impact Eng 18(6):611–631
58. Addessio FL, Johnson JN (1990) A constitutive model for the dynamic response of brittle materials. J Appl Phys 67(7):3275–3286
59. Espinosa HD, Zavattieri PD, Dwivedi SK (1998) A finite deformation continuum /discrete model for the description of fragmentation and damage in brittle materials. J Mech Phys Solids 46(10):1909–1942
60. Margolin LG (1984) Microphysical models for inelastic material response. Int J Eng Sci 22(8–10):1171–1179
61. Krajcinovic D (1996) Damage mechanics. North-Holland, Elsevier, Amsterdam, The Netherlands
62. Nemat-Nasser S, Hori M (1999) Micro-mechanics: overall properties of heterogeneous materials, 2nd edn. NorthHolland, Elsevier, Amsterdam, The Netherlands
63. Margolin LG (1983) Elastic moduli of a cracked body. Int J Fract 22:65–79
64. Margolin LG (1983) Numerical simulation of fracture. In: Desai CS, Gallagher RH (eds) Proceedings of international conference on constitutive laws for engineering materials. Tucson, pp 567–572
65. Horii H, Nemat-Nasser S (1986) Brittle failure in compression: splitting, faulting, and ductile-brittle transition. Philosoph Trans R Soc A319:337–374
66. Ilankamban R, Krajcinovic D (1987) A constitutive theory for progressively deteriorating brittle solids. Int J Solids Struct 23(11):1521–1534
67. Krajcinovic D, Fanella D (1986) A micro-mechanical model for concrete. Eng Fract Mech 25(5/6):585–596
68. Costin LS, Stone CM (1987) A finite element material model for micro-fracture - damaged brittle rock. Sandia Report, SAND87-1227, UC-13
69. Costin LS, Stone CM (1987) Implementation of a finite element damage model for rock. In: Desi CS et al (eds) Constitutive laws for engineering materials. theory & applications. Elsevier Science Publishing Co., Inc., pp 829–841
70. Zuo QH, Addessio FL, Dienes JK, Lewis MW (2006) A rate-dependent damage model for brittle materials based on the dominant crack. Int J Solids Struct 43:3350–3380
71. Riou P, Cottenot CE, Boussauge M (1996) Steel rod impact on silicon carbide beams: experiments and anisotropic model of damage. In: Jones N, Brebbia CA, Watson AJ (eds) iStructures under Shock and Impact. Computational Mechanics Publications, pp 533–541
72. Rajendran AM (1994) Modelling the impact behavior of AD85 ceramic under multi-axial loading. Int J Impact Eng 15(6):749–768
73. Rajendran AM, Grove DJ (2002) Computational modelling of shock and impact response of alumina. Comput Modell Eng Sci 3(3):367–380
74. Dienes JK (1986) Comments on "A generalized griffith criterion for crack propagation", by L.G. Margolin. Eng Fract Mech 23(3):615–617
75. Gurson AL (1977) Continuum theory of ductile rupture by void nucleation & growth; part i: yield criteria and flow rules for porous ductile materials. J Eng Mater Technol 99:2–15
76. Mackenzie JK (1950) The elastic constants of a solid containing spherical holes. Proc Phys Soc (Sect B) 2:63
77. Johnson JN (1981) Dynamic fracture and spallation in ductile solids. J Appl Phys 52(4):2812
78. Cook WH, Rajendran AM, Grove DJ (1992) An efficient numerical implementation of Bodner-Partom model in EPIC-code. J Eng Fract Mech 41(5):607–623
79. CTH Release Information, Sandia National Laboratory on CTH code - Version 7.0 - April 2005
80. **ABAQUS**: Abaqus/Explicit, a special-purpose Finite-Element analyzer that employs explicit integration scheme to solve highly nonlinear systems with many complex contacts under transient loads. https://www.3ds.com/products-services/simulia/products/abaqus/
81. Rajendran AM (2003) Critical measurements for validation of constitutive equations under shock and impact loading conditions. J Opt Lasers Eng 40:249–262

82. Chhabildas, Furnish MD, Reinhart WD, Grady DE (1997) Impact of AD995 alumina rods, shock compression of condensed matter, Sandia National Laboratory report SAND-97-2012C; CONF-970707
83. Woolsey P (1991) Residual penetration ballistic testing of Armour ceramics. U. S. Army Materials Technology Laboratory, Watertown, MA, (Unpublished Work)
84. Grove DJ, Rajendran AM (2002) Overview of the Rajendran-Grove ceramic failure model. In: McCauley J, Crowson A, Gooch W, Rajendran AM, Bless SJ, Logan KV, Normandia M, Wax S (eds) Ceramic Armour materials by design. Ceramic Transactions, pp 134, 371-384
85. Grady DE (1996) Dynamic properties of ceramic materials. Chapter 1 "Physics of shock and impact". Sandia Report: SAND94-3266 • UC-704, IOP Publishing
86. Kondo N, Shimamura A, Hotta M, Shimizu H, Ujihar T, Ohnish Y (2021) Control of microstructure and mechanical properties of sintered aluminium nitride through addition of aluminium nitride whiskers. J Asian Ceramic Soc 9(3):1248–1254. https://doi.org/10.1080/21870764.2021.1953760
87. Mashimo T, Uchino M, Nakamura A, Kobayashi T, Takasawa E, Sekine T, Noguchi Y, Hikosaka H, Fukuoka K, Syono Y (1999) Yield properties, phase transition, and equation of state of aluminum nitride (AlN) under shock compression up to 150 GPa. J Appl Phys 86(12)
88. Kaur S, Cutler RA, Shetty DK (2009) Short-crack fracture toughness of silicon carbide. J Amer Ceramic Soc 92(1):179–185
89. Bhat HS, Rosakis AJ, Sammis CG (2012) A micro-mechanics based constitutive model for brittle failure at high strain rates. J Appl Mech 79:031016–12
90. Kolari K (2017) A complete three-dimensional continuum model of wing-crack growth in granular brittle solids. Int J Solids Struct 115–116:27–42
91. Ashby MF, Sammis CG (1990) The damage mechanics of brittle solids in compression. Pure Appl Geophys 133(3):489–521
92. Deshpande VS, Evans AG (2008) Inelastic deformation and energy dissipation in ceramics: a mechanism-based constitutive model. J Mech Phys Solids 56(10):3077–3100
93. Press WH, Teukolski SA, Vetterling WT, Flannery BP (1993) Numerical recipes in Fortran 77: the art of scientific computing 1. Cambridge University Press, Cambridge
94. Barham MI, King MJ, Mseis G, Faux DR (2013) Hyperelastic fiber-reinforced composite model with damage. Lawrence Livermore National Laboratory
95. Brazil M, Ashrafi H, Jafari A, Zhao XL, Singh Raman RK, Bai Y (2019) Effect of fibers configuration and thickness on tensile behavior of GFRP laminates exposed to harsh environment. Polymers Bol 11(9). https://doi.org/10.3390/polym11091401
96. Scott NR, Rajendran AM, Nelms MD, Prakash V (2022) A new woven composite constitutive model validated by shock wave experiments. J Appl Phys 131:165905
97. Holzapfel GA (2005) Nonlinear solid mechanics: a continuum approach for engineering. Wiley, New York
98. Chen W, Ravichandran G (1997) Dynamic compressive failure of a glass ceramic under lateral confinement. J Mech Phys Solids 45:1303–1328
99. Yuan F, Prakash V, Tullis T (2011) Origin of pulverized rocks during earthquake fault rupture. J Geophys Res 116(B06309). https://doi.org/10.1029/2010JB007721
100. Fraser J (2022) ABAQUS implementation of a hyperelastic damage model for glass-reinforced polymers under shock and impact loading. Master of Science Thesis, Department of Mechanical Engineering, The University of Mississippi
101. Tsai L, Yuan F, Prakash V, Dandekar DP (2009) Shock compression behavior of a S2-glass fiber reinforced polymer composite. J Appl Phys 105(9):093526
102. ALE3D Computational Manual Material Model, Lawrence Livermore National Laboratory 7000 East Avenue • Livermore, CA 94550
103. Prifti J, DeLuca E, Bethoney W, Chou S-C (2022) Dynamic response of S-2 glass reinforced plastic structural armour. Army Research Lab Watertown MA Materials Directorate. https://apps.dtic.mil/sti/citations/ADA273868
104. McCauley J, Crowson A, Gooch W, Rajendran AM, Bless SJ, Logan KV, Normandia M, Wax S (eds) (2002) Ceramic Trans 134:371–384

Chapter 6
Computational Modelling

Prologue

Solving nonlinear shock and impact problems is a formidable challenge. In the absence of analytical solutions, these complex problems are solved through advanced finite element/finite difference computer codes (hydrocodes). Many of these codes are widely accessible across the globe. This chapter presents essential factors of obtaining realistic and precise solutions from three-dimensional numerical hydrocodes, addressing a wide array of problem types.

In-depth discussions on the mathematical formulations of governing equations, encompassing concepts like shape functions, weight functions, trial functions, and test functions used in discretizing a continuum, exceed the scope of this chapter. However, to help orientate the reader, brief discussions on different discretization schemes are presented. For follow-up, mathematical developments associated with a broad range of algorithms are readily available in the open literature through journals and books.

Becoming proficient in the use of any high-velocity impact code demands substantial investments in terms of training and time. To ensure that these codes yield realistic solutions, users, be they graduate students or practicing engineers, must be well-versed in the myriad of physical and numerical parameters that govern the simulation. Modelling and simulating impact problems using finite element codes hinges on the accurate determination of forces acting on the mesh at each time step. Three critical modelling aspects come into play: Newtonian laws of motion, the equation of state, and the constitutive/damage model. Furthermore, the incorporation of specialized algorithms to address solid-to-solid contact, artificial viscosity, crack propagation, interfaces, element-to-particle conversion, and more is often necessary to enforce appropriate boundary and interface conditions. These algorithms often introduce additional model parameters or inputs to the code. Users must understand the limitations and constraints imposed by the code in order to obtain realistic

A. M. Rajendran et al., *Failure of Brittle Materials Under Shock and Impact*, Materials Horizons: From Nature to Nanomaterials, https://doi.org/10.1007/978-981-97-7026-7_6

solutions. In practice, it is sometimes necessary to rely on default values for many numerical parameters, such as the artificial viscosity coefficients.

In the past, computer programs describing shock wave propagation and calculating velocities, strains, and more, as functions of time and position, were commonly referred to as "Hydrocode". These hydrocodes primarily found utility in gas dynamics and hypervelocity impact problems, with material strength being disregarded in the solutions, assuming that the applied shock pressure vastly exceeded material strength. The term "hydro" alluded to the absence of shape change in the solid, and the material flow was attributed solely to volume change. However, when describing shape changes resulting from shear loading, constitutive or strength models became necessary. Additionally, the evolution of high-strength metallic and ceramic materials, such as steel alloys, tantalum, and ceramics over the past five decades, mandated the incorporation of material strength into impact analyses, especially at low- and high-impact velocities. Material strength could no longer be treated as negligible; it was now comparable to the shock pressure. Consequently, modern impact dynamics codes encompass both equation of state and strength models. For simplicity, we often refer to these as hydrocodes throughout this chapter unless a specific code is mentioned during discussions.

6.1 Exploring Impact Analysis Codes

Significant number of finite element method (FEM), finite difference method, and particle-based codes are available to practicing engineers and researchers for performing impact analyses. These versatile codes, such as Abaqus [1], LS-Dyna [2], Ansys-Autodyn [3], EPIC [4], CTH [5], ALEGRA [6], SIERRA [7], ALE3D [8], and others [9, 10], cater to a wide spectrum of applications. They empower analysts to tackle challenges ranging from crashworthiness assessments, foreign object impacts on structures, and blast explosions on buildings to oil shale explosions and debris impacts on space structures. These codes seamlessly handle an array of materials, including metals, ceramics, layered polymeric composites, rocks, concrete, and soils.

While the utility of these tools is evident, direct referencing to certain restricted codes, often lacking user manuals and source code accessibility, poses a challenge. Some hydrocodes, particularly those developed within US National laboratories, exhibit highly restrictive access policies. An illustrative example is the EPIC3D code [11], a cutting-edge design software for warheads and armours, which remains inaccessible to both US citizens and non-citizens.

To harness the power of hydrocodes, be they FEM or mesh-free, for realistic solutions to two- or three-dimensional impact problems, users must cultivate an acute awareness of how numerical artefacts can significantly influence solution accuracy. These artefacts encompass a wide range of factors, including contact modelling, artificial viscosity, mesh size, mesh architecture, mesh distortion, mesh conversion to particles, and the imposition of boundary and interfacial conditions.

Spatial discretization is a cornerstone of these codes, entailing the representation of fields and structures within the problem domain using computational points in space, typically interconnected through computational grids. As a rule of thumb, finer grids yield more accurate solutions. Common spatial discretization methods include Lagrange, Euler, ALE (a blend of Lagrange and Euler), and mesh-free techniques like Smooth Particle Hydrodynamics (SPH). Additionally, specialized codes exist based on the "*Material Point Method (MPM)*" and "*Discrete Element Method (DEM)*". Users of hydrocodes must grasp the nuances and trade-offs associated with each framework, as each method carries its own set of advantages and limitations.

6.1.1 Lagrangian

A large number of highly sophisticated Lagrangian finite element methods are available, both in the open-source and commercial domains, to tackle complex nonlinear dynamics problems. Lagrangian formulations are particularly adept at tracking the motion of material points embedded within solid structures, providing time histories of critical parameters like displacements (strains), velocities (strain rates), and stresses (forces) at grid or material points. Importantly, the material coordinates of these points remain constant over time, with no material flow between elements.

To address contact and collision scenarios, Lagrangian codes employ contact algorithms to prevent material surfaces, such as a projectile and target, from interpenetrating. These contact surfaces are often referred to as "*slide lines*", wherein one surface is designated as the master, and the other as the slave, while the corresponding nodes are labelled as master and slave nodes. Numerous Lagrangian contact algorithms exist, categorized by their approach to contact search, handling of contact and release conditions, and determination of contact forces. These codes offer extensive libraries of contact algorithms, encompassing automatic and manual surface-to-surface, node-to-surface, and single-surface contact conditions.

However, a limitation of Lagrangian methods emerges when large deformations occur, as the computational grid is tied to the material. This leads to a reduction in the time step size as deformations progress, ultimately impacting computational efficiency. To ensure a realistic solution for impact problems, the time step must satisfy the Courant criterion given by

$$\Delta t \leq \eta \frac{\Delta x}{C_L} \tag{6.1}$$

where Δx represents the smallest mesh size, C_L stands for the longitudinal wave speed, and η is a user-defined fraction, typically around 0.1. This criterion ensures that the time step accommodates the transit of a shock wave across the smallest element in the finite element mesh.

In impact problems involving large deformations, it becomes advantageous to move the material relative to a fixed mesh in space, an algorithm accomplished through Eulerian and Arbitrary Lagrangian-Eulerian (ALE) methods. In such cases, element quadrature or integration points remain coincident with material points, simplifying the application of boundary and interface conditions throughout the solution process in ALE codes.

Generally, Lagrangian methods offer efficient and accurate solutions for tracking state variables over time, with relatively lower CPU and memory requirements compared to Eulerian and ALE techniques. However, they encounter challenges in simulating scenarios with free surfaces generated by solid fracture and fragmentation. Additionally, mesh development for complex, irregular 3D geometries can be time-consuming and challenging.

For Lagrangian codes to yield realistic results, it's crucial to maintain a reasonable element aspect ratio and ensure elements maintain positive volume. Elements with large aspect ratios or inverted contact geometry can introduce errors or hinder solver convergence, posing challenges for users tackling high-impact problems.

To address collisions and sliding between different solids, contact descriptions and algorithms, both smooth and non-smooth, play a pivotal role. A wealth of mathematical frameworks and review articles are available in the open literature for reference. Commercial hydrocodes typically offer various contact options to simulate a range of impact scenarios, emphasizing the importance of selecting the appropriate option for specific situations, such as sharp corners impacting smooth surfaces or the interaction of fragments generated by fracture.

Recent advancements, as exemplified by Gerlach and Johnson's work [12] on the Combined Particle-Element Method (CPEM), introduce innovative solutions for impact problems with severe distortions. The CPEM combines finite element and meshless particle approaches to handle lower strained and higher strained particles efficiently, without the tensile instabilities sometimes associated with other particle methods. Their research highlights the challenges associated with existing contact and sliding algorithms and presents an improved algorithm, showcasing its capabilities and accuracy through practical examples.

In summary, recent decades have seen significant progress in addressing the limitations of Lagrangian codes, particularly in simulating large deformations and mesh distortions. Solutions such as element-particle conversion algorithms and re-meshing techniques have mitigated these issues, leading to more accurate predictions, even in scenarios involving fracture and fragmentation, as demonstrated by simulations involving a Lagrangian code like EPIC3D.

6.1.2 Eulerian

Eulerian methods initially emerged as tools for simulating fluid dynamics. Nevertheless, Eulerian and Arbitrary Lagrangian-Eulerian (ALE) hydrocodes differ from contemporary Computational Fluid Dynamics (CFD) codes by their incorporation of material strength characteristics.

Take, for example, the CTH Eulerian code [5], designed to simulate large deformations within solids subjected to high-velocity and hypervelocity impacts, like metallic projectiles penetrating thick steel or ceramic plates. In CTH, a two-step Eulerian solution method prevails: first, a Lagrangian step causes cell distortion to track material motion and the second step is a re-meshing process where the distorted cells are mapped back to the Eulerian mesh.

Eulerian methods permit arbitrarily large deformations and automatically allow formation of new free surfaces. The fixed spatial Eulerian mesh eliminates the challenges associated with mesh distortion that often plague Lagrangian formulations.

Nonetheless, since materials can move within the mesh, Eulerian methods necessitate additional data to characterize the content of each element and update ongoing calculations. This inclusion of material transport and advection computations renders Eulerian methods more computationally intensive compared to Lagrangian methods.

Furthermore, in scenarios involving surface contact and separation, it becomes imperative to consider contact forces between interacting materials. Eulerian hydrocodes conventionally employ mixture theory to handle subelement thermodynamics and stresses.

In this context, "*mixture theory*" refers to a technique employed to approximate stress and pressure field discontinuities encountered at interface boundaries within an element, effectively consolidating them into a single pressure and stress field.

Distinct materials possess their unique sets of properties, with each material assumed to occupy a portion of the element's volume. This necessitates the incorporation of specific thermodynamic conditions, defining a nonlinear system of equations wherein the volume fraction of each material within a cell acts as the unknown variable.

Numerous mixture theories are available, yet it's important to note that none of these methods fundamentally address the physics occurring at interface boundaries. Consequently, they are regarded as ad hoc solutions, potentially impacting the accuracy of the overall solution.

Solid mechanics researchers have embraced the Eulerian approach to mitigate mesh distortions arising from substantial deformations, a concern often encountered with Lagrangian-based discretization techniques. Eulerian methods exhibit computational robustness, particularly in the realm of large deformation material mechanics. Additionally, they afford the capability to introduce or remove materials as needed. In contrast to mesh-based Lagrangian approaches, mesh development is notably less cumbersome.

Robust algorithms exist within the Eulerian framework for generating free surfaces due to fracture and fragmentation. Nevertheless, challenges emerge when handling material flow within the spatial grid, making the tracking of boundary fronts and material interfaces more intricate. In shock wave propagation problems, the numerical algorithms employed for front or interface tracking may introduce excessive shock diffusion.

Consider, for instance, a scenario where a projectile and a target consisting of different materials coexist within a cell, effectively converting it into a mixed cell.

In such instances, managing the conservation of mass can pose challenges. It's worth noting that the Eulerian method frequently demands substantial computational resources, both in terms of CPU utilization and memory requirements.

6.1.3 Arbitrary Lagrangian-Eulerian

In 2017, Wall's Ph.D. dissertation [13] provided a comprehensive examination of the strengths and weaknesses inherent to diverse ALE methods. This section encapsulates many of Wall's insights into Eulerian and ALE hydrocodes. The ALE method of space discretization is a hybrid of the Lagrange and Euler methods. It allows redefining the grid continuously in arbitrary and predefined ways as the calculation proceeds, which effectively provides a continuous rezoning facility. Various predefined grid motions can be specified, such as free (Lagrange), fixed (Euler), equipotential, equal spacing, and others. The ALE method can model solids as well as liquids. The advantage of ALE is the ability to reduce and sometimes eliminate difficulties caused by severe mesh distortions encountered by the Lagrange method, thus allowing a calculation to continue efficiently. One of the best example applications is fluid–structure interaction, such as the bomb blast on a building.

The ALE method has been developed to take advantage of the strengths of both the Lagrangian and Eulerian coordinate systems. In this method, the mesh is permitted to deform as it would in a Lagrangian formulation as shown in Fig. 6.1.

ALE methods allow for improved handling of distortions compared to Lagrangian methods while also providing more flexibility for handling material interfaces. ALE formulation has two steps, a Lagrangian step in which the solution is advanced in time and the mesh is distorted, and an advection, or remap, step in which the mesh is relaxed and material is transferred across element boundaries. This step

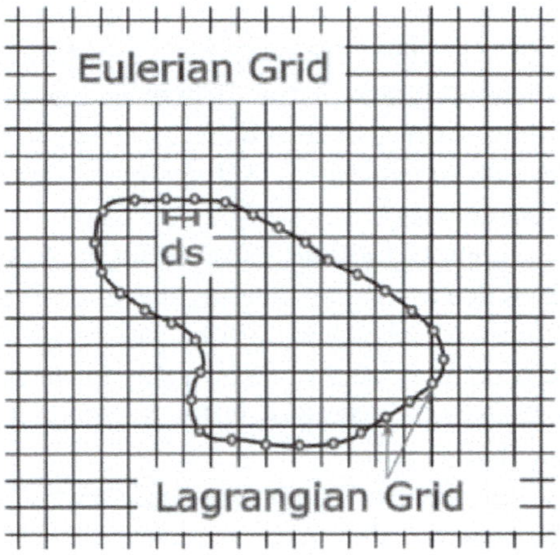

Fig. 6.1 Deforming Lagrangian mesh of a plastically deforming or distorted solid across a fixed Eulerian cells in space

can be simplified to restore the mesh to its original position and create the Eulerian formulation that was used in the validation of the contact approach developed by Walls for his work [13]. While ALE methods provide more flexibility in performing the calculation they still suffer from the same drawbacks as Eulerian methods when multiple materials are permitted to occupy the same element. These multi-material ALE formulations (MALE) uses the same mixture theory approach to mixed cells as Eulerian formulations.

A variety of other numerical techniques have also been developed to alleviate the difficulties associated with Lagrangian methods. These include r-adaptive or Single material Arbitrary Lagrangian-Eulerian (SALE), Coupled Lagrangian-Eulerian (CLE), and Smoothed-Particle Hydrodynamics (SPH). r-adaptive and SALE techniques have been successfully used to improve Lagrangian calculations that involve moderate amounts of deformation. Since they do not allow multiple materials within a computational element they do not suffer from the drawbacks of mixture theory seen in traditional Eulerian and ALE formulations. These techniques allow for the independent movement of nodes, allowing element distortions to be controlled and, thus, dramatically increasing the time step required for numerical stability. SALE and r-adaptive methods, however, are not able to overcome the difficulties associated with problems that involve large deformations and penetrations found in Lagrangian methods. Therefore, these techniques alone are insufficient to prevent severe mesh distortions [14].

It is possible to combine Eulerian and Lagrangian mesh systems into a single formulation. This method is referred to as the Coupled Eulerian-Lagrangian (CLE) method, and is implemented in several commercially available codes. These codes facilitate the modelling of fluid–structure and gas–structure interaction by allowing some of the materials or sub-grids to be described using an Eulerian coordinate system while others are simultaneously modelled using a Lagrangian construction. CLE codes require special attention to be given to the interaction of sub-grids, especially those comprising differing coordinate systems.

Benson's extensive work [15] provides a comprehensive exploration of various algorithms utilizing explicit finite element and finite difference methods. These methods are employed to address transient, large deformation problems in solid mechanics. Benson discussed several key developments in numerical techniques, encompassing both Lagrangian and Eulerian methods, operator splitting, and interface tracking. He also addressed the issues associated with stable time step estimation for multi-material Eulerian hydrocodes [16], along with his implicit multi-material Eulerian formulation [17]. These concepts hold significant relevance for ALE-type hydrocodes.

One of the most critical challenges in Eulerian and ALE codes is the accurate tracking of boundaries and interfaces within multi-material cells. For example, consider a scenario where a tungsten rod penetrates a ceramic plate. How can we identify the interface between these two different materials within a single cell? Several code developers have focussed on achieving precise treatment of contact in multi-material Eulerian and ALE methods. Furthermore, Zhong's book [18] provides an in-depth exploration of finite element procedures related to the contact-impact problem. For a more detailed understanding of the ALE method, work of Donea et al. [19] offers an exhaustive description.

Vitali and Benson's work [20] employed XFEM, which involves "*enriching*" nodes to introduce additional degrees of freedom. In this context, enrichment enabled the determination of independent velocity fields for each material within a mixed cell. Separate elements were created for each material, overlaying one another. During the Lagrangian step, these overlaying elements deformed independently, and a contact enforcement algorithm was activated if the combined volume fractions of the cells exceeded 1.0.

While promising, this approach necessitates the use of extended finite element algorithms, which are not readily available in many production Eulerian codes. To the best of the author's knowledge, these methods are currently limited to two dimensions. Apart from the research code RAVEN discussed in Vitali and Benson's work, Sandia National Laboratories hydrocode ALEGRA has recently incorporated an XFEM Eulerian contact algorithm in two dimensions [21]. However, this implementation has proven to be computationally expensive, resulting in runtimes three to four times longer than ALEGRA's traditional Eulerian approach. Nevertheless, it's worth noting that little optimization has been applied to the code at this stage, and this speed penalty is expected to diminish over time.

6.1.4 Smooth Particle Hydrodynamics

Smooth Particle Hydrodynamics (SPH) is a "*meshless*" Lagrangian, particle-based method that holds substantial promise for simulating multiphase flow problems. SPH, as a particle-based method, proves to be a potent tool for solving problems involving the fragmentation of brittle materials into granular or pulverized forms upon impact. This efficacy arises from its ability to accurately track the total energy throughout each time step of the particle-based hydrocode solution, thereby reasonably ensuring the conservation of mass, momentum, and energy.

This section discusses further the conversion from finite element to particles within the Lagrangian framework. The origins of the SPH method can be traced back to the pioneering work of Lucy [22] and Gingold and Monaghan [23]. In 1993, Wingate and Stellingwerf [10] developed a research code named "*SPHINX*" at the Los Alamos National Laboratory, primarily for modelling scenarios involving large deformations and impact. Wang Zhi-Bin et al. [24] conducted a comprehensive review of recent advancements in SPH, with a particular focus on addressing challenges related to pressure handling in incompressible flow, boundary conditions, and surface tension within the SPH framework.

Particle-based numerical methods for simulating impacts and multiphase phenomena often exhibit superior robustness compared to traditional mesh-based techniques such as finite elements and finite differences.

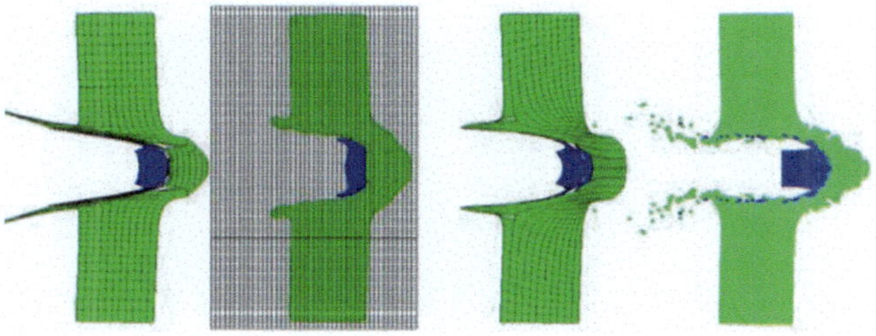

Fig. 6.2 Examples of Lagrange, Euler, ALE, and SPH simulations on an impact problem (Wang et al. [25])

Figure 6.2 provides a visual comparison of simulations conducted using various numerical frameworks, including Lagrange, Euler, ALE, and SPH, in the context of an impact problem. Wang et al. [25] reported the results of simulations performed using these four methods, and Fig. 6.2 schematically illustrates the key differences among them.

In SPH, particles are embedded within the material, serving both as mass points and interpolation points used to calculate physical variables based on data from neighbouring SPH particles. These calculations are scaled by a weighting function. Notably, SPH does not rely on a predefined grid, which helps prevent issues related to grid distortion and tangling. In contrast to the Euler method, SPH offers well-defined material boundaries and interfaces, making it particularly suited for problems involving extensive material damage and separation, such as micro-cracking. These responses are often observed in brittle materials and hypervelocity impacts. However, it is important to note that mesh-free methods, including SPH, can sometimes be less computationally efficient compared to mesh-based Lagrangian methods with similar resolution. Figure 6.2 clearly shows the material deformations observed under different numerical frameworks, with each method exhibiting unique characteristics despite achieving similar deformed configurations. For instance, ALE simulations predict thicker segments with pronounced bulging on the target's back surface and stretched elements around the cavity, while Lagrangian simulations reveal crushed elements beneath the projectile. The cavity volumes also vary among the four methods.

In the year 2000, Randles and Libersky [26] introduced a significant advancement known as "*normalized SPH with stress points*". They extended SPH to a normalized, staggered particle formulation that effectively addressed boundary conditions and instability issues in tension. This extension incorporated a companion set of interpolation points to store stress, velocity gradients, and other derived field variables.

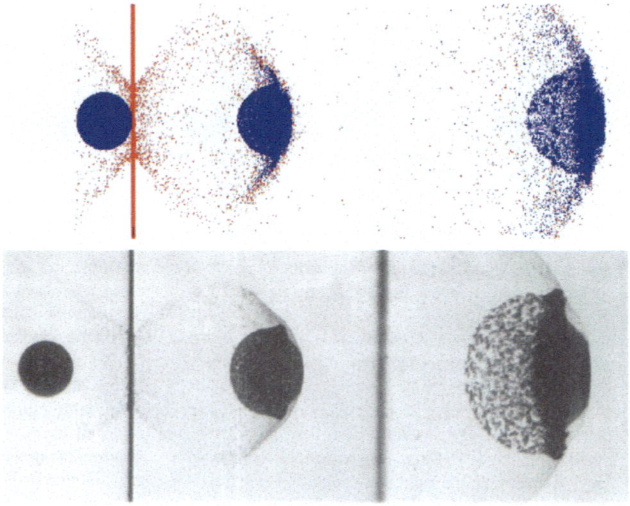

Fig. 6.3 Computed (top) and experimental (bottom) debris clouds from an aluminium sphere impacting an aluminium plate at 6.15 km/s. Images taken 0.0, 8.1, and 23.2 μs after impact [27]

The resulting method demonstrated stability, linear consistency, and an explicit treatment of boundary conditions. Furthermore, it introduced a novel approach for finding neighbouring particles, which selected a minimal and robust set while remaining insensitive to anisotropy in particle arrangements. Test problems confirmed that these improvements led to enhanced accuracy and stability.

Readers can find simulations of a debris cloud impact problem by different types of hydrocodes, including the SPH code in Beissel et al. [27], who performed computations with the EPIC3D code utilizing both finite elements and mesh-free particles as shown in Fig. 6.3. They implemented an algorithm for the automatic conversion of distorted finite elements into mesh-free particles to simulate this hypervelocity impact problem. The algorithm in EPIC3D (a Lagrangian Code) tracks material boundaries and properties without the errors typically associated with Eulerian formulations as the material traverses large distances. In Fig. 6.3, the simulation successfully reproduced the three regions in the debris cloud observed in X-ray pictures: a front region composed of droplets of melted projectile and target, a middle region of fragmented projectile, and a back region of spalled projectile. This example demonstrates the capability of the element-to-particle conversion algorithm in Lagrangian codes to replicate experimental results.

Several aspects related to SPH particle generation are elaborated upon in the Sandia Report authored by Beckwith et al., titled "*Sierra/Solid Mechanics 4.56 User's Guide*" [7]. According to this report, if two SPH particles are created in extremely close proximity (e.g. within 10% of a particle's radius of each other), these particles may violently separate, introducing significant artificial energy into the system. Similar instability concerns can arise in many Lagrangian hydrocodes when using element-to-particle conversion or embedded particle algorithms (e.g. material point method). This can happen when particles are placed very close to elements or other particles. Zheng et al. [28] presented this method to solve large deformation problems in two-phase porous media such as soils. Additionally, in high-velocity impact scenarios, contact or particle repulsion may not always function perfectly, allowing bodies to temporarily occupy the same space. If a newly converted particle coincides with an existing one, it can lead to a substantial influx of energy, causing instability. To mitigate such issues, potential solutions include

1. Reducing the time step.
2. Minimizing contact or particle repulsion errors.
3. Implementing a "failsafe" death criterion to detect and remove unstable particles (e.g. particles with velocities exceeding a certain threshold).
4. Enhancing the enforcement of contact mechanics (e.g. increasing the number of contact momentum balance iterations).

These measures can help prevent stability errors from propagating throughout the analysis. The Material Point Method (MPM) is further discussed in the following section.

6.1.5 Material Point Method

The origin of the "*Materials Point Method (MPM)*" can be traced to the "*Particle-in-Cell*" method by Sulsky, Chen, and Schreyer [29]. The MPM is a hybrid Lagrangian/Eulerian discretization method designed to address the challenges posed by large mesh distortions in Lagrangian and front (and interface) tracking in Eulerian cells. However, this method remains relatively unfamiliar as compared to finite element methods among engineers and research students involved in computational analysis for several reasons:

1. Lack of fully developed commercial codes.
2. Method is still evolving.
3. The need for careful handling of artificial viscosity when modelling shock wave propagation in MPM.
4. Limited studies demonstrating the impact of particle resolution and background grid on results.
5. MPM is considered a "weak" solution compared to Finite Element Method (FEM).
6. Lack of fully developed algorithms to describe fracture and fragmentation of solids.
7. Insufficient data demonstrating robust scalable algorithms.

In this section, we highlight key features of MPM and its application in simulating high-velocity projectile penetration problems. Zhen and Brannon [30] traced the origins of MPM techniques for solving impact problems in their Sandia National Laboratory report and summarized through Table 6.1.

Table 6.1 MPM solution steps [30]

For each material point (particle), perform the mapping operation from particles to cell nodes
Map the mass from the particles to the nodes of the cell containing these particles
Map the momentum from the particles to the nodes of the cell containing these particles
Find the internal force vector at the grid nodes
Apply essential and natural boundary conditions to the grid nodes, and compute the nodal force vector
Update the momenta at the grid nodes
For each material particle, perform the mapping operation from the nodes of the cell containing the particle to the particle
Map the nodal accelerations back to the particle
Map the current nodal velocities back to the particle
Compute the current particle velocity for strain calculations
Compute the current particle position
Compute the particle displacement vector
This unique feature of the MPM allows simulations of impact and penetration problems without the need for a special contact algorithm
Map the updated particle momenta back to the nodes of the cell containing these particles
Find the updated nodal velocities
Apply the essential boundary conditions to the grid nodes of the cells containing the boundary particles. For the essential boundary conditions, this treatment is consistent with the weak form of the governing equations because are assumed to be zero on the essential boundary
If needed for a constitutive model, find the current gradient of particle velocity the particle strain increment and the updated deformation gradient tensor
Find the stress increment from the constitutive model for the given strain increment and update the particle stress tensor
Identify which grid cell each particle belongs to and update the natural coordinates of the particle. This is the convective phase for the next time increment
Repeat the above steps for the next time increment if the required termination time has not been reached

Despite utilizing a background mesh for gradient and divergence calculations, MPM maintains a meshless approach, continuously adjusting spatial discretization as the body deforms. Notably, the use of the same set of nodal basis functions for both particle-to-cell and cell-to-particle mappings ensures a natural no-slip contact/impact scheme, eliminating interpenetration issues during penetration problems.

When compared to other meshless or particle-based methods, the MPM stands out as an excellent choice for addressing penetration and perforation problems. Its implementation is straightforward, and MPM is notably less complex and cost-efficient compared to finite elements. This efficiency is attributed to the use of the same set of continuous shape functions for both mapping—from material points to cell nodes and from cell nodes to material points. Consequently, this naturally leads to a no-slip contact and impact scheme, effectively preventing interpenetration [31].

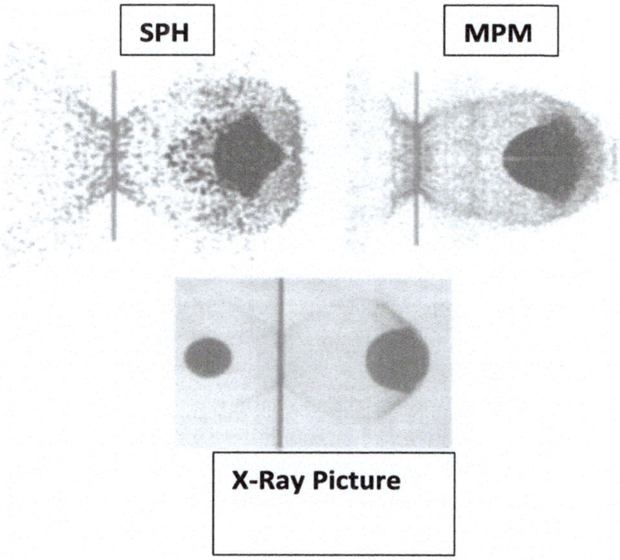

Fig. 6.4 Debris cloud obtained by SPH (top left), MPM (top right), and experiment [32] (bottom) at 8 ms after impact

Experimental validation is crucial for both SPH and MPM, as demonstrated by Ma et al. [32] in their simulations of hypervelocity impact problems. The time-consuming neighbour searching, which is required in both the SPH and element-free Galerkin methods, is not required in MPM. The shape functions used in MPM exactly satisfy the constant and linear consistency.

In addition, boundary conditions and contact algorithms can be implemented in MPM as easily as in FEM. Since a regular background mesh or computational grid is used in all time steps, the time step size remains the same in MPM simulation. Figure 6.4 shows simulations of debris cloud impact problem using an in-house research MPM3D code and the SPH option in the LS-DYNA code, respectively.

Though both methods showed promising results, a comparison with the X-RAY picture revealed that the MPM predictions matched the experimental data more closely. In these simulations, both codes employed the Johnson–Cook material model and Mie–Gruneisen equation of state with the same model constants. The material point method seems to be an efficient and promising method for simulating the hypervelocity impact problems.

There are a few open-source MPM codes [31, 33] available, including the *"Karamelo" C++ code*, designed for research purposes and easy modification. These codes facilitate the study of impact and penetration problems involving large deformations and fragmentations.

A book by Zhang et al. [34] provides theoretical insights and practical applications of MPM for high-speed impact, ballistic, and penetration problems. They also offer a Fortran code open to the public, aiding understanding of MPM calculations for various engineering problems. Additionally, a repository (https://github.com/topics/material-point-method) lists 30 sets of free MPM codes for researchers to explore.

Several review articles delve into MPM's development, including its current status and future challenges [34–36]. This method continues to mature, necessitating further developments, including multi-physics modelling, contact laws, and complex boundary conditions. However, MPM has already demonstrated its effectiveness in simulating high and hypervelocity impact scenarios.

Key challenges in MPM modelling include incorporating realistic artificial viscosity terms, tensile failure criteria and algorithms, noise control, and mitigating numerical oscillations during high-speed impact [36]. Recent work by Zhang et al. [37] explored shock wave propagation in solids using the "*Generalized Interpolation Material Point (GIMP) Method*", demonstrating its potential in solving dynamic problems involving shock compression and spallation.

MPM has also found application in modelling complex materials like glass fibre-reinforced plastics (GRP or FRP), where traditional Finite Element Method (FEM) meshes would be cumbersome. Recently, Chen et al. [38] demonstrated the effectiveness of MPM in analysing shock wave propagation in fibre-reinforced plastics by enabling detailed modelling of woven composites and various constituents/interfaces. FEM is not a user-friendly approach in the detailed modelling of the fibre bundles and the matrix as separate materials or structural components. The MPM can model the interfaces between the polymer or glass fibre and the brittle matrix (epoxy, polyester, etc.).

There are several types of MPM available in open literature. For instance, the MLS-MPM is a moving least squares MPM that does not require a structured or regular background mesh. According to Hu et al. [39], this method works well on a variety of problems, including large-scale flow of particles. In large-scale simulations of wide range of static or quasi-static or dynamic applications, visualization of the MPM results (such as sand flows) becomes essential. When several millions of particles are employed in a high-resolution modelling, not only a scalable three-dimensional MPM code is required, but also high-fidelity visualization capability. In his Ph.D. thesis at Massachusetts Institute of Technology, Yuanming Hu [40] reported the development of a programming language, called "*Taichi*", for visual computing.

Wang et al. [41] simulated a "*crushing concrete*" problem, using a scalable MPM for high-performance parallel computing on a single workstation with 4 NVIDIA Quadro P6000 GPUs. This simulation contained 93.8 million particles on a 10243 grid, achieving a 3.9 min/frame performance. This demonstrates the capability of a scalable MPM in simulating large-scale deformation, damage, and fracture without bottle necked by any mesh distortion. The high-resolution graphic rendering enhances our ability to identify particle-level failure mechanisms leading to complete failure of a brittle material under quasi-static compression.

In summary, MPM has evolved significantly and continues to be a promising method for simulating various high-velocity impact scenarios. Researchers can access a range of open-source MPM codes and resources to advance their work in this field. The application of massively parallel multi-GPU MPM by Wang et al. [41] has potential to simulate missile penetration into soil, projectile penetration into thick concrete structure, as well as to extract damage mechanisms under compressive loading at high strain rates. There are also other applications of scalable MPM, such as physics-based simulation to facilitate robot learning in navigation.

6.1.6 Discrete Element Method

For the sake of completeness, we present some discussions on the Discrete Element Method (DEM), which is considered a versatile tool for conducting fundamental research on soil and rock behaviours at both the micro- and macro-scales. Its applications span a wide variety of fields, including hydraulic fracturing, soil–tool interactions, brittle rock fracturing, slope and excavation stability, rock cutting, pavement design, material handling, bulk material flow/mixing, and cave mining.

Reference [42] states, "*The Discrete Element Model (DEM) is an alternative to the classical continuum mechanics approach for problems with flow-like deformations of solids. DEM is especially well-suited for problems such as soil ploughing where the media is subject to large, discontinuous deformations*". DEM is extensively employed in modelling vehicle–terrain interactions [43]. The DEM has inherent capability to model soil deformation mechanisms, such as localized shear deformation, friction-dependent resistance, and soil dilatancy. According to Peters et al. [43], the principal limitation of DEM is the large number of particles required to achieve a reasonably small ratio of discrete element radius to tire size within sufficiently large test bed dimensions. This necessitates scalable and parallel high-performance computing to solve large dimension problems using DEM.

There are several commercial codes available for analysing applications involving particle dynamics, including YADE [44] and PFC™ [45], which is based on the "*Discrete Element Method*". In this method, particles (blocks) are assumed to be rigid, and the behaviour of contacts between blocks is characterized using relative displacement–force laws. According to the PFC website https://www.itascacg.com/software/pfc, it can be used for fundamental research on soil and rock behaviour at the micro-scale to a wide variety of large-scale applications, as previously mentioned.

Simulation of systems with real particle shapes and sizes is a challenge, allowing for the specification of both spherical and non-spherical particle shapes. An example of a high-fidelity particle simulation software, The Ansys Rocky DEM (https://www.ansys.com/products/fluids/ansys-rocky), is a DEM code capable of simulating the flow behaviour or motion of particles in granular and discontinuous material particles with varying shapes, sizes, flexibility, and adhesion combinations to create unique particle sets. Also, numerous commercial DEM codes have been uniquely developed for pharmaceutical process simulations.

The application of the DEM method to projectile penetration into solid targets (ballistics) or shock wave propagation problems has been explored to a limited extent by in-house code developers in private and government organizations or laboratories. However, most of these developmental efforts remain undocumented or unpublished technical write-ups. Interested readers can find applications of DEM to large projectile penetration into soils or granular systems in open-access journals (https://www.intechopen.com/chapters/61821), as reported by Alshanti [46].

DEM has the capability to model large deformations, fracture, and fragmentation of solids, as seen in hypervelocity impact scenarios involving space shield problems. Steinhauser et al. [47] presented results of high-speed impact experiments on aluminium oxide and silicon carbide ceramics. They also modelled the fracture behaviour of ceramics based on discrete particles at meso-scale. Using a DEM-type particle method, high-speed pictures of crack nucleation and propagation in a ceramic plate due to an impact of a metallic cylinder on the edge of the plate (often called as "Edge-on-Impact test") were simulated. Steinhauser et al. [47] attempted to capture the experimentally observed crack patterns through two-dimensional DEM simulations in which the random distribution of particles could describe local variation of shear strength due to damage or cracks. This reference is an example of the DEM application to describe damage in brittle materials under multi-axial stress.

To model the dynamics of material fragmentation upon impact, Watson and Steinhauser [48] developed a computational model based on the DEM. While DEM is more commonly used in simulating industrial granular flow, the discrete point method developed by Watson and Steinhauser [48] extends DEM to model debris cloud formation under hypervelocity impact. DEM has also been employed in simulating brittle failure and impact into granular materials by Steinhauser et al. [47–49].

6.2 Numerical Artefacts

In most cases, solutions to impact problems are obtained using the "*Explicit*" method or solver. With an explicit algorithm, the new state or configuration can be calculated directly from the values of current state variables, such as stress, strain, strain rate, damage, etc., or data available in the current state. The explicit algorithm essentially involves an extrapolation operation. Since there is no need for matrix assembly or inversions in the explicit scheme, the computational cost of a single explicit increment is relatively small. All the required information is readily available, making the calculation straightforward and fast. However, several parameters, including the time step, artificial viscosity, contact, and mesh type, can influence the solutions of impact problems.

Many of the more advanced constraint types are not fully compatible with each other. Therefore, if a single node is involved in multiple types of constraints, the resultant behaviour may become unstable, particularly if the constraints are not fully orthogonal. The currently supported types of constraints include kinematic (such as prescribed velocity), rigid body, periodic, MPC, contact, and spot weld. Kinematic constraints are compatible with contact, while MPC constraints have limited compatibility with contact. However, most other combinations exhibit limited compatibility. For instance, a node with both a contact constraint and a periodic constraint may lead to model instability.

A blog by Christine Obbink-Huizer [50] discusses the appropriate usage of either of these two solvers (implicit versus explicit) in finite element applications. The choice of solver type influences the set of equations that are solved, the availability of certain features, the runtime, and even whether a solution can be obtained. This blog further explains the differences between the two solvers available in ABAQUS, their advantages and disadvantages, and when to choose each.

6.2.1 Time Increment

The determination of the time increment (Δt) follows the Courant criterion as in Eq. (6.1) ($\Delta t \leq \eta\, \Delta x / C_L$), where Δx is the smallest mesh size in the finite mesh, C_L is the wave speed in bars under a uniaxial stress state or plates under a one-dimensional strain state, such as in the plate impact configuration discussed in Chap. 5. The parameter η is a fraction (≤ 1). The maximum time increment is determined for $\eta = 1$. Most codes use 0.1 as the default value.

For all practical purposes, the stable time increment decreases with smaller element size, lower density, and higher stiffness (E—Young's modulus, K—bulk modulus, and G—shear modulus). Figure 6.5 is from reference and illustrates how the numerical solution can deviate from the analytical solution for large time increments. However, solutions for smaller time steps tend to converge to accurate and correct results.

Under severe or large deformation of a solid, such as in a projectile penetrating a thick target problem, the smallest mesh in the FEA determines the stable time at each time step. A highly distorted single element in the analysis can significantly impact the simulation time and could lead to a breakdown in the solution. In Ref. [50], the influence of the time increment in the explicit scheme is demonstrated through ABAQUS code simulations.

In addition to element distortion problems, the constitutive relationships describing damage evolution in materials can lead to solving extremely stiff differential equations. The damage-induced stress softening can cause serious time step issues. In general, the stable time increment can be further divided into smaller time steps to solve constitutive equations within the user-defined material model subroutines, such as VUMAT, when none of the existing material models in the ABAQUS/Explicit material library accurately represents the behaviour of the material to be modelled. Specialized numerical integration schemes, such as the iterative second-order Runge–Kutta method [51], and sub-cycling of the time-marching schemes can be used to solve stiff differential equations. A recent proposal by Soares [52] introduces an adaptive hybrid explicit formulation for wave propagation analysis as an example of handling global time-stepping algorithms. When seeking solutions to impact problems using any FEM code, it is important to conduct a time step sensitivity study to ensure accurate and realistic solutions. This also applies to many particle problems, whether or not they involve a mesh.

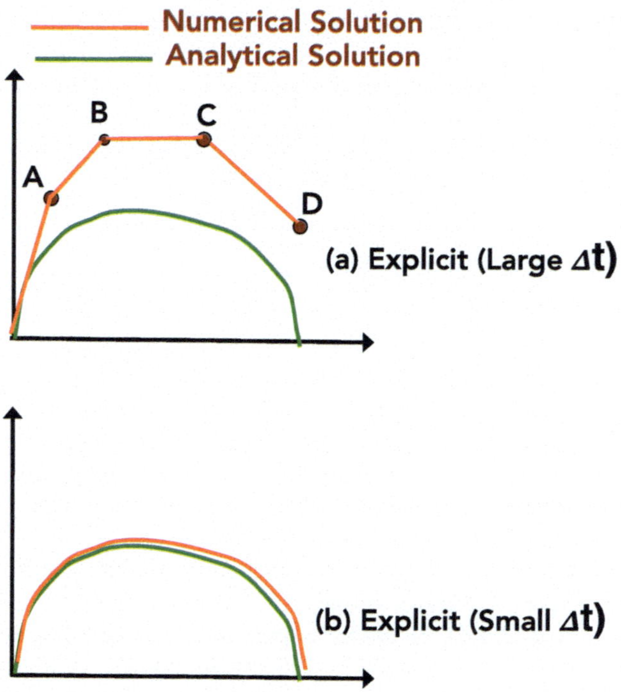

Fig. 6.5 Time incrementation with explicit solver. **a** Larger time increments lead to larger errors on the solution; **b** Smaller time converges to analytical solution

In hydrocode simulations, such as SPH or MPM, many of the numerical problems related to particle generation or closely spaced particles that are unstable at the default time step can be made stable by running them at a manually reduced time step [7]. The SIERRA code manual further suggests that there are several related issues: SPH time step calculation is not very robust, especially in high volumetric compression. If, over a single time step, two particles can move more than about one-half particle radius length, it can cause instability, effectively leading to the inversion of the SPH material. This can be problematic in high-velocity impact scenarios. The critical time step calculation does not consider the relative particle velocities, so the code may not detect when this instability is occurring. If this problem arises, it may be resolved by reducing the global time step scaling factor (e.g. from 1.0 to 0.75 or 0.5) and evaluating whether stability can be achieved. Generally, if stability cannot be obtained at a factor of 0.5, there may be factors other than the time step contributing to the instability.

6.2.2 *Artificial Viscosity*

The mathematical foundation for all hydrocode algorithms is the assumption that we are dealing with a continuum. This assumption, however, excludes the presence of shock waves, as in this formalism, shocks are considered mathematical discontinuities. Although it is possible to devise mathematical methods that establish internal floating boundaries connecting regions of continuous flow, the complexity increases significantly for three-dimensional (3D) flows, making it numerically challenging. Moreover, shocks often interact in intricate ways, and their typical narrow widths require extremely small cell dimensions for proper resolution. This, in turn, imposes significant computational memory and time requirements.

A solution to this dilemma was discovered by von Neumann and Richtmyer [53] in 1950. They introduced the concept of artificial viscous pressure (often imprecisely referred to as artificial viscosity), which, when added to the pressure, effectively spreads the shock wave over several cells, transforming the discontinuity into a steep gradient. The artificial viscosity serves to convert kinetic energy into internal energy within the shock. For accurate shock wave calculations, an artificial viscosity is essential in the difference equations to represent shock discontinuities correctly. The classical quadratic artificial viscosity, denoted as Q2, is calculated as follows [53]:

$$Q_2 = \mathrm{C}_2\ \mathrm{L}^2\tilde{\rho}\begin{cases}[\nabla\cdot\mathbf{u}]^2 & \text{if } \nabla\cdot\mathbf{u}\le 0\\ 0 & \text{if } \nabla\cdot\mathbf{u}>0\end{cases} \tag{6.2}$$

where C_2 is a constant (~ 2.0 for many Eulerian hydrodynamic codes), L is a length appropriate to the cell in which the artificial viscosity is calculated, and $\tilde{\rho}$ is the cell average density. The default value of 2 for C_2 spreads the shock over 3 or 4 cells. Note that $\nabla.\mathbf{u}$ is the volumetric term of material velocity, u. When the deformation is under volume expansion (or tension), no artificial viscosity is added to the pressure.

The linear artificial viscosity is given by

$$Q_1 = -\mathrm{C}_1\ \mathrm{L}\tilde{\rho}\tilde{c}\begin{cases}\nabla\cdot\mathbf{u} & \text{if } \nabla\cdot\mathbf{u}\le 0\\ 0 & \text{if } \nabla\cdot\mathbf{u}>0\end{cases} \tag{6.3}$$

Here, C_1 is a constant and $\tilde{c}$ denotes the cell average sound speed. The linear artificial viscosity is used to dampen oscillations behind shock waves. Users have the flexibility to select both coefficients (C_1 and C_2) and the length scale (L). The total artificial viscosity, Q, can be expressed as

$$Q = Q_1 + Q_2 \tag{6.4}$$

or it can be written in terms of bulk strain rates as

$$Q = C_2L^2\tilde{\rho}\,(\dot{\varepsilon}_B)^2 - C_1L\tilde{\rho}\tilde{c}\dot{\varepsilon}_B \tag{6.5}$$

where $\dot{\varepsilon}_B$ represents the bulk or volumetric strain rate.

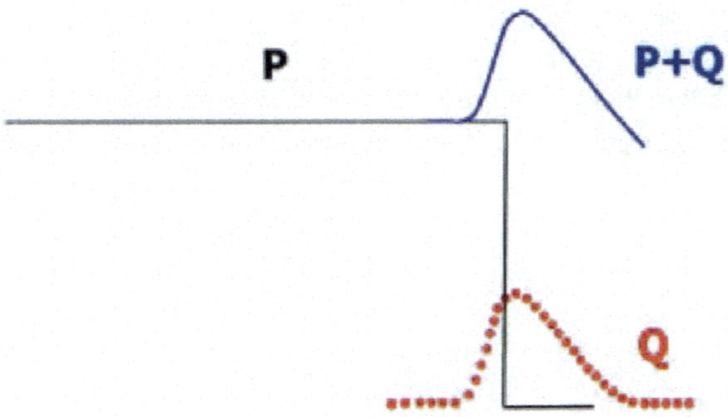

Fig. 6.6 A typical pressure and artificial viscosity in the region of a shock [9]

Artificial viscosity works by spreading the shock over several zones, ensuring that the variables vary continuously through the shock region and satisfy the Rankine–Hugoniot conservation relations. Regardless of cell size, the shock will be spread over multiple zones, and a few cells away from the shock, artificial viscosity becomes zero.

The influence of artificial viscosity depends significantly on cell size. Simulations conducted at very fine mesh resolutions may not require any artificial viscosity. Moreover, in simulations where extreme gradients are absent, artificial viscosity may need to be suppressed entirely. Excessive use of artificial viscosity can often dampen the solution.

For an ideal shock, the pressure forms a square wave. When the artificial viscosity is computed and added to the pressure, the result is shown in Fig. 6.6. The shock is spread out over several cells, and the artificial viscosity is a fraction of the pressure amplitude in the shock region, becoming nonexistent away from the shock.

6.2.3 Mesh Size and Time Increment Dependency

In general, commercial and research hydrocodes exhibit robustness, and solutions to nonlinear dynamics problems using uniform fine meshes tend to be largely independent of mesh size. These codes yield realistic solutions, especially when deformation and damage processes exhibit strain rate dependence, indicative of dissipative behaviour. For metallic materials, adopting viscoplastic constitutive equations and rate-dependent cumulative damage models, implemented through robust algorithms, goes a long way in mitigating mesh dependency. Nevertheless, achieving identical solutions across all meshes and time increments can be challenging, primarily because material models operate in a "*local*" fashion. Stresses are determined using information from element nodes or integral (quadrature) points.

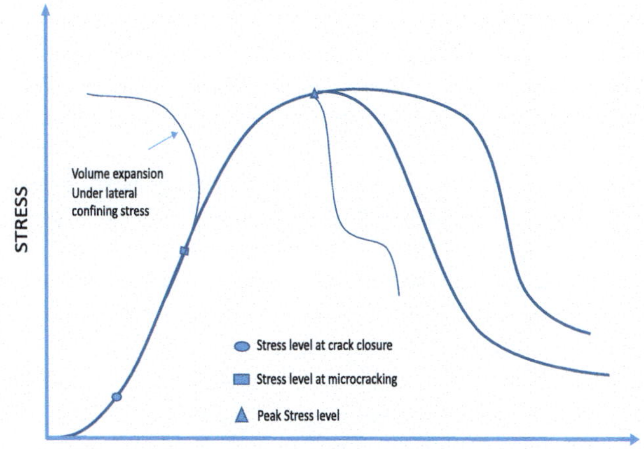

Fig. 6.7 A schematic of stress–strain curve for a typical rock exhibiting stress softening after reaching a maximum stress (failure initiation point). (Brady and Brown)

To attain a mesh-independent solution, constitutive relationships should adopt a "*non-local*" character that takes into account length scales like strain gradients. Conventional constitutive/damage models that rely on classical rate-independent plasticity theory or local theory for solving dynamic problems lack an inherent length scale. Furthermore, when brittle materials, like rocks, exhibit strain (or stress) softening behaviour, as illustrated in Fig. 6.7, resulting in declining stresses with increasing strains (or strain localization), the solution tends to become highly mesh dependent. The shape of strain softening portion of the stress–strain curve is highly dependent on the type of rock as well as loading conditions. Also, the softening response of rocks is highly strain rate and lateral confinement pressure dependent. For instance, marble, shale, and red sandstone exhibit different post-peak behaviours [54]. To capture accurately the complex response of rocks in the finite element analysis, use of high-resolution meshes is essential. However, further refinement of the mesh often worsens issues, such as numerical stability problems linked to mesh size and mesh alignment sensitivities. Notably, the anisotropic constitutive model by Costin and Stone modelled volume expansion (bulking) behaviours in Westerly Granite as shown in Fig. 5.13 in Chap. 5. Illankamban and Kracinovitch modelled the volume expansion and softening in Norite rock using their micro-mechanical model as was shown earlier in Fig. 5.11 of Chap. 5.

Several regularization approaches in constitutive modelling can address this problem. These approaches encompass viscoplastic models, thermal dissipation models, non-local models, and strain-gradient models. Numerous non-local plasticity and damage models proposed in the literature introduce intrinsic length scale measures into the constitutive equations.

A review paper by Murakami and Liu [55] discusses the reasons for mesh dependency in Continuum Damage Mechanics (CDM) models in hydrocodes. They observe that in brittle materials, material instabilities, particularly those leading to bifurcation and strain localizations, are difficult to capture accurately without adopting appropriate mesh sizes. Their paper outlines various regularization methods to mitigate mesh dependencies, including the use of a minimum admissible mesh size or employing mesh-dependent material properties. While a detailed discussion of these numerical methods is beyond the scope of this section, it is clear that these methods have their limitations. Determining the mesh size that aligns with material properties and discontinuities can be challenging, and the use of mesh-dependent material properties may not always be practical. Nonetheless, material model parameters can be calibrated by selecting a realistic or reasonable mesh size that captures features in a composite material, such as woven glass fibre-reinforced plastic (GRP).

In general, realistic solutions are achieved in hydrocodes when the mesh remains uniformly consistent throughout the body. The design of the mesh hinges on the type of impact problem under investigation. For example, in Chap. 4, a variety of experimental configurations were presented to evaluate shock and high strain rate properties of brittle materials. Stress and strain states vary across each configuration, and depending on the impact geometries, one-, two-, or three-dimensional simulations may be carried out using hydrocodes. For instance, the plate impact test is conducted under one-dimensional strain conditions at different pressures and strain rates. Details of plate impact experiments, particularly the characteristics of shock wave propagation and the velocity profiles measured by VISAR on the free surface, can be found in Chaps. 3 and 4.

In computational simulations, a single row of three-dimensional elements (such as the 8-node brick element) with periodic or roller boundary conditions applied to all nodes along the four lateral surfaces can generate one-dimensional strain states. This is achieved by allowing all nodes to move in only one direction. The time history of velocity (often referred to as profiles or signals) resulting from the flyer impact can be used to assess the impact of time step and mesh aspect ratio.

The ensuing discussions and results, as reported in the Masters Theses of James Fraser [56] and Nicholas Scott [57] at the University of Mississippi, serve to illustrate how significantly the time increment can influence the computed velocity profiles on the free surface in a plate impact simulation. They conducted simulations of a plate impact experiment [58], where a 25-mm-thick elastic flyer plate collided with a thin 6.8-mm-thick GRP at relatively low velocities. The flyer was modelled as an elastic solid (either aluminium or steel), while GRP was modelled using a continuum damage mechanics (CDM)-based hyperelastic solid [59]. Fraser examined mesh dependency on shock wave profiles using the Lagrangian (ABAQUS) code, while Scott utilized the Arbitrary Lagrangian-Eulerian code, ALE3D. Both studies employed the CDM/hyperelastic model for the GRP and the same plate impact configuration.

6.2.3.1 ABAQUS Simulations

Fraser [56] implemented a weakly strain rate-dependent hyperelasticity/CDM (continuum damage mechanics) model in the ABAQUS/explicit code using the VUMAT

Table 6.2 Abaqus plate impact model mesh details

Flyer plate				GRP			
Total dimension mm	Element dimension mm	Elements	Nodes	Total dimension mm	Element dimension mm	Element	Nodes
0.1 × 0.1 × 25.4	0.05 × 0.05 × 0.05	2032	4581	0.1 × 0.1 × 6.8	0.05 × 0.05 × 0.05	544	1233

subroutine. The ABAQUS simulations represented the GRP as a 6.8-mm-thick plate impacted by a thick steel or aluminium flyer plate. Table 6.2 presents some details of the simulations.

Preliminary analyses suggested that better results could be achieved by using a finer mesh size and different mesh geometries. A comparison of results for an array of mesh configurations is presented in the sensitivity study. The best quality velocity profiles were attained using cubic 0.05 mm × 0.05 mm × 0.05 mm elements, so this was chosen instead of the original 0.1 mm × 0.1 mm × 0.1 mm element size. The overall dimensions of the GRP and flyer were kept the same. Table 6.2 summarizes the mesh geometries and element amounts. Furthermore, a trade-off between velocity profile quality and computational time led to the decision to use a 0.4 ns time step, meaning 15,000 increments were used for the analysis. The maximum time window of interest was 6 μs, which was 4 μs after the rear surface velocity profiles commenced.

To study the effect of a mesh on the velocity profiles, Fraser [56] presented results for six different mesh sizes. Figure 6.8 shows the effect of mesh size on the velocity profile from a plate impact test configuration. The finest 8-node cubic element size was 0.02 mm × 0.02 mm × 0.2 mm with an aspect ratio of 10, while the finest volume element tested was 0.025 mm × 0.025 mm wide × 0.01 mm with an aspect ratio of 0.4. Fraser found the velocity profile from a mesh size of 0.05 mm × 0.05 mm × 0.05 mm cubic elements matched reasonably well with the plate impact experimental data.

As the thickness (size) of the elements decreased from 0.1 mm (blue line) to 0.01 mm (green line), the arrival time of the shockwave from the impact plane to the back surface of GRP increased with increasing fineness of the mesh. Extremely fine or small mesh size causes features, like the spikes in Fig. 6.8 that are not observed in plate impact VISAR measurements. Since the conservation of mass, momentum, and energy is "*nearly*" preserved each time step or, in other words, the Rankine–Hugoniot condition is met, the peak velocity is not influenced by the mesh size. It should be noted though the transient dynamic behaviour (ramping of the shock wave) is severely mesh dependent, the shock response or the solution (peak velocity) is however very accurate.

Table 6.3 shows when the fineness increased by 10-fold (from 0.1 to 0.01), the arrival time increased only by 10%; however, the wave rise time became significantly

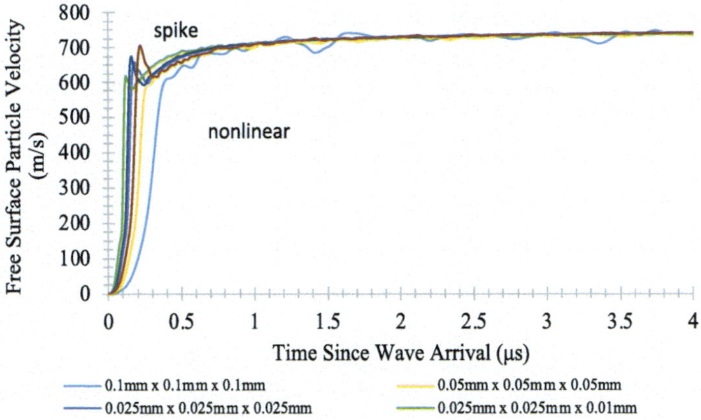

Fig. 6.8 Element size particle velocity profile comparisons

Table 6.3 Element thickness effects on shock wave arrival time

Element thickness (mm)	Shock wave arrival time (μs)
0.10	1.87
0.05	1.97
0.04	2.00
0.025	2.02
0.02	2.03
0.01	2.06

steeper as seen in Fig. 6.8. For larger mesh size, the shock artificially spreads across the element introducing "*numerical*" viscosity. The opposite case was also true; using a too coarse mesh, the wave propagated too quickly as the discrete jumps between elements were too large. Fraser chose 0.05 mm element in the plate impact simulations by matching the arrival time to the experimentally determined 1.978 μs [58]. In addition to shock wave arrival time, the point of nonlinearity initiation (seen at the top of the velocity profiles in Fig. 6.8) and development of the nonlinearity in the velocity profile until the peak particle velocity was reached were important in element size choice. The velocity profiles for finer mesh sizes were characterized by steeper rise times, a more pronounced Elastic–Elastic Cracking (EEC) transition point, and a smoother tapering up to the peak velocity. The coarser element sizes possessed a more gradual rise time yet achieved a more accurate representation of the oscillations in the transition between the EEC initiation and the peak velocity. These findings further motivated the choice of the 0.05 mm cubic element.

6.2.3.2 ALE3D Simulations

This section discusses the mesh-dependent solution from a robust and well-developed Eulerian code, ALE3D [8], due to a weakly strain rate-dependent elastic damage model. Mesh dependency is inherent in damage models, rendering true convergence unattainable due to their intrinsic nature. Nicholas Scott [57] conducted simulations of plate impact experiments described by Tsai et al. [58] using the ALE3D. He demonstrated that excessively fine resolution could trigger numerical instabilities, such as severe mesh distortion and inversion of rectangular meshes resulting in negative volumes. These issues necessitated remedies like re-meshing, adaptive meshing, artificial viscosity, and particle conversion methods. Given the unidirectional motion experienced by the target in a plate impact test, Scott devised a rectangular mesh in ALE3D with an aspect ratio of 8:5. This configuration allowed for contraction in the shock direction, maintaining elements' shape close to cubes (1:1) in high-velocity impact simulations.

The mesh design considered the physical dimensions of the GRP, particularly the thickness of the fibre bundle, ensuring that each finite element encapsulated a portion of the 0.68-mm-thick fibre bundle or ply. This approach aimed for at least 25 elements to model the fibre bundle. Consequently, the length of each element in the shock direction amounted to 27.2 μm with 25 layers of elements for each ply, totalling 250 elements in the thickness direction for 10 plies. To model the GRP target in ALE3D, simulations utilized a total of $10 \times 10 \times 250$ parallelepiped elements.

For handling the excessive applied compressive strains in the longitudinal (shock) direction, a moderately refined rectangular mesh with an aspect ratio of 1.6 was chosen. This selection allowed for a significant reduction in mesh dimensions in the shock direction. Additionally, a nearly uniform mesh was adopted in the lateral directions, with element dimensions of 0.017 mm in height and width. This choice facilitated exact contact between the target and flyer faces (node to node), ensuring a robust contact response. An auto-contact algorithm was employed to model the impact-contact process.

Figure 6.9 illustrates a mesh resolution study indicating the absence of convergence for a plate impact test at an impact velocity of about 112 m/s with a large mesh size, such as 100 μm. The solution presented in this figure is unrealistic and breaks down. As the uniform mesh size decreases, both the linear and nonlinear portions of the velocity profile shift leftward, leading to a significant decrease in the rise time of the wave, as indicated by the arrow. The slope of the elastic line for a finer mesh becomes steeper, aligning with observed experimental data. Convergence is achieved, and mesh dependency minimized when employing a uniform mesh size of 5 μm [57].

LS-DYNA-SPH Simulation

To illustrate the impact of zone or mesh size on free surface velocity profiles in a ballistic test at very high-impact velocities (around 5 km/s), we present findings from simulations using the SPH option in LS-DYNA [2]. Shawoon Roy et al. [60] conducted a comparative study employing two different finite element

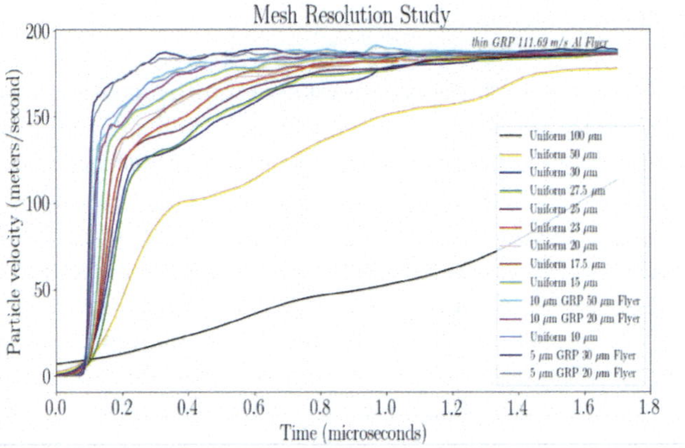

Fig. 6.9 Mesh size dependency on the wave profiles in a plate impact test simulation using ALE3D. The "elastic" damage model which is very weakly rate dependent tends to cause the severe mesh dependency [57]

methods: Lagrangian-based Smooth Particle Hydrodynamics (SPH) and Eulerian-based hydrocode, CTH, for simulating a hypervelocity ballistic experiment. They utilized the Johnson–Cook material model for strength and the Mie–Gruneisen equation of state in both codes. Figure 6.10 shows the influence of varying particle spacing on the free surface velocity profiles obtained from these simulations.

The study revealed that increasing particle density led to a more accurate representation of the elastic precursor wave. However, it was noted that higher particle density resulted in peak velocities higher than those observed experimentally. Moreover, augmenting the particle count in the SPH model substantially increased computational time. This study underscores how particle spacing significantly impacts computational results [60].

When applying the same ballistic configuration in CTH (an Eulerian code), the computed velocity profiles exhibited notable zone size effects, as depicted in Fig. 6.10. The elastic wave and the sharp rise of the plastic wave varied with different zone sizes. Despite employing a strain rate-dependent strength model and artificial viscosities, the zone dependency persisted in the computations. It's worth mentioning that our focus isn't on the code's ability to match experimental data, but rather on understanding the effects of mesh size on computational results, influenced by various factors. As discussed earlier, several numerical artefacts and the nature of deformation and failure models can induce significant zone size effects even in a Eulerian code like CTH, as shown in Fig. 6.11.

In general, solutions obtained from hydrocodes tend to exhibit some degree of mesh dependency due to the inherent nature of continuum constitutive and failure models, especially in the absence of strong strain rate-sensitive ("*viscoplastic*") terms. Mesh size and time-step-dependent solutions can also be influenced by steep stress and strain gradients induced by damage. Various numerical algorithms are available to mitigate mesh dependency issues. For instance, one approach involves

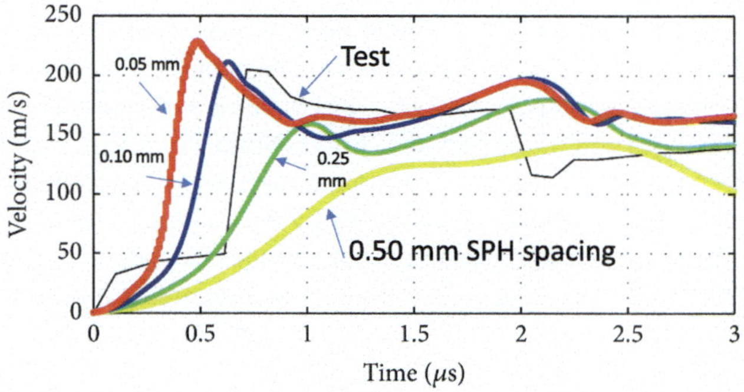

Fig. 6.10 Typical SPH particle sensitivity study [60]

Fig. 6.11 CTH simulations of grid or zone size effects on free surface velocity profiles in a ballistic test

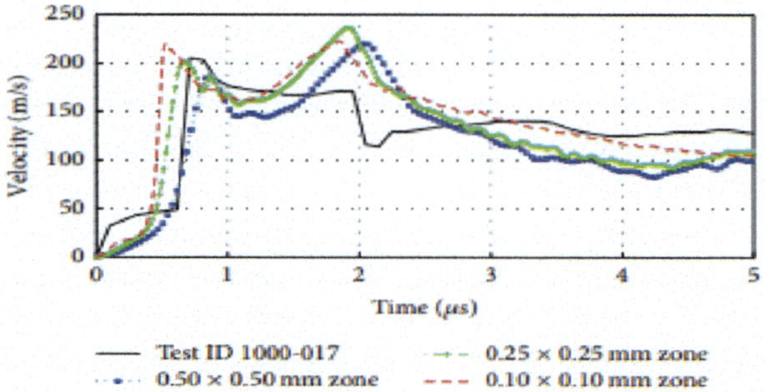

employing a statistically seeded mesh that captures the size-dependent Weibull distribution of strengths observed in experimental data [61]. According to the developers of ALEGRA [6], simulations conducted with ALEGRA have shown a significant reduction in mesh size dependency, a crucial step towards enhancing the model's predictive capabilities.

Chapter 5 addresses these issues by introducing what are commonly referred to as *"Local"* strain rate-independent constitutive (or strength) and damage models, which are implemented in various hydrocodes. The open literature offers numerous numerical remedies to mitigate the mesh dependency in these solutions.

6.2.4 Boundary and Contact Constraints

Handling boundary and contact constraints in material deformation modelling using Lagrangian finite element codes is typically straightforward. However, when it comes to addressing fracture and fragmentation in brittle materials, most traditional FEM codes rely on several algorithms. These include element-to-particle conversion, cohesive zones, as well as both smooth and non-smooth contact methodologies. The challenge arises as many of these advanced constraint types are not entirely compatible with each other. Consequently, if a single node is subjected to multiple types of constraints, the resulting behaviour may become unstable.

These constraints encompass various types such as kinematic constraints (e.g. prescribed velocity), rigid body, periodic, MPC (Multi-Point Constraints), contact, and spot weld constraints. Kinematic constraints are generally compatible with contact, and MPC exhibit some level of compatibility with contact. However, most other combinations have limited compatibility. For instance, a node subjected to both a contact constraint and a periodic constraint might cause model instability.

6.2.5 *Conversion of Finite Elements into Particles*

Currently, most Lagrangian codes, such as ABAQUS, incorporate element-to-particle conversion algorithms to mitigate the loss of conservation of mass caused by erosion of high aspect ratio and negative volume elements. However, this section is primarily influenced by the Generalized Particle Algorithm (GPA) and the Combined Particle-Element Method (CPEM), pioneered and developed by Gordon Johnson and his collaborators [62–64]. In 2002, Johnson et al. [64] introduced the GPA to address challenges related to large deformations, interfaces, and boundaries. In this algorithm, all variables (position, velocity, strain, stress, etc.) are carried at the nodes, which, similar to SPH codes, renders it susceptible to tensile instabilities. Gerlach and Johnson [12] introduced a novel contact and sliding interface algorithm for CPEM, specifically tailored for intense loading computations involving severe distortions.

CPEM represents a recent advancement enabling the conversion of distorted elements into meshless particles. The method offers the advantage of computing lower strain particles (stress points) using a fast and accurate finite element formulation, while higher strain particles are computed using a meshless particle formulation capable of handling severe distortions. Moreover, the meshless particle algorithm, employing Moving Least Squares (MLS) and Element-Free Galerkin (EFG) weak-form forces, remains consistent and free from tensile instabilities. Its adaptability extends to the conversion of finite elements into variable-connectivity meshless particles without the need for element deletion and particle addition; instead, a branch point based on equivalent plastic strain suffices. This basic approach can be applied solely to the element algorithm or the particle algorithm. However, the CPEM may introduce significant inaccuracies when used with existing contact/sliding algorithms.

In Ref. [64], Johnson et al. compared the capabilities and efficiency of three meshless particle algorithms in EPIC: the Generalized Particle Algorithm (GPA), the Nodal Element-Free Galerkin algorithm (Nodal EFG), and the Combined Particle-Element Method (CPEM). They evaluated various problems to analyse the effects of these algorithms with and without element-to-particle conversion, conversion from hexahedral and tetrahedral elements, and various sliding/contact interfaces. The subsequent section on "*Hydrocodes*" presents simulations of fracture and fragmentation of layered ceramic plates using the particle algorithms within the EPIC code.

6.2.6 *Length and Scale Effects*

Many constitutive relationships and damage models, often referred to as "*local*" models, are based on the determination of state variables, such as stress and damage, using values at specific nodal or integration points. However, it is well recognized that these local constitutive models can introduce mesh dependencies in finite element solutions.

To address this issue, "*non-local*" models incorporate length scale parameters, like stress gradients ($\partial\sigma/\partial x$) and strain gradients ($\partial\varepsilon/\partial x$), to capture the influence of stress and strain from neighbouring regions on the state variables at a node or particle. Non-local micro-mechanical models tend to be theoretically rigorous and may involve different length scales.

Another approach to mitigate mesh dependency, particularly in continuum damage mechanics-based models where stiffness linearly degrades with scalar damage variables, is to use techniques like the crack band model (CBM). Mesh dependency becomes apparent when stress decreases with increasing strain after reaching a critical level due to damage nucleation, growth, and coalescence, as depicted in Fig. 6.12.

Figure 6.12 illustrates the process of stress softening. At the peak stress, micro-cracks begin to activate, leading to a degradation in stiffness. Strain softening accelerates as micro-cracks coalesce. Beyond a certain point (indicated by the green solid circle in the figure), macro-cracks form, leading to material failure. The differential equations describing strain softening tend to be highly stiff, requiring special numerical treatment. The CBM approach based on fracture energy provides a numerical scheme to minimize the influence of finite element mesh size on solutions.

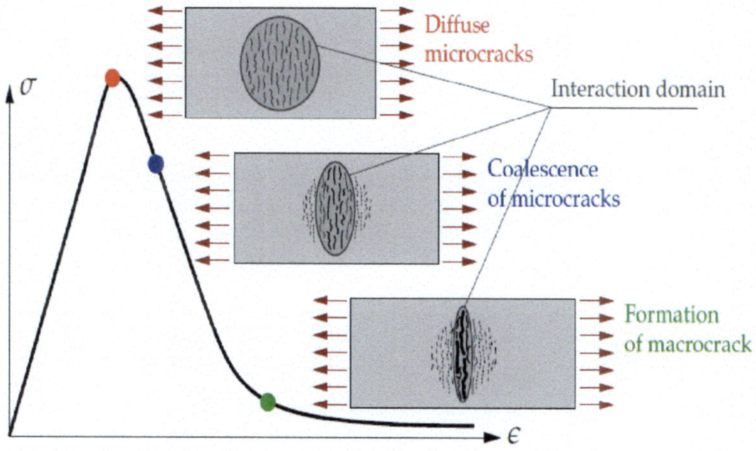

Fig. 6.12 Stress softening as described by the crack band model [65]

A recent article by Abdullah and Kirane [66] in 2022 suggests that the CBM is one of the most commonly used energy regularization techniques in continuum-scale finite element fracture analyses of brittle materials. They demonstrated the effectiveness of CBM in eliminating or reducing element size effects by simulating Mode-I dynamic crack propagation in a cracked PMMA plate subjected to uniform lateral edge displacements.

Their simple linear tensile-damage model assumes that damage, represented by "D", is zero when the applied effective strain (ε_{app}) is below the strain (ε_0) at maximum stress (F_t), and "D" equals 1 when ε_{app} equals or exceeds the fracture strain ε_f. In this model, the damage is computed from the following empirical expression:

$$D = \frac{\varepsilon_f}{\varepsilon_f - \varepsilon_0}\left[1 - \left(\frac{\varepsilon_0}{\varepsilon_{app}}\right)\right]; \varepsilon_0 \le \varepsilon_{app} \le \varepsilon_f \tag{6.6}$$

The elastic modulus, E, degrades using the CDM approach, where $E = E_0(1 - D)$, with E_0 being the intact solid's modulus before damage initiation. As shown in Fig. 6.13a, E decreases beyond the maximum stress, F_t with increasing strain (or damage). The effect of mesh or element size on the slope of the softening portion is illustrated in Fig. 6.13b.

CBM regularization is achieved by equating the crack width "h" equal the element size "h_e". For linear softening, this is incorporated into the definition of the failure strain:

$$\varepsilon_f = \frac{2G_F}{F_t h} \tag{6.7}$$

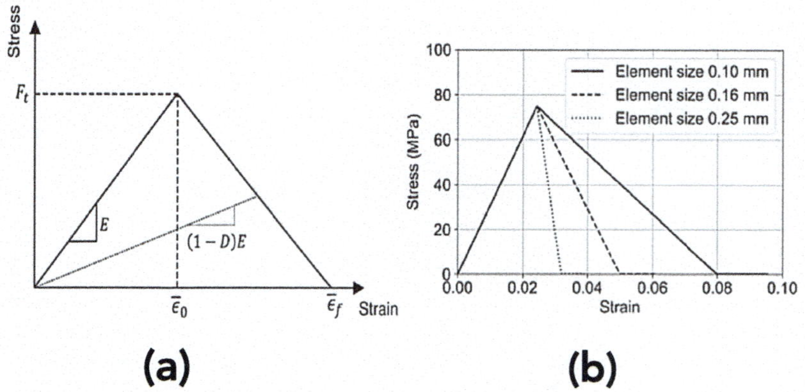

Fig. 6.13 Elastic stress–strain curve with linear softening. **a** The current slope after damage initiation is represented by the degraded modulus, $(1D)E$. **b** Mesh size effect on the softening path

where G_F is the material's fracture energy, also called strain energy release rate. The area under the stress–strain curve, $\varepsilon_f F_t/2$, representing the internal or fracture energy is scaled as $\frac{G_F}{h}$. According to Abdullah and Kirane [66], CBM regularization involves rescaling the post-peak slope by recalculating the failure strain based on the element size, as shown in Fig. 6.13b for three different mesh sizes. It's evident that the softening branch becomes steeper for larger element sizes, suggesting an upper limit to the usable mesh size to maintain model stability.

The simulation of dynamic Mode-I fracture in a thin plate under various strain rates produced results with localized, slightly branched, highly branched, and extensively branched diffused cracks. Their study revealed that CBM provides mesh-independent results only for localized or slightly branched cracks. For diffused or highly branched cracks, results depend on mesh size when CBM is used, but more realistic results are obtained if CBM is bypassed. CBM works when energy dissipation scales linearly with element size, which typically occurs for localized or slightly branched cracks. According to Abdullah and Kirane [66], this finding has significant implications for finite element modelling of dynamic fracture applications.

Figure 6.14 illustrates CBM's capability to minimize the mesh effect on various variables, such as crack speed. Without CBM regularization, the simple CDM model is expected to show significant mesh size dependency in ABAQUS results. Abdullah and Kirane's article [66] provides detailed simulations with ABAQUS code, emphasizing the usefulness of CBM-based regularization, especially for dynamic problems.

In 2018, Nelms [67] developed a damage model to describe deformation in a fish scale resulting from nanoindentations. This fish scale damage model follows the RG ceramic model discussed in Chap. 5. In this model, a characteristic volume for a scalar damage variable "γ" defines the micro-crack density as $N\psi^3$, where N is the average number of flaws and ψ is the average flaw (or micro-crack) length in the brittle layer of a fish scale. A damage parameter, D, ranging from 0 to 1, is further defined as follows:

$$D = \frac{\gamma}{\gamma_{crit}}; \qquad \gamma_{crit} = l_c^3 \tag{6.8}$$

The characteristic length l_c represents the critical flaw size and is dependent on the size of the finite element. Consequently, this crack length parameter makes solutions highly mesh dependent. The value of l_c is used as a limiter for crack growth and to reduce the common mesh dependency observed in damage-softening constitutive models. To further address mesh sensitivity issues, a viscous regularization employed in ABAQUS is also applied. The damage accumulation, which is a generalization of the Duvaut–Lions viscous regularization [68], is described by

$$\dot{D}_I^v = \frac{1}{\eta_I}\left(D_I - D_I^v\right) \tag{6.9}$$

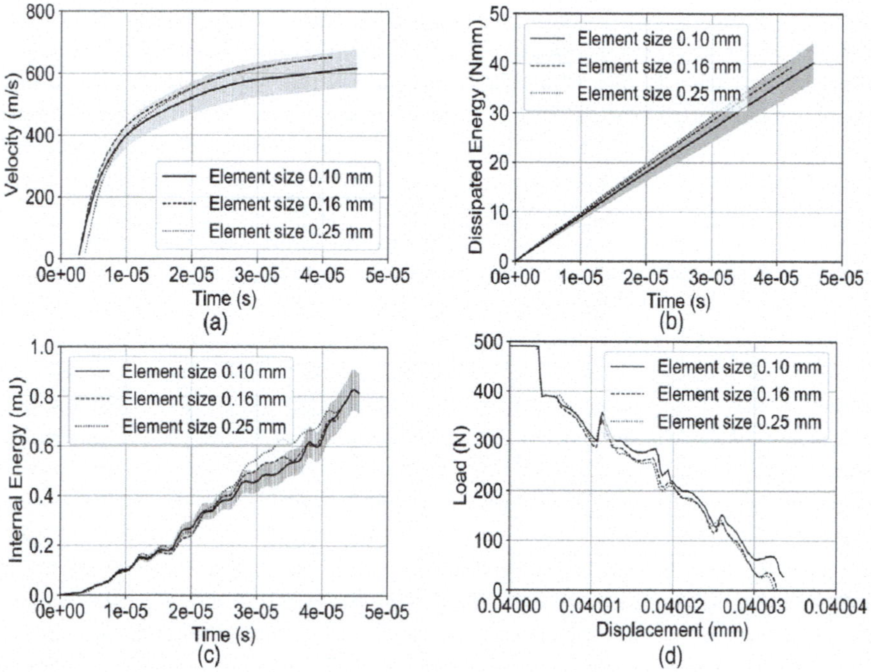

Fig. 6.14 **a** Leading crack velocity evolution, **b** dissipated energy, **c** internal energy, and **d** structural softening curves for load case A (localized crack) predicted with the CBM regularization with three different mesh sizes [66]

where η_I is an arbitrary parameter that should be significantly smaller than the simulation time step. Lapczyk and Hurtado [69] used this evolution equation in their regularization scheme for the regularized damage variable for various damage modes, denoted by the subscript "I" in fibre composites. D_I^v represents the regularized damage variable for mode I. While complete elimination of mesh sensitivity is challenging in finite element analysis with a local constitutive description, it is possible to reduce its effects to an acceptable level through the implementation of viscosity, regularization methods, inclusion of characteristic lengths, and damage growth viscous regularization.

Nelms [67] utilized the regularization options in ABAQUS to study how mesh dependency due to damage or stress softening can be minimized or eliminated in a nanoindentation experiment. His work demonstrated the effectiveness of viscous regularization and the use of a length scale damage parameter (l_c) linked to the finite element mesh size for individual elements in a finite element model.

To mitigate mesh dependency resulting from stiffness degradation or strain softening, the ABAQUS code introduces a characteristic length associated with the element size. It expresses the softening portion of the constitutive law as a stress–displacement relation. In this context, the energy dissipated during the damage process is specified per unit area, rather than per unit volume. The dissipated energy is treated as an additional material parameter and is used to compute the displacement at which full material damage occurs. This approach aligns with the concept of the critical energy release rate as a material parameter for fracture mechanics. This formulation ensures that the correct amount of energy is dissipated and significantly reduces mesh dependency, as suggested in the ABAQUS manual [70].

6.3 Hydrocodes

Many hydrocodes were initially developed to address impact and penetration problems primarily involving ductile metallic systems. However, over time, these codes have evolved to encompass the deformation and fracture of brittle materials such as rocks, concrete, and ceramics. Specialized treatments were introduced to handle fracture and fragmentation in these brittle materials. Despite the fact that brittle materials do not exhibit plastic deformation like metals, plasticity theories were adapted to describe permanent or "inelastic" strains. Special numerical schemes, including cohesive zones for delamination, element splitting for crack propagation, erosion to account for material distortion, and element-to-particle conversion for pulverization, were developed to simulate dynamic failure in brittle solids.

A key feature of hydrocodes is their ability to incorporate material models, strength properties, and equations of state through user-defined subroutines. For example, the VUMAT option in ABAQUS (explicit) allows users to implement advanced models to achieve accurate solutions for various impact-related problems. In recent decades, the use of lightweight, layered fabric composites with epoxy or polyester-type matrices in impact-resistant structural components has driven the need for the development of constitutive theories and continuum damage models.

The computational results obtained from hydrocodes are highly sensitive to the choice of constitutive models, as discussed in a previous section. It is well known that so-called "*local*" constitutive models, particularly those with damage descriptions, tend to produce mesh-dependent solutions compared to "non-local" models. We discussed this aspect of code applications in Chap. 5. Any stress or strain softening model will yield a mesh-dependent solution unless some strain $(\partial\varepsilon/\partial x)$ or stress gradient $(\partial\sigma/\partial x)$ is included in the constitutive equations, as described in Sect. 6.2.6. When a strong dissipative or viscous description, such as strain rate-dependent inelastic strains, is included in the constitutive/damage models, the mesh and solution time increment effects on the solution tend to be less pronounced.

It's essential to understand that the mere generation of a solution by the code does not guarantee its accuracy. As mentioned previously, the solution often depends on a wide range of factors, including mesh size, mesh shape, time increment, material model parameters, and appropriate boundary conditions. Boundary conditions should be carefully selected, especially when dealing with simplified problems, such as taking advantage of the geometrical symmetry of the structure or target being modelled. For instance, a 3D problem can be effectively represented using 2D axisymmetric, plane stress, or plane strain approaches or even reduced to 1D with periodic boundary conditions, as in the case of modelling plate impact geometry under 1D strain.

In the following section, we will present sample simulation results from various hydrocodes applied to the modelling of metal projectile penetration into a layered target composed of anisotropic fibre-reinforced plastics (FRP or GRP) and isotropic ceramic layers. A significant focus of these simulations lies in accurately describing deformation and damage using a wide array of material continuum mechanics damage (CDM) models.

Several notable hydrocodes include LS-DYNA, ABAQUS-EXPLICIT, ANSYS-AUTODYN, EPIC3D, CTH, ALE3D, DYNA3D, ALEGRA, SIERRA, PRONTO3D, and more. Developers of these codes, whether in the commercial, research, or open-source domains, continually enhance them with robust algorithms to address numerical challenges stemming from mesh and time dependencies, contact mechanics, cohesive zones, and stress–strain softening.

An example of this ongoing development is the ALE3D code, which remains actively evolving at Lawrence Livermore National Laboratory (LLNL). Besides advancements in physics and methodologies, the underlying computer science framework is continually adapted to ensure optimal performance on the latest High-Performance Computing (HPC) platforms. It's essential to note that certain codes, such as EPIC3D, ALE3D, and CTH, are export-controlled and restricted to official US Government use only.

ALEGRA: ALEGRA is an arbitrary Lagrangian-Eulerian (multi-physics) computer code developed at Sandia National Laboratories since 1990. This versatile code encompasses a wide range of physics options, including magnetics, radiation, and multi-material flow. Although it has been in development for nearly two decades, recent work has led to substantial enhancements in accuracy and robustness. These improvements encompass techniques applied to basic Lagrangian differencing, artificial viscosity, and the remap step of the method, notably improving the fundamental conservation of energy in the scheme. We will delve into these algorithmic advancements and their impact on critical applications, including magnetic implosions, ceramic fracture modelling, and electromagnetic launches.

Sierra/SolidMechanics: Sierra/SolidMechanics (Sierra/SM) is a Lagrangian, three-dimensional code designed for finite element analysis of solids and structures. It boasts capabilities for explicit dynamic, implicit quasi-static, and dynamic analyses. Explicit dynamics features facilitate efficient and robust modelling of solids and structures subject to extensive contact and large, suddenly applied loads. For implicit problems, Sierra/SM employs a multi-level iterative solver, enabling effective solutions for problems involving significant deformations, nonlinear material behaviour, and contact. The code offers a versatile library of continuum and structural elements, alongside a comprehensive selection of material models. Sierra/SM is designed for parallel computing environments, enabling scalable solutions for extremely large problems in both implicit and explicit analyses. It is built upon the SIERRA Framework, which streamlines integration with other SIERRA mechanics codes. This document outlines Sierra/SM's functionality and input syntax.

PRONTO3D: PRONTO3D is a time-dependent explicit finite element structural dynamics code, developed and maintained by Sandia National Labs. It is employed to model the transient responses of structures to time-varying loads. PRONTO3D serves as a component of the coupled Eulerian-Lagrangian code ZAPOTEC but can also operate independently. It supports most features found in other structural dynamics codes, such as DYNA3D, including automatic contact handling and the capability to model the elastic–plastic response of various engineering materials.

Simulation of Explosive Effects: Simulating the effects of explosives on structures presents a challenge. This is because the explosive response is best simulated using Eulerian computational techniques, while structural behaviour is more accurately modelled using Lagrangian methods. Due to differences in methodology and code architecture requirements, these techniques are typically implemented in separate computer programs. When explosive and structural modelling are performed using different codes, conducting coupled explosive/structure interaction simulations becomes challenging, if not nearly impossible. To address this issue, Sandia National Laboratories has developed two techniques.

The first technique employs Smoothed-Particle Hydrodynamics (SPH), a relatively new gridless method, akin to Eulerian, ideal for treating liquids and gases produced by explosives. The SPH capability has been fully integrated into the transient dynamics finite element (Lagrangian) codes PRONTO2D and -3D. This paper presents a PRONTO-3D/SPH simulation demonstrating the effects of an explosive blast on a protective-wall barrier.

The second technique leverages a new code called Zapotec, which combines the 3D Eulerian code CTH with the Lagrangian code PRONTO3D, requiring minimal changes to either code. Currently, CTH and PRONTO3D are operating on the Sandia Terraflops machine, and plans are underway for them to be executed in a coupled fashion on this machine.

6.3.1 EPIC-3D—Lagrangian Code (Restricted Code)

Dr. Gordon Johnson wrote a historical article [62] titled, "*Numerical Algorithms and Material Models for High-Velocity Impact Computations*", and showed the effectiveness of the EPIC code in simulating damage and fracture in several ballistic applications. Further, the readers can find a detailed history of this code's development in Refs. [11, 12, 63, 64], EPIC-3D, which stands for Elastic–Plastic Impact Computations in three dimensions, is based on an explicit Lagrangian finite element formulation. It incorporates nonlinear material strength and compressibility effects to account for elastic–plastic flow and wave propagation. In the earlier versions of the EPIC code [4], two-dimensional triangular elements and three-dimensional tetrahedral elements were used, as they handle severe distortions well. However, these elements introduced inaccuracies in certain conditions.

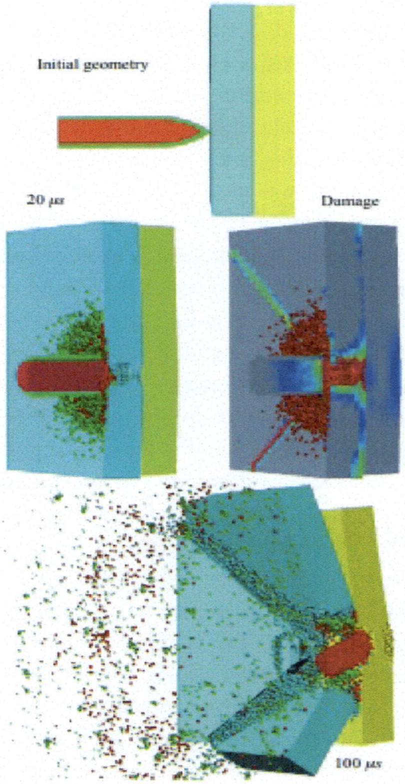

Fig. 6.15 Computation of a small-calibre projectile impacting a body armour component using element-to-particle conversion algorithms in the EPIC code

To enhance accuracy, Johnson's team adopted elements arranged in a crossed triangle pattern for two-dimensional elements and utilized tetrahedral elements with a new symmetric arrangement for three-dimensional simulations. While primarily designed for solving projectile-target impact problems, the fundamental formulation of the code is applicable to a wide range of dynamic response problems involving continuous media.

The example in Fig. 6.15 illustrates a small-calibre projectile (8.62 mm diameter) impacting a layered ceramic/metallic component (silicon carbide over aluminium) commonly found in some body armours. The projectile impacts 5.1 mm below the centre of a square tile (51 mm × 51 mm) at a normal velocity of 800 m/s.

The top part of Fig. 6.15 displays the initial geometry, featuring a copper jacket (in green) around the hard steel core. At 20 μs after impact, the ceramic plate shows significant damage, particularly under the projectile and along two radial cracks, despite no significant penetration into the ceramic. However, in this case, the finite thickness of the ceramic plate fails to resist the projectile effectively. By 100 μs after impact, the steel core has penetrated the ceramic and some of the aluminium. The ceramic has fragmented into three pieces (for the half of the problem shown), and the copper jacket has moved forward along the core. Due to its lower strength, the copper jacket cannot decelerate as effectively as the stronger steel core. Consequently, the steel core and aluminium move at almost the same low velocity, preventing core penetration through the aluminium.

The simulation depicted in Fig. 6.15 incorporates various algorithms, including those related to contact, contact interface, element-to-particle conversion, boundary and interface conditions, and more. This complex simulation showcases the capabilities of the EPIC code in modelling significant deformations, the initiation of various types of cracks (ring, cone, and radial) in the ceramic plate, followed by penetration and perforation of the remaining projectile into the aluminium substrate.

In the following sections, we present several example simulations of penetration or ballistic problems, primarily to highlight the effectiveness of hydrocodes in modelling extreme deformation and fracture due to impact loading. Our focus is on demonstrating the types of models used, rather than providing in-depth details of constitutive/damage/fracture models.

6.3.2 LS-DYNA—Lagrangian Code (Commercial)

This section discusses the application of the Lagrangian code LS-DYNA for ballistic design through modelling and simulation. To obtain meaningful solutions for high-velocity impact problems using hydrocodes, it is crucial to establish reasonable values for various numerical parameters. These parameters include erosion strain, mesh size, and density for a zero-thickness layer.

Referencing the work of Krishnan et al. [71], this section emphasizes the complexities associated with selecting input parameters for simulations. The study presents results for a ballistic problem involving the penetration of an APM2 projectile into three different targets: (1) a thick aluminium block, (2) a two-layer ceramic/aluminium armour, and (3) a two-layer ceramic/composite armour. We can employ Johnson–Cook and John–Holmquist models for the aluminium and ceramic, respectively, and the Mie–Gruneisen (equation of state) for both materials. The challenge is in the modelling of any layered composite. Krishnan et al. [71] used the ultra-high molecular weight polyethylene, UHMWPE laminate for the composite.

In the LS-DYNA simulations, Krishnan et al. considered four different meshes for the target and bullet configurations, respectively, to demonstrate the effects of mesh resolution on the depth of penetration of an APM2 bullet. In the finite element model, the bullet with tetrahedral elements and the target (ceramic/aluminium) with eight-node solid elements were modelled. LS-DYNA input requires parameters like IBULK (bulk modulus) and IG (shear modulus) for user-defined materials. Krishnan et al. set these values conservatively to the highest possible stiffness expected in the simulation. They also adjusted the time increment scale factor (η) and activated second-order stress updates for improved results. Attention to various energy checks by LS-DYNA in large deformation, high-impact ballistic problems is crucial for stable and physically meaningful solutions.

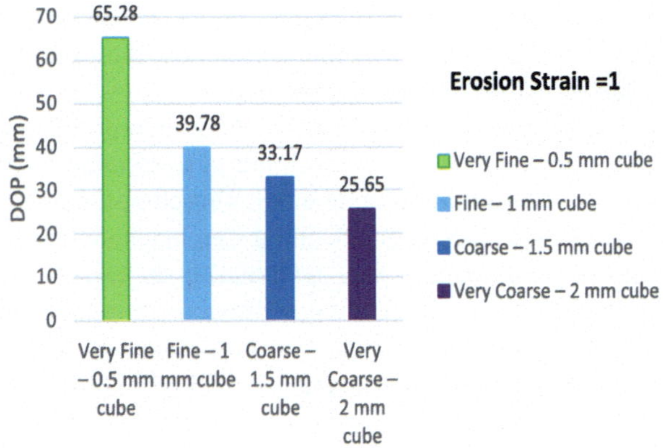

Fig. 6.16 Effect of mesh resolution on the depth of penetration of APM2 bullet into a thick block of aluminium

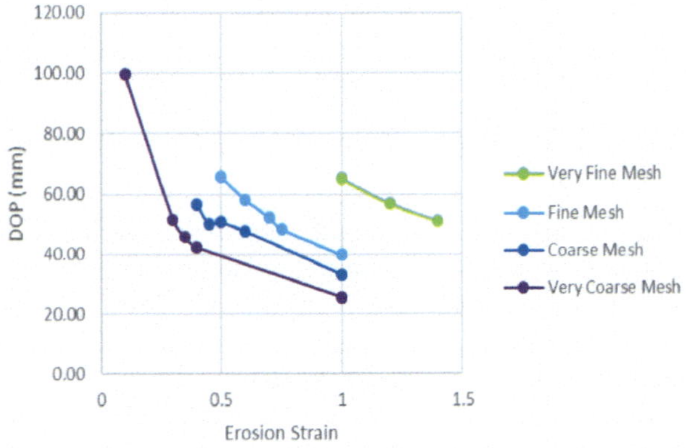

Fig. 6.17 The depth of penetration (DOP) versus erosion strain for four different mesh designs for the 6061-T6 Aluminium target

Figure 6.16 shows LS-DYNA simulation results for projectile penetration into a thick aluminium block for four different meshes with an erosion strain of 1. Element distortion, inevitable in large deformation and damage problems, necessitates the use of the element erosion option in LS-DYNA. Krishnan et al. examined the influence of mesh size on element erosion by matching experimental DOP test values with various erosion levels across four distinct meshes (Fig. 6.17). It's important to note that erosion is a non-physical parameter, and optimal erosion strain values may lack inherent physical significance.

In Fig. 6.17, the finest mesh with a 0.5 mm size projected a greater depth of penetration compared to coarser meshes. When erosion strain was set to 1, penetration depth decreased with increasing mesh coarseness, potentially indicating an artificial increase in target penetration resistance.

However, the calibration of erosion strain may be necessary in Lagrangian codes to achieve realistic solutions that closely match experimental measurements or data. In Fig. 6.17, the finest mesh with a 0.5 mm size projected a greater depth of penetration compared to meshes with sizes of 1, 1.5, and 2 mm. Notably, when the erosion strain

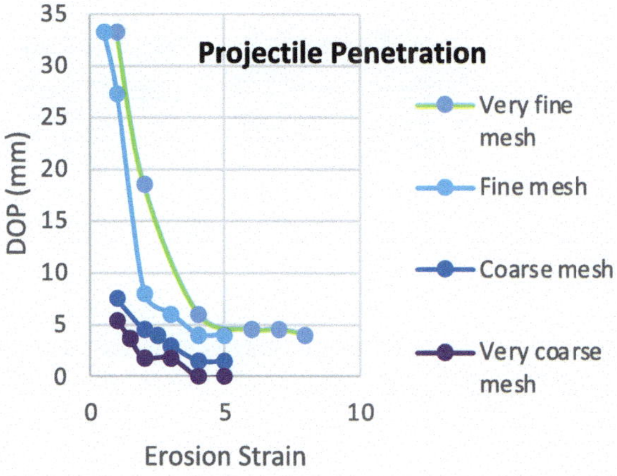

Fig. 6.18 The depth of penetration (DOP) versus erosion strain for four different mesh designs for the two-layer ceramic/aluminium target

was set to 1, the depth of penetration decreased as the mesh coarseness increased. This trend could be attributed to substantial differences in stress (or stress gradient) distributions near the penetration site for varying mesh coarseness levels. We can speculate that the penetration resistance of the target artificially increases with the coarseness of the mesh.

Krishnan et al. proceeded with their investigation of the combined effects of mesh and erosion strain on the DOP in the second target configuration: two-layer ceramic/aluminium armour. Figure 6.18 summarizes the effects of these two parameters on DOP results from LS-DYNA for the two-layer target. The DOP variation to mesh size and erosion strain follows the same trend observed in the aluminium block target. The erosion strains were applied to both the target and the bullet materials, coupled with a highly refined mesh. This further enabled the calibration of JH-2 strength and damage parameters for the ceramic, as outlined in Chap. 5 of this book. The calibration process was designed to bring the depth of penetration (DOP) and the extent of damage experienced by the ceramic in line with experimental results. As a result of this calibration method, the computational outcomes from LS-DYNA exhibited a reasonable alignment with ballistic data.

Following the simulation methods of a monolithic aluminium and ceramic/aluminium targets, Krishnan et al. extended their investigation to a two-layer ceramic armour target with a UHMWPE laminate instead of aluminium as the substrate. They considered different ratios of ceramic thickness to laminate thickness, modelling delamination of the substrate under combined shear/tension loading. The simulation effectively captured the influence of ceramic thickness on target resistance, as indicated by projectile penetration depth (Fig. 6.19).

Figure 6.19 state of bullet at 150 μs in a two-layer configuration with ceramic as top layer and layered composite (ultra-high molecular weight polyethylene, UHMWPE) as the substrate [70].

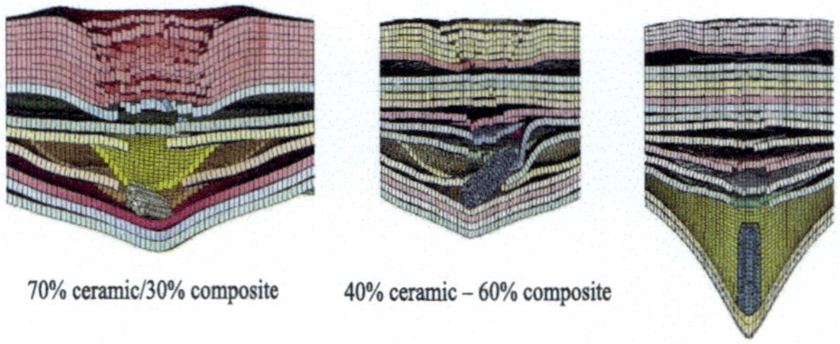

Fig. 6.19 State of bullet at 150 μs in a two-layer configuration with ceramic as top layer and layered composite (ultra-high molecular weight polyethylene, UHMWPE) as the substrate [72]

It's noteworthy that bonding between sheets is numerically represented using zero-thickness cohesive elements in LS-DYNA. Krishnan et al. utilized the LS-DYNA model -MAT-COHESIVE-GENERAL to describe delamination processes. This modelling effort involves specifying parameters for cohesive elements, such as density per unit area and peak traction, as detailed by Krishnan et al. [71].

In summary, Krishnan et al. conducted a study on modelling and simulating ballistic impact using LS-DYNA. To address excessively distorted elements, erosion strain served as a failure model parameter. Their calibration method involved selecting an appropriate combination of mesh resolution and erosion strain value to match DOP values from ballistic tests. This approach has the potential to optimize armour systems by reducing the need for trial-and-error ballistic tests.

Ballistic Response of Carbon Composite: Yen et al. [72] investigated the ballistic response of carbon composite systems used in aerospace and aircraft structures. They employed a user-defined computational constitutive model within LS-DYNA to characterize the strain rate-dependent progressive failure process in carbon laminated composite plates subjected to ballistic impact and blast loading conditions. Their simulations aimed to validate the user-defined failure model for two different target plates with varying ply configurations.

Validation through Simulations: In the LS-DYNA simulations (as depicted in Fig. 6.20), two distinct target plates were considered. One was a thin 7-ply configuration, while the other was a thick 36-ply with different ply arrangements. These targets were modelled using 8-node 3D brick elements, each with a single integration point. To accurately represent each ply, a single row of elements was employed through the thickness. The projectile used in the simulations was an AISI 4340 steel, 0.30 calibre fragment simulating projectile (FSP). The primary objective of these simulations was to validate a user-defined failure model by predicting the ballistic limit velocity (V50) for both target configurations under a striker velocity of 300 m/s.

Modelling Contact and Penetration: The LS-DYNA simulations utilized the eroding contact algorithm in conjunction with a user-supplied CDM (Continuous Damage Mechanics) damage/failure model to simulate the contact and penetration between

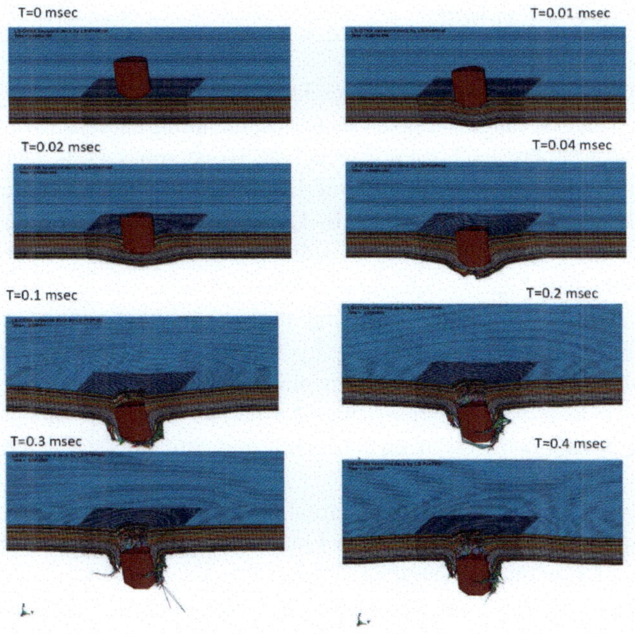

Fig. 6.20 LS-DYNA results. Plots of deformed mesh for a 36-Ply laminate subjected to 300 m/s impact at 0, 0.01, 0.02, 0.04, 0.1, 0.2, 0.3, and 0.4 ms [72]

the projectile and the impact area of the target plate. When complete fibre failure occurred in the unidirectional axial direction, the corresponding elements were systematically removed, and the contact surfaces were automatically updated to accurately represent the newly exposed layers of material. Figure 6.20 presents typical results of deformed mesh with damage for the 36-ply composite laminate subjected to a 300 m/s impact.

Effective Utilization of Through Thickness Element: According to Yen [73], a striking correlation between computational and experimental results was established. This correlation suggests that the ballistic resistance of various composite laminates can be effectively captured using only a single through-thickness element per layer. Remarkably, the coarse one-element-per-layer finite element model (FEM) used for the 36-ply thick composite target provided satisfactory predictions for overall energy dissipation, primarily dominated by fibre failure modes. This capability to accurately depict and forecast the penetration and perforation processes (or V50) in highly heterogeneous, thick targets using a relatively coarse mesh is an encouraging development in terms of computational efficiency.

6.3.3 ABAQUS Explicit

ABAQUS Explicit, a widely used commercial code, has been successfully employed in simulating a diverse range of impact scenarios, including car crashworthiness, bullet penetration, and blast loads on infrastructure. It encompasses an array of numerical algorithms designed to address numerical instabilities, mesh distortions, erosions, and other complex aspects. The ABAQUS user manual [1] offers an extensive explanation of the code's capabilities for solving highly nonlinear impact problems.

When dealing with highly nonlinear problems, both in material and geometry, it is crucial to perform simulations with varying levels of mesh refinement and time step sizes to validate results generated by the hydrocode. A comparison study between numerical solutions and available experimental data is a valuable step in ensuring the accuracy and realism of the problem setup. Numerous articles in the open literature showcase the impact response of various solids and structures using ABAQUS-like hydrocodes.

In this section, we draw upon the journal article by Goda and Girardot [74] as an example to illustrate the application of hydrocodes like ABAQUS in modelling projectile penetration into a layered target plate. The primary objective of this example is to demonstrate ABAQUS' capabilities in modelling brittle failure in ceramics, damaged fibre/epoxy plies, and the use of various features such as contact, erosion, boundary conditions, and the cohesive zone method. Goda and Girardot [74] also conducted numerical simulations of ballistic impacts, as shown in Fig. 6.21, to comprehensively investigate the damage behaviours of ceramic/composite materials caused by a rigid projectile.

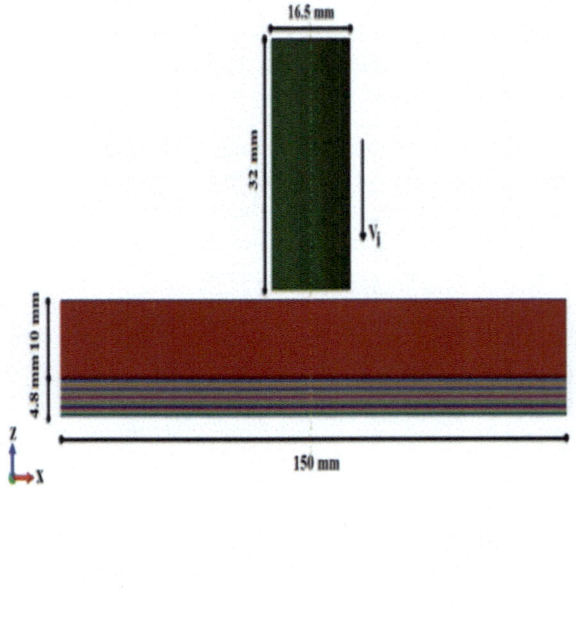

Fig. 6.21 The ceramic/composite configuration showing a thick ceramic backed by layers of plies. Plies are highlighted with different colours [74]

The top layer consists of a thick ceramic plate, while the substrate comprises layered woven fibre-reinforced plastic (WFRP). ABAQUS simulations use two different WFRP composites: carbon fibre fabric/epoxy and glass fabric/epoxy. The finite element model (FEM) employs brick element-type C3D8R for ceramics and shell element SC8R for the plies with a uniform mesh of 1.25 mm × 1.25 mm × 1.25 mm. In the thickness direction of the composite layer, the mesh size corresponds to the ply thickness of 0.3 mm, rather than 1.25 mm. The projectile, treated as rigid, is described using element-type R3D4 to represent a non-deformable solid. Boundary conditions dictate that all four edges of the target plate are fixed to prevent rigid body rotation. Goda and Girardot [74] provide a detailed description of the discretization of the target configuration, with their mesh sensitivity study revealing an optimal mesh size that balances accuracy and computational efficiency.

Within the JH-2 failure model described in Chap. 5, fracture occurs either when damage reaches a critical value of 1.0 or when negative pressure reaches the maximum tensile pressure. Element deletions in the ceramic layer are executed when elements satisfy the failure criterion. For the composite layers, elements are removed when any damage or deformation parameter in the CDM-based model reaches the specified maximum value.

Various types of contact occur within the ballistic configuration: (1) projectile and ceramic, (2) projectile and composite, (3) ceramic and composite, (4) individual plies of the composite laminate, and (5) newly created fragmented surfaces. ABAQUS employs its general contact algorithm to model these contacts, preventing element interpenetration. Friction between contact surfaces can also dissipate energy. Goda and Girardot [74] set the coefficient of friction between fabric plies at 0.3, and 0.28 for projectile-ceramic and ceramic-composite surfaces.

Delamination between interfaces plays a significant role in energy dissipation. A cohesive zone-based constitutive model is adopted to describe interface deformation and delamination. While a cohesive element method would require fine mesh and extremely small time increments, Goda and Girardot [74] employ a surface-based contact approach to avoid significant computational demands.

The final configurations of three different target assemblies, shown in Fig. 6.22 for varying ceramic thicknesses, demonstrate the effectiveness of this approach. The rigid projectile's post-perforation positions with different residual velocities are not depicted in the figure.

When describing the various modes of progressive damage and failure, it is crucial to monitor the total energy of the system over time. The majority of the projectile's kinetic energy ($mv^{2/}/2$) is transformed into internal energy through deformation. As noted by Goda and Girardot [74], energy balance analysis reveals that the primary form of energy absorption is inelastic deformation, followed by energy from vibrations, elastic strain energy stored in panels, energy dissipated viscously, artificial strain energy, and other energy forms, including that lost to delamination, friction, and contact penalties. For instance, in a high-impact scenario with 3520 J (corresponding to a projectile velocity of 800 m/s), inelastic deformation accounts for approximately 51% of the total energy dissipation. The exact distribution of energy estimates may vary depending on the specific damage models used in ABAQUS.

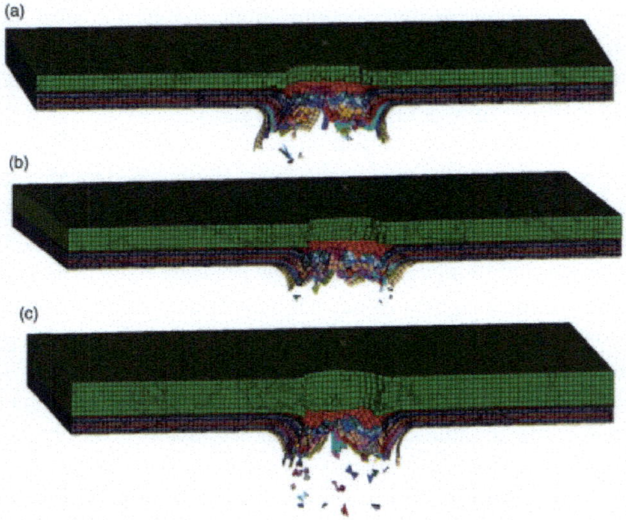

Fig. 6.22 Deformed configurations of three different layered target of ceramic and fabric composite material after the rigid projectile perforation at 80 μs. **a** Ceramic 4 mm, composite 5 mm, **b** Ceramic 6 mm, composite 5 mm, and **c** Ceramic 9 mm, composite 5 mm; Goda and Girardot [74]

Nevertheless, maintaining energy balance at each time step is a crucial factor in obtaining accurate solutions for the design analysis of impact-resistant structures.

In general, research students or engineers who are implementing their own empirical damage models involving material stiffness degradation within commercial hydrocodes should consider special numerical or modelling techniques. Techniques such as the crack band method or iterative sub-cycling time steps can be highly beneficial to reduce mesh size effects. We provided a few examples from open literature that directly addressed mesh size-dependent solutions, especially to shock and impact problems.

6.4 Summary

Hydrocodes serve as effective design tools for computing the impact response of solids and structures under shock and impact. Over the past five decades, hydrocodes have made remarkable progress, integrating advanced material models and numerical algorithms to accurately describe fracture and fragmentation of layered heterogeneous structures. In general, well-developed codes yield realistic and predictive solutions for multi-axial impact problems. Consequently, when employed correctly, whether through commercial- or government-controlled hydrocodes, they are expected to provide accurate results in designing impact-resistant structures.

This chapter primarily underscores the significance of utilizing diverse numerical algorithms within hydrocodes, including element-to-particle conversion, the utilization of meshless method codes, and the element erosions, among others. These techniques play a pivotal role in addressing mesh distortion, ensuring stable and robust solutions. Moreover, achieving precise results from FEM or meshless codes demands expertise in setting up the appropriate mesh, selecting mesh size, special numerical schemes to minimize mesh-dependent solutions, and suitable time increments.

The response of brittle materials to high-velocity impact or high strain rate loading is complex, necessitating realistic strength models, equations of state, descriptions of damage evolution, and models for failure (or fracture). One of the challenges lies in determining reasonable and physically meaningful values for model parameters. Chapters 4 and 5 provided specific methodologies for assigning realistic values to these parameters and for determining fundamental material properties, including strain rate-dependent compressive strength, spall strength, and the Hugoniot elastic limit.

References

1. **ABAQUS**: Abaqus/Explicit, a special-purpose Finite-Element analyzer that employs explicit integration scheme to solve highly nonlinear systems with many complex contacts under transient loads. https://www.3ds.com/products-services/simulia/products/abaqus/
2. Livermore Software Technology Corporation (LSTC) (2007) **LSDYNA** Keyword User's Manual, vol 1, 971st edn. Livermore Software Technology Corporation (LSTC)
3. ANSYS Inc. Available Solution Methods, https://www.ansys.com/en-in/products/structures/ansys-autodyn; **ANSYS AUTODYN** User's Manual Theoretical manual 2005, ANSYS Inc. Southpointe, 2600 ANSYS Drive, Canonsburg, PA, 15317 USA
4. Johnson GR (1978) **EPIC-2**, A Computer program for Elastic-Plastic Impact computations in Two Dimensions Plus Spin. ARBRL-CR-00373
5. Scalable Shock Physics Code (2020) **CTH 12.2**, https://www.sandia.gov/cth/
6. **ALEGRA**: Shock hydrodynamics and Multiphysics: https://www.sandia.gov/alegra/
7. Frank NB, **SIERRA/**Solid Mechanics 4.56 User's Guide. SANDIA REPORT SAND2020-3547. https://www.osti.gov/servlets/purl/, 1608404
8. **ALE3D**, Computational Manual Material Model. Lawrence Livermore National Laboratory 7000 East Avenue • Livermore, CA 94550
9. **PAGOSA**, Weseloh W, Clancy S, Painter J (2010) PAGOSA physics manual. Technical Report (LAUR-14425- M). Los Alamos National Laboratory
10. Wingate CA, Stellingwerf RF (1993) Smooth Particle Dynamics - The SPHINX and SPHC codes. Technical Report LA-UR-93-1938, Los Alamos National Laboratory
11. Johnson GR, Stryk RA (2003) Conversion of 3D distorted elements into meshless particles during dynamic deformation. Int J Impact Eng 28(9):947–966
12. Gerlach CA, Johnson GR (2018) A contact and sliding interface algorithm for the combined particle-element method. Int J Impact Eng 113:21–28
13. Walls KC, III (2017) An improved contact method for multi-material Eulerian hydrocodes. A PhD, Dissertation, University of Alabama, Birmingham, AL
14. Littlefield D (2001) the use of r-adaptivity with local, intermittent Remesh for modelling hypervelocity impact and penetration. Int J Impact Eng 26(1):433–442
15. Benson D (1992) Computational methods in Lagrangian and Eulerian hydrocodes. Comput Methods Appl Mech Eng 99(2–3):235–394

16. Benson D (1998) Stable time step estimation for multi-material Eulerian hydrocodes. Comput Methods Appl Mech Eng 167:191–205
17. Benson D (2000) An implicit multi-material Eulerian formulation. Int J Numer Meth Eng 48:475–499
18. Zhong Z (1993) Finite element procedures for contact-impact problems. Oxford University Press, New York
19. Donea J, Huerta A, Ponthot J, Rodriguez-Ferran A (2004) Arbitrary Lagrangian-Eulerian methods. In: Stein E, de Borst R, Hughes J (eds) Encyclopedia of computational mechanics. Wiley, pp 1–25
20. Vitali E, Benson D (2006) An extended finite element formulation for contact in multi-material arbitrary Lagrangian-Eulerian calculations. Int J Numer Meth Eng 67:1420–1444
21. Park B, Leavy R, Niederhaus J (2013) Penetration of rod projectiles in semi-infinite targets: a validation test for Eulerian XFEM in ALEGRA. Sandia National Laboratories Report SAND2013-1863
22. Lucy LB (1977) A numerical approach to the testing of Fission hypothesis. Astron J 82:1013–1024
23. Gingold RA, Monaghan JJ (1975) Smoothed particle hydrodynamics: theory and application to non-spherical stars. Mon Not R Astron Soc 181:375–389
24. Zhi-Bin W, Rong C, Hong W, Qiang L, Xun Z, Shu-Zhe L (2016) An overview of smoothed particle hydrodynamics for simulating multiphase flow. Appl Math Modell 40(23–24):9625–9655
25. Wang G, Cheng AH-D, Ostoja-Starzewski M, Al-Ostaz A, Radziszewski P (2010) Hybrid lattice particle modelling approach for polymeric materials subject to high strain rate loads. Polymers 2(1):3–30. https://doi.org/10.3390/polym2010003
26. Randles PW, Libersky LD (2000) Normalized SPH with stress points. Int J Numer Meth Eng 48(10):1445–1462
27. Beissel SR, Gerlach CA, Johnson GR (2006) Hypervelocity impact computations with finite elements and meshfree particles. Int J Impact Eng 33(1–4):80–90
28. Zheng X, Pisano F, Vardon PJ, Hicks MA (2021) An explicit stabilised material point method for coupled hydromechanical problems in two-phase porous media. Comput Geotech 135:104112
29. Sulsky D, Chen Z, Schreyer HL (1994) A particle method for history-dependent materials. Comput Methods Appl Mech Eng 18(1–2):179–196
30. Chen Z, Brannon R (2002) An evaluation of the material point method. Sandia National Laboratory Technical report, SAND2002-0482, 2002
31. de Vaucorbeil A, Nguyen VP, Nguyen-Thanh C (2021) Karamelo: an open source parallel C++ package for the material point method. Comput Particle Mech 8(4):767–789
32. Ma S, Zhang X, Qiu XM (2009) Comparison study of MPM and SPH in modelling hypervelocity impact problems. Int J Impact Eng 36(2):272–282
33. Karamelo opensource MPM code: https://github.com/adevaucorbeil/karamelo/
34. Zhang X, Chen Z, Liu Y (2016) The material point method - a continuum-based particle method for extreme loading cases. Academic, Elsevier/AP
35. de Vaucorbeil A, Nguyen VP, Sinaie S, Wu JY (2019) Material point method after 25 years: theory, implementation, and applications. Adv Appl Mech 53:185–398
36. Wojciech TS, Sołowski, Berzins M, Coombs WM, Guilkey JE, Möller M, Anh Tran Q, Adibaskoro T, Seyedan SM, Tielen R, Soga K (2021) Material point method overview and challenges ahead. Adv Appl Mech 54:113–204
37. Zhang H, Shukla MK, Rajendran AM, Jiang S (2023) Simulations of single and double shock experiments using generalized interpolation material point method with a noise control strategy. Comput Particle Mech 29. https://doi.org/10.1007/s40571-023-00590-1
38. Chen Z, Su Y-C, Rajendran AM, Su H, Liu Y, Jiang S (2020) A new woven composite constitutive model validated by shock wave experiments. Compos Struct 252:112751
39. Hu Y, Fang Y, Ge Z, Qu Z, Zhu V, Pradhana A, Jiang C (2018) A moving least squares material point method with displacement discontinuity and two-way rigid body coupling. ACM Trans Graph 37(4):146

40. Hu Y (2021) The Taichi high-performance and differentiable programming language for sparse and quantized visual computing. PhD Thesis, Massachusetts Institute of Technology
41. Wang X, Qiu Y, Slattery YR, Fang Y, Li M, Zhu SC, Zhu Y, Tang M, Manocha D, Jiang C (2020) A massively parallel and scalable multi-GPU material point method. ACM Trans Graph 39(4):30, 30:1-30:15
42. Carrillo AR, Horner DA, Peters JF, West JE (1996) Design of a large scale discrete element soil model for high performance computing systems. In: Supercomputing '96: proceedings of the 1996 ACM/IEEE conference on Supercomputing. https://doi.org/10.1145/369028.36913
43. Peters J, Vahedifard F, Jelinek B, Mason G, Priddy J (2019) The discrete element method for vehicle-terrain analysis. EasyChair Preprint No 1304
44. Haustein M, Gladkyy A, Schwarze R (2017) Discrete element modeling of deformable particles in YADE. TU Bergakademie Freiberg, Institut für Mechanik und Fluiddynamik, Lampadiusstr. 4, 09596 Freiberg, Germany, SoftwareX, Elsevier Publications, vol 6, pp 118–123
45. PFC | US Minneapolis - Itasca Consulting Group, Inc. https://www.itascacg.com/software/pfc
46. Alshanti WG (2018) Discrete element modelling of a projectile impacting and penetrating into granular systems. In: Osheku C (ed) https://doi.org/10.5772/intechopen.75550
47. Steinhauser MO, Grass K, Strassburger E, Blumen A (2009) Impact failure of granular materials-non-equilibrium multi-scale simulations and high-speed experiments. Int J Plast 25:161–162
48. Watson E, Steinhauser MO (2017) Discrete particle method for simulating hypervelocity impact phenomena. Materials 10(4):379
49. Steinhauser MO, Kühn M, Grass K (2006) Numerical simulation of fracture and failure dynamics in Brittle solids. In: Proceedings of the 12th international symposium on plasticity and its current applications, Halifax, NS, Canada
50. Obbink-Huizer C (2021) Implicit vs explicit finite element analysis: when to use which? A Simuleon FEA Blog. https://info.simuleon.com/blog/implicit-vs-explicit-finite-element-analysis, on https://www.msn.com/en-us/feed
51. Cook WH, Rajendran AM, Grove DJ (1992) An efficient numerical implementation of Bodner-Partom model in EPIC-code. J Eng Fract Mech 41(5):607–623
52. Soares Jr (2020) A novel time-marching formulation for wave propagation analysis: the adaptive hybrid explicit method. Comput Methods Appl Mech Eng 366:113095
53. von Neumann J, Richtmyer RD (1950) A method for the numerical calculation of hydrodynamic shocks. J Appl Phys 21:232–237
54. Brady BHG, Brown ET (2004) Rock mechanics: for underground mining, 3rd edn. Springer Science and Business Media, Dordrecht
55. Murkami S, Liu Y (1996) Local approach of fracture based on continuum damage mechanics and related problems. Mater Sci Res Int 2(3):131–142
56. Fraser J (2022) ABAQUS implementation of a hyperelastic damage model for glass-reinforced polymers under shock and impact loading. AMS Thesis, Department of Mechanical Engineering, University of Mississippi
57. Scott N (2020) Computational modelling of shock wave propagation in a layered composites. MS Thesis, Department of Mechanical Engineering, The University of Mississippi
58. Yuan LTF, Prakash V, Dandekar DP (2009) Shock compression behavior of a S2-glass Fibre reinforced polymer composite. J Appl Phys 105:093526. https://doi.org/10.1063/1.3124622
59. Barham MI, King MJ, Mseis G, Faux DR (2013) Hyperelastic fibre-reinforced composite model with damage. Technical report, Lawrence Livermore National Laboratory, LLNL-MI-644243
60. Roy SK, Trabia M, O'Toole B, Hixson R, Becker S, Pena M, Jennings R, Somasoundaram D, Matthes M, Daykin E, Machorro E (2016) Study of hypervelocity projectile impact on thick metal plates. Shock Vib 2016:4313480. https://doi.org/10.1155/2016/4313480
61. Brannon RM, Wells JM, Strack OE (2007) Validating theories for brittle damage. Metall Mater Trans A 38a(12):2861–2868
62. Johnson GR (2002) Numerical algorithms and material models for high-velocity impact computations. Int J Impact Eng 28:456–472

63. Johnson GR, Stryk RA, Beissel SR (1996) SPH for high velocity impact computations. Comput Methods Appl Mech Eng 139:347–373
64. Johnson GR, Gerlach CA, Beissel SR (2019) Comparisons of meshless particle algorithms and improvements in parallelization for the EPIC code. Technical report no: 18.17637/09, Southwest Research Institute
65. Bazant ZP, Oh BH (1983) Crack band theory for fracture of concrete. Mater Struct 16:155–177
66. Abdullah T, Kirane K (2022) Strain rate dependence of the mesh objectivity in dynamic fracture analyses with the crack band model. Eng Fract Mech 269:108501
67. Nelms M (2018) Computational modelling of damage mechanisms in mineralized biological tissue. PhD Dissertation, Department of Mechanical Engineering, University of Mississippi
68. Duvaut G, Lions JL (1976) Inequalities in mechanics and physics. Springer, Berlin
69. Lapczyk I, Hurtado JA (2007) Progressive damage modelling in fibre-reinforced materials. Compos A Appl Sci Manuf 38(11):2333–2341
70. Hibbit KAS (2014) Abaqus theory and user manuals. Version 6.14. USA: ABAQUS Inc
71. Krishnan K, Sockalingam S, Bansal S, Rajan SD (2010) Numerical simulation of ceramic composite armour subjected to ballistic impact. Compos: Part B 41:583–593
72. Yen CF, Kaste B, Chen CCT, Carey N (2020) Modelling and simulation of carbon composite ballistic and blast behavior. J Compos Mater 54(4):485–499
73. Yen CF (2012) A ballistic material model for continuous-fibre reinforced composites. Int J Impact Eng 46:11–21
74. Goda I, Girardot J (2021) Numerical modeling and analysis of the ballistic impact response of ceramic/composite targets and the influence of cohesive material parameters. Int J Damage Mech 30(7):1079–1122

Appendix

Fraser (see reference [101] in Chap. 5) developed the HECDM model subroutines in Fortran for VUMAT of explicit version of ABAQUS. The listing of this code is given below:

```
C
      SUBROUTINE VUMAT(
     !     nblock, ndir, nshr, nstatev, nfieldv, nprops, lanneal,
     !     stepTime, totalTime, dt, cmname, coordMp, charLength,
     !     props, density, strainInc, relSpinInc,
     !     tempOld, stretchOld, defgradOld, fieldOld,
     !     stressOld, stateOld, enerInternOld, enerInelasOld,
     !     tempNew, stretchNew, defgradNew, fieldNew,
     !     stressNew, stateNew, enerInternNew, enerInelasNew )
C
      INCLUDE 'vaba_param.inc'
C
      DIMENSION props(nprops), density(nblock), coordMp(nblock,*),
     !          charLength(nblock), strainInc(nblock,ndir+nshr),
     !          relSpinInc(nblock,nshr), tempOld(nblock),
     !          stretchOld(nblock,ndir+nshr),
     !          defgradOld(nblock,ndir+nshr),
     !          fieldOld(nblock,nfieldv), stressOld(nblock,ndir+nshr),
     !          stateOld(nblock,nstatev), enerInternOld(nblock),
     !          enerInelasOld(nblock), tempNew(nblock),
     !          stretchNew(nblock,ndir+nshr),
     !          defgradNew(nblock,ndir+nshr),
     !          fieldNew(nblock,nfieldv),
     !          stressNew(nblock,ndir+nshr), stateNew(nblock,nstatev),
     !          enerInternNew(nblock), enerInelasNew(nblock)
C
      CHARACTER*80 cmname
C
C     NUMBER DEFINITIONS
C     ------------------
      REAL*8 zero, one, two, three, four, ten, half, third, twothirds,
     !  deetee
      PARAMETER(zero = 0.d0, one = 1.d0, two = 2.d0, three = 3.d0,
     !  four = 4.d0, ten = 10.d0, half = 0.5d0, third = 1.d0/3.d0,
```


A. M. Rajendran et al., *Failure of Brittle Materials Under Shock and Impact*, Materials Horizons: From Nature to Nanomaterials, https://doi.org/10.1007/978-981-97-7026-7

```
      !  twothirds = 2.d0/3.d0, deetee = 1.d0*(10.d0**-10.d0))
C
C     LOCAL VARIABLES
C     ---------------
      REAL*8 G, Kay, beta, k1, k2, M11, M12, M13, M21, M22, M23,
     !       c1ems, Jec, Jet, iota, ems, tcrit, ccrit, C11, C22, C33,
     !       C44, C55, C66, C12, C13, C23, eta, alpha, c1alpha, trace,
     !       J, Cxx, Cyy, Czz, Cxy, Cyz, Cxz, Ubarxx, Ubaryy, Ubarzz,
     !       Ubarxy, Ubaryz, Ubarxz, Ubar2xx, Ubar2yy, Ubar2zz, Ubar2xy,
     !       Ubar2yz, Ubar2xz, I1, p0, Gfac, facxx, facyy, faczz, facxy,
     !       facyz, facxz, fac2, m1dpm1, m2dpm2, m1outxx, m1outyy,
     !       m1outzz, m2outxx, m2outyy, m2outzz, fac3xx1, fac3xx2,
     !       fac3yy1, fac3yy2, fac3zz1, fac3zz2, exx, eyy, ezz, exy,
     !       eyz, exz, epeqold, epeq, strrate, Rmeold, Rmsold, Rms, Rme,
     !       lograte, emsadj, alphaadj, Rm, matdamCurrent, matdamOld,
     !       matdamNew, Rfit1, Rfit2, Rfic1, Rfic2, fibdam1, fibdam2,
     !       fibdam1Old, fibdam2Old, fibdam1New, fibdam2New, Totaldam,
     !       Kayd, Kdterm, uliniso, active, inactive
C
C     MATERIAL PROPERTIES
C     ---------------
      G      = props(1)
      Kay    = props(2)
      beta   = props(3)
      k1     = props(4)
      k2     = props(5)
      M11    = props(6)
      M12    = props(7)
      M13    = props(8)
      M21    = props(9)
      M22    = props(10)
      M23    = props(11)
      c1ems  = props(12)
      Jec    = props(13)
      Jet    = props(14)
      iota   = props(15)
      ems    = props(16)
      tcrit  = props(17)
      ccrit  = props(18)
      C11    = props(19)
      C22    = props(20)
      C33    = props(21)
      C44    = props(22)
      C55    = props(23)
      C66    = props(24)
      C12     = props(25)
      C13     = props(26)
      C23     = props(27)
      eta     = props(28)
      alpha   = props(29)
      c1alpha= props(30)
C
C     *****************************************************************
C     ------------ INITIALIZE MATERIAL AS LINEARLY ELASTIC -------------
C     *****************************************************************
C
      IF (totalTime.EQ.0.0) THEN
```

```
      DO i = 1,nblock
              trace = strainInc(i,1) + strainInc(i,2) +
     !                strainInc(i,3)
              stressNew(i,1) = stressOld(i,1) +
     !                         two * G * strainInc(i,1) +
     !                         (Kay - (twothirds * G)) * trace
              stressNew(i,2) = stressOld(i,2) +
     !                         two * G * strainInc(i,2) +
     !                         (Kay - (twothirds * G)) * trace
              stressNew(i,3) = stressOld(i,3) +
     !                         two * G * strainInc(i,3) +
     !                         (Kay - (twothirds * G)) * trace
              stressNew(i,4) = stressOld(i,4) +
     !                         two * G * strainInc(i,4)
              stressNew(i,5) = stressOld(i,5) +
     !                         two * G * strainInc(i,5)
              stressNew(i,6) = stressOld(i,6) +
     !                         two * G * strainInc(i,6)
      END DO
      RETURN
      END IF
C
C     *******************************************************
C     ----------- START LOOP FOR USER CALCULATIONS -----------
C     *******************************************************
C
      DO k = 1,nblock
C
C        CALCULATE J = |F| = |U|
C        ----------------------
         J =    stretchNew(k,1) *
     !        ( stretchNew(k,2) * stretchNew(k,3)   -
     !          stretchNew(k,5) * stretchNew(k,5) ) +
     !          stretchNew(k,4) *
     !        ( stretchNew(k,5) * stretchNew(k,6)   -
     !          stretchNew(k,3) * stretchNew(k,4) ) +
     !          stretchNew(k,6) *
     !        ( stretchNew(k,4) * stretchNew(k,5)   -
     !          stretchNew(k,2) * stretchNew(k,6) )
C
C        CALCULATE RIGHT CAUCHY TENSOR, C = U*U
C        --------------------------------------------------
         Cxx = stretchNew(k,1) * stretchNew(k,1) +
     !   stretchNew(k,4) * stretchNew(k,4) +
     !   stretchNew(k,6) * stretchNew(k,6)
         Cyy = stretchNew(k,2) * stretchNew(k,2) +
     !   stretchNew(k,4) * stretchNew(k,4) +
     !   stretchNew(k,5) * stretchNew(k,5)
         Czz = stretchNew(k,3) * stretchNew(k,3) +
     !   stretchNew(k,6) * stretchNew(k,6) +
     !   stretchNew(k,5) * stretchNew(k,5)
         Cxy = stretchNew(k,1) * stretchNew(k,4) +
     !   stretchNew(k,4) * stretchNew(k,2) +
     !   stretchNew(k,6) * stretchNew(k,5)
         Cyz = stretchNew(k,4) * stretchNew(k,6) +
     !   stretchNew(k,2) * stretchNew(k,5) +
     !   stretchNew(k,5) * stretchNew(k,3)
```

```
      Cxz = stretchNew(k,1) * stretchNew(k,6) +
     !  stretchNew(k,4) * stretchNew(k,5) +
     !  stretchNew(k,6) * stretchNew(k,3)
C
C     CALCULATE MODIFIED STRETCH TENSOR, Ubar = U/(J^1/3)
C     ------------------------------------------------
      Ubarxx = stretchNew(k,1) / (J**third)
      Ubaryy = stretchNew(k,2) / (J**third)
      Ubarzz = stretchNew(k,3) / (J**third)
      Ubarxy = stretchNew(k,4) / (J**third)
      Ubaryz = stretchNew(k,5) / (J**third)
      Ubarxz = stretchNew(k,6) / (J**third)
C
C     CALCULATE MODIFIED STRETCH TENSOR SQUARED, Ubar^2
C     -------------------------------------------------
      Ubar2xx = Ubarxx * Ubarxx
      Ubar2yy = Ubaryy * Ubaryy
      Ubar2zz = Ubarzz * Ubarzz
      Ubar2xy = Ubarxy * Ubarxy
      Ubar2yz = Ubaryz * Ubaryz
      Ubar2xz = Ubarxz * Ubarxz
C
C     FIRST INVARIANT OF Ubar^2 = tr(Ubar2)
C     ---------------------------------
      I1 = Ubar2xx + Ubar2yy + Ubar2zz
C
C     ISOCHORIC TERM
C     ----------------------
      p0 = third * I1
      Gfac = G / J
C
      facxx = Gfac * (Ubar2xx - p0)
      facyy = Gfac * (Ubar2yy - p0)
      faczz = Gfac * (Ubar2zz - p0)
      facxy = Gfac * Ubar2xy
      facyz = Gfac * Ubar2yz
      facxz = Gfac * Ubar2xz
C
C     VOLUMETRIC CAUCHY STRESS TERM
C     ----------------------------------
      fac2 = (Kay / (J * beta)) * (one - (one / (J**beta)))
C
C     MATRIX DIRECTION DEFINITIONS
C     ----------------------------------
      m1dpm1 = M11 * (M11 * Cxx + M12 * Cxy + M13 * Cxz) +
     !         M12 * (M11 * Cxy + M12 * Cyy + M13 * Cyz) +
     !         M13 * (M11 * Cxz + M12 * Cyz + M13 * Czz)
      m2dpm2 = M21 * (M21 * Cxx + M22 * Cxy + M23 * Cxz) +
     !         M22 * (M21 * Cxy + M22 * Cyy + M23 * Cyz) +
     !         M23 * (M21 * Cxz + M22 * Cyz + M23 * Czz)
C
C     MATRIX OUTER PRODUCTS
C     ----------------------------------
      m1outxx = (stretchNew(k,1) * M11 + stretchNew(k,4) * M12 +
     !  stretchNew(k,6) * M13)**two
      m1outyy = (stretchNew(k,4) * M11 + stretchNew(k,2) * M12 +
     !  stretchNew(k,5) * M13)**two
```

```
      m1outzz = (stretchNew(k,6) * M11 + stretchNew(k,5) * M12 +
     !   stretchNew(k,3) * M13)**two
      m2outxx = (stretchNew(k,1) * M21 + stretchNew(k,4) * M22 +
     !   stretchNew(k,6) * M23)**two
      m2outyy = (stretchNew(k,4) * M21 + stretchNew(k,2) * M22 +
     !   stretchNew(k,5) * M13)**two
      m2outzz = (stretchNew(k,6) * M21 + stretchNew(k,5) * M22 +
     !   stretchNew(k,3) * M23)**two
C
C     ANISOTROPIC STRESSES
C     ----------------------------------
      fac3xx1 = (two * k1 / J) * (m1dpm1 - one) * m1outxx
      fac3xx2 = (two * k2 / J) * (m2dpm2 - one) * m2outxx
      fac3yy1 = (two * k1 / J) * (m1dpm1 - one) * m1outyy
      fac3yy2 = (two * k2 / J) * (m2dpm2 - one) * m2outyy
      fac3zz1 = (two * k1 / J) * (m1dpm1 - one) * m1outzz
      fac3zz2 = (two * k2 / J) * (m2dpm2 - one) * m2outzz
C
C     MATRIX DAMAGE CRITERIA
C     ----------------------------------
      exx = twothirds * (stretchNew(k,1) - one) -
     !   third * (stretchNew(k,2) - one) -
     !   third * (stretchNew(k,3) - one)
      eyy = twothirds * (stretchNew(k,2) - one) -
     !   third * (stretchNew(k,1) - one) -
     !   third * (stretchNew(k,3) - one)
      ezz = twothirds * (stretchNew(k,3) - one) -
     !   third * (stretchNew(k,1) - one) -
     !   third * (stretchNew(k,2) - one)
      exy = two * stretchNew(k,4)
      eyz = two * stretchNew(k,5)
      exz = two * stretchNew(k,6)
      epeqold = stateOld(k,1)
      epeq = twothirds * sqrt((three / two) * (exx**two + eyy**two +
     !   ezz**two) + (three / four) * (exy**two + eyz**two + exz**two))
      stateNew(k,1) = epeq
      strrate = (epeq - epeqold) / deetee
      Rmsold = stateOld(k,2)
      Rmeold = stateOld(k,3)
      IF (strrate.LT.one) THEN
          Rms = Rmsold
          Rme = Rmeold
          lograte = zero
          emsadj = ems
          alphaadj = alpha
      ELSE
          lograte = log(abs(strrate))
          emsadj = ems * (one + c1ems * lograte)
          Rms = epeq / emsadj
          IF (J.GT.one) THEN
              Rme = (J - one) / (Jet - one)
          ELSE
              alphaadj = alpha * (one + c1alpha * lograte)
              Rme = alphaadj * (one - J) * exp(J / Jec)
          END IF
      END IF
      stateNew(k,2) = Rms
```

```
      stateNew(k,3)= Rme
      Rm = sqrt((Rms**two) + (Rme**two))
      IF (Rm.LT.one) THEN
          matdamCurrent = zero
      ELSE
          matdamCurrent = one - exp((one - Rm) / iota)
      END IF
      matdamOld = stateOld(k,4)
      IF (matdamCurrent.GE.matdamOld) THEN
          matdamNew = matdamCurrent
      ELSE
          matdamNew = matdamOld
      END IF
      stateNew(k,4) = matdamNew
C
C     FIBER DAMAGE CRITERIA
C     ----------------------------------
      Rfit1 = (m1dpm1 - one) / tcrit
      Rfit2 = (m2dpm2 - one) / tcrit
      Rfic1 = (m1dpm1 - one) / ccrit
      Rfic2 = (m2dpm2 - one) / ccrit
      stateNew(k,7) = Rfit1
      stateNew(k,8) = Rfic1
      IF (Rfit1.GT.one) THEN
          fibdam1 = one
      ELSE IF (Rfic1.GT.one .AND. matdamNew.EQ.zero) THEN
          fibdam1 = one - eta
      ELSE IF (Rfic1.GT.one .AND. matdamNew.GT.zero) THEN
          fibdam1 = one
      ELSE
          fibdam1 = zero
      END IF
      IF (Rfit2.GT.one) THEN
          fibdam2 = one
      ELSE IF (Rfic2.GT.one .AND. matdamNew.EQ.zero) THEN
          fibdam2 = one - eta
      ELSE IF (Rfic2.GT.one .AND. matdamNew.GT.zero) THEN
          fibdam2 = one
      ELSE
          fibdam2 = zero
      END IF
      fibdam1Old = stateOld(k,5)
      fibdam2Old = stateOld(k,6)
      IF (fibdam1.GE.fibdam1Old) THEN
          fibdam1New = fibdam1
      ELSE
          fibdam1New = fibdam1Old
      END IF
      IF (fibdam2.GE.fibdam2Old) THEN
          fibdam2New = fibdam2
      ELSE
          fibdam2New = fibdam2Old
      END IF
      stateNew(k,5) = fibdam1New
      stateNew(k,6) = fibdam2New
      stateNew(k,9) = strrate
C
```

```
C         COROTATIONAL CAUCHY STRESSES
C         ----------------------------------
          Totaldam = matdamNew + fibdam1New + fibdam2New
          IF (Totaldam.EQ.zero) THEN
              stressNew(k,1) = C11 * (stretchNew(k,1) - one) +
     !     C12 * (stretchNew(k,2) - one) + C13 * (stretchNew(k,3) -
     !     one)
              stressNew(k,2) = C12 * (stretchNew(k,1) - one) +
     !     C22 * (stretchNew(k,2) - one) + C23 * (stretchNew(k,3) -
     !     one)
              stressNew(k,3) = C13 * (stretchNew(k,1) - one) +
     !     C23 * (stretchNew(k,2) - one) + C33 * (stretchNew(k,3) -
     !     one)
              stressNew(k,4) = C66 * two * stretchNew(k,4)
              stressNew(k,5) = C55 * two * stretchNew(k,5)
              stressNew(k,6) = C44 * two * stretchNew(k,6)
          ELSE
              IF (matdamNew.EQ.one .AND. J.GE.one .AND.
     !            fibdam1New.EQ.one .AND. fibdam2New.EQ.one) THEN
              ELSE
                  Kayd = Kay / ten
              END IF
              Kdterm = (Kayd / (J * beta)) * (one -
     !        (one / (J**beta)))
              stressNew(k,1) = (one - matdamNew) * facxx +
     !    (one - matdamNew) * fac2 + matdamNew * Kdterm +
     !    (one - fibdam1New) * fac3xx1 + (one - fibdam2New) * fac3xx2
              stressNew(k,2) = (one - matdamNew) * facyy +
     !    (one - matdamNew) * fac2 + matdamNew * Kdterm +
     !    (one - fibdam1New) * fac3yy1 + (one - fibdam2New) * fac3yy2
              stressNew(k,3) = (one - matdamNew) * faczz +
     !    (one - matdamNew) * fac2 + matdamNew * Kdterm +
     !    (one - fibdam1New) * fac3zz1 + (one - fibdam2New) * fac3zz2
              stressNew(k,4) = (one - matdamNew) * facxy
              stressNew(k,5) = (one - matdamNew) * facyz
              stressNew(k,6) = (one - matdamNew) * facxz
          END IF
C
C         UPDATE SPECIFIC INTERNAL ENERGY
C         ------------------------------
          uliniso = half*((stressOld(k,1)+stressNew(k,1))*strainInc(k,1)+
     !               (stressOld(k,2)+stressNew(k,2))*strainInc(k,2) +
     !               (stressOld(k,3)+stressNew(k,3))*strainInc(k,3) +
     !                two * ( (stressOld(k,4) + stressNew(k,4))*
     !                         strainInc(k,4) +
     !                        (stressOld(k,5) + stressNew(k,5))*
     !                         strainInc(k,5) +
     !                        (stressOld(k,6) + stressNew(k,6))*
     !                         strainInc(k,6) ) )
C
          enerInternNew(k) = enerInternOld(k) + uliniso / density(k)
       END DO
C
       RETURN
C
       END SUBROUTINE VUMAT
```

Index

A. M. Rajendran et al., *Failure of Brittle Materials Under Shock and Impact*, Materials Horizons: From Nature to Nanomaterials, https://doi.org/10.1007/978-981-97-7026-7

The manufacturer's authorised representative in the EU is Springer Nature Customer Service Centre GmbH, Europaplatz 3, 69115 Heidelberg, Germany. If you have any concerns regarding our products, please contact ProductSafety@springernature.com

Printed and bound by CPI Group (UK) Ltd, Croydon, CR0 4YY
01/10/2026
02225962-0007